BIOMATERIALS
An Introduction
SECOND EDITION

BIOMATERIALS

An Introduction

SECOND EDITION

Joon B. Park
and
Roderic S. Lakes
The University of Iowa
Iowa City, Iowa

PLENUM PRESS • NEW YORK AND LONDON

Library of Congress Cataloging-in-Publication Data

Park, Joon Bu.
Biomaterials : an introduction / Joon B. Park and Roderic S. Lakes. -- 2nd ed.
p. cm.
Includes bibliographical references and index.
ISBN 0-306-43992-1
1. Biomedical materials. I. Lakes, Roderic S. II. Title.
R857.M3P37 1992
610'.28--dc20 92-23214
CIP

ISBN 0-306-43992-1

A Division of Plenum Publishing Corporation
233 Spring Street, New York, N.Y. 10013

Printed in the United States of America

PREFACE

This book is intended as a general introduction to the uses of artificial materials in the human body for the purposes of aiding healing, correcting deformities, and restoring lost function. It is an outgrowth of an undergraduate course for senior students in biomedical engineering, and it is offered as a text to be used in such courses. Topics include biocompatibility, techniques to minimize corrosion or other degradation of implant materials, principles of materials science as it relates to the use of materials in the body, and specific uses of materials in various tissues and organs. It is expected that the student will have successively completed elementary courses in the mechanics of deformable bodies and in anatomy and physiology, and preferably also an introductory course in materials science prior to undertaking a course in biomaterials.

Many quantitative examples are included as exercises for the engineering student. We recognize that many of these involve unrealistic simplifications and are limited to simple mechanical or chemical aspects of the implant problem. We offer as an apology the fact that biomaterials engineering is still to a great extent an empirical discipline that is complicated by many unknowns associated with the human body. In recognition of that fact, we have endeavored to describe both the successes and the failures in the use of materials in the human body. Also included are many photographs and illustrations of implants and devices as an aid to visualization.

Any errors of commission or omission that have remained in spite of our efforts at correction are our responsibility alone.

We thank Bea Park and Diana Lakes for their patience and support during a lengthy undertaking.

Joon B. Park
Roderic S. Lakes

Iowa City, Iowa

CONTENTS

CHAPTER 1

INTRODUCTION TO BIOMATERIALS

In the treatment of disease or injury it has been found that a variety of nonliving materials are of use. Commonplace examples include sutures and tooth fillings. A *biomaterial* is a synthetic material used to replace part of a living system or to function in intimate contact with living tissue. The Clemson University Advisory Board for Biomaterials has formally defined a biomaterial to be "a systemically and pharmacologically inert substance designed for implantation within or incorporation with living systems." By contrast, a *biological material* is a material such as bone matrix or tooth enamel, produced by a biological system. Artificial materials that simply are in contact with the skin, such as hearing aids and wearable artificial limbs are not biomaterials since the skin acts as a barrier with the external world. The uses of biomaterials, as indicated in Table 1-1, include replacement of a body part that has lost function due to disease or trauma, to assist in healing, to improve function, and to correct abnormalities. The role of biomaterials has been influenced considerably by advances in many areas of medicine. For example, with the advent of antibiotics, infectious disease is less of a threat than in former times, so that degenerative disease assumes a greater importance. Moreover, advances in surgical technique have permitted materials to be used in ways that were not possible previously.

This book is intended to develop in the reader a familiarity with the uses of materials in medicine and with the rational basis for these applications. It should be suitable as a text for a junior or senior level course in biomaterials for students in biomedical engineering.

The performance of materials in the body can be viewed from several conceptual perspectives. First, we may consider biomaterials from the point of view of the problem area that is to be solved, as in Table 1-1. Second, we may consider the body on a tissue level, an organ level (Table 1-2), or a system level (Table 1-3). Third, we may consider the classification of materials as metals, polymers, ceramics, and composites as is done in Table 1-4. In that vein, the role of such materials as biomaterials is governed by the interaction between the

Table 1-1. Uses of Biomaterials

Problem area	Examples
Replacement of diseased or damaged part	Artificial hip joint, kidney dialysis machine
Assist in healing	Sutures, bone plates and screws
Improve function	Cardiac pacemaker, contact lens
Correct functional abnormality	Harrington spinal rod
Correct cosmetic problem	Augmentation mammoplasty, chin augmentation
Aid to diagnosis	Probes and catheters
Aid to treatment	Catheters, drains

Table 1-2. Biomaterials in Organs

Organ	Examples
Heart	Cardiac pacemaker, artificial heart valve
Lung	Oxygenator machine
Eye	Contact lens, eye lens replacement
Ear	Artificial stapes, cosmetic reconstruction of outer ear
Bone	Bone plate
Kidney	Kidney dialysis machine
Bladder	Catheter

material and the body, specifically, the effect of the body environment on the material, and the effect of the material on the body.

It should be evident in any of these perspectives that most current applications of biomaterials involve structural functions even in those organs and systems that are not primarily structural in their nature, or very simple chemical or electrical functions. Complex chemical functions such as those of the liver, and complex electrical or electrochemical functions such as those of the brain and sense organs, cannot be carried out by biomaterials. For the sake of completeness

Table 1-3. Biomaterials in Body Systems

System	Examples
Skeletal	Bone plate, total joint replacements
Muscular	Sutures
Digestive	Sutures
Circulatory	Artificial heart valves, blood vessels
Respiratory	Oxygenator machine
Integumentary	Sutures, burn dressings, artificial skin
Urinary	Catheters, kidney dialysis machine
Nervous	Hydrocephalus drain, cardiac pacemaker
Endocrine	Microencapsulated pancreatic islet cells
Reproductive	Augmentation mammoplasty, other cosmetic replacements

Table 1-4. Materials for Use in the Body

Materials	Advantages	Disadvantages	Examples
Polymers			
Nylon	Resilient	Not strong	Sutures, blood
Silicones	Easy to fabricate	Deform with time	vessels, hip socket,
Teflon®		May degrade	ear, nose, other
Dacron®			soft tissues
Metals			
Titanium	Strong, tough	May corrode	Joint replacement,
Stainless steels	Ductile	Dense	bone plates and
Co–Cr alloys			screws, dental
Gold			root implants
Ceramics			
Aluminum oxide	Very biocompatible,	Brittle	Dental; hip socket
Carbon	inert	Difficult to make	
Hydroxyapatite	Strong in	Not resilient	
	compression		
Composites			
Carbon–carbon	Strong, tailor-made	Difficult to make	Joint implants;
			heart valves

in these areas, a chapter dealing with transplantation of organs and tissues has been included.

1.1. HISTORICAL BACKGROUND

The use of biomaterials did not become practical until the advent of aseptic surgical technique as developed by Lister in the 1860s. Earlier surgical procedures, whether they involved biomaterials or not, were generally unsuccessful as a result of infection. Problems of infection tend to be exacerbated in the presence of biomaterials, since the implant can provide a region inaccessible to the body's immunologically competent cells. The earliest successful implants, as well as a large fraction of modern ones, were in the skeletal system. Bone plates were introduced in the early 1900s to aid in the fixation of fractures. Many of these early plates broke as a result of unsophisticated mechanical design: they were too thin and had stress-concentrating corners. It was also discovered that materials such as vanadium steel, which were chosen for good mechanical properties, corroded rapidly in the body. Better designs and materials soon followed. Following the introduction of stainless steels and cobalt chromium alloys in the 1930s, greater success was achieved in fracture fixation, and the first joint replacement surgeries were performed. As for polymers, it was found that warplane pilots in World War II who were injured by fragments of plastic [polymethyl methacrylate (PMMA)] aircraft canopy did not suffer adverse chronic reactions from the presence of the fragments in the body. PMMA became widely used after that time for corneal replacement and for replacements of sections of damaged skull bones. Following further advances in materials and in surgical technique, blood vessel replacements were tried in the 1950s; and heart valve replacements and cemented joint replacements in the 1960s. Recent years have seen many further advances.

1.2. PERFORMANCE OF BIOMATERIALS

The success of a biomaterial in the body depends on factors such as the material properties, design, and *biocompatibility* of the material used, as well as other factors not under control of the engineer, including the technique used by the surgeon, the health and condition of the patient, and the activities of the patient. If we can assign a numerical value f to the probability of failure of an implant, then the *reliability* can be expressed as

$$r = 1 - f \tag{1-1}$$

If, as is usually the case, there are multiple modes of failure, the total reliability r_t is given by the product of the individual reliabilities $r_1 = (1 - f_1)$, etc.:

$$r_t = r_1 r_2 \cdots r_n \tag{1-2}$$

Consequently, even if one failure mode such as implant fracture is perfectly controlled so that the corresponding reliability is unity, other failure modes such as infection could severely limit the utility represented by the total reliability of the implant.

One mode of failure that can occur in a biomaterial but not in engineering materials used in other contexts is an attack by the body's immune system on the implant. Another such failure mode is an unwanted effect of the implant upon the body—e.g., toxicity, inducing an inflammation, or causing cancer. Consequently, biocompatibility is included as a material requirement in addition to those requirements associated directly with the function of the implant. *Biocompatibility* involves the acceptance of an artificial implant by the surrounding tissues and by the body as a whole. Biocompatible materials do not irritate the surrounding structures, do not provoke an inflammatory response, do not incite allergic reactions, and do not cause cancer. Other characteristics that may be important in the function of an implant device made of biomaterials include adequate mechanical properties such as strength, stiffness, and fatigue properties; appropriate optical properties if the material is to be used in the eye, skin, or tooth; appropriate density; manufacturability; and appropriate engineering design.

The failure modes may differ in importance as time passes following the implant surgery. For example, consider the case of a total joint replacement in which infection is most likely soon after surgery, while loosening and implant

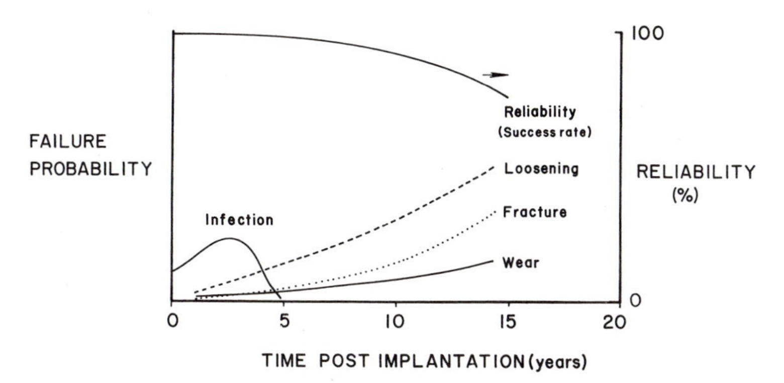

Figure 1-1. Schematic diagram of role of various failure modes as they depend on time for a joint replacement prosthesis. Not to scale. Failure modes of small probability, such as surgical error, allergic reaction to metal, not shown.

fracture become progressively more important as time goes on, as shown in Figure 1-1. Failure modes also depend on the type of implant and its location and function in the body. For example, an artificial blood vessel is more likely to cause problems by inducing a clot or becoming clogged with thrombus than by breaking or tearing mechanically.

Example 1-1

Suppose the probabilities of failure of a knee replacement in the first year are 5% for infection, 3% for wear, 2% for loosening, 1% for surgical complication, and 4% for fracture. Calculate the reliability of the implant during the first year. Suppose in addition 10% of the patients complain of excessive pain. Again calculate the reliability.

Answer

$r = (1 - 0.05)(1 - 0.03)(1 - 0.02)(1 - 0.01) = \underline{0.89}$, so that 11% or about one of ten procedures will not be satisfactory. If we have, in addition, a 10% pain problem, then $r = 0.89(1 - 0.10) = \underline{0.80}$. We remark that the foregoing is under the assumption that the failure processes are essentially independent. In reality, pain may follow from loosening or it may have no identifiable cause.

PROBLEMS

1-1. Determine the probability of failure of a hip replacement arthroplasty after 1 year and after 10 years, assuming the following. t is in years.

Infection:	$f_{\mathrm{i}} = 0.05 \exp(-t)$
Loosening:	$f_{\mathrm{lo}} = 0.01 \exp(+0.2t)$
Fracture:	$f_{\mathrm{fr}} = 0.01 \exp(+0.12t)$
Wear:	$f_{\mathrm{w}} = 0.01 \exp(+0.1t)$
Surgical error:	$f_{\mathrm{su}} = 0.001$

1-2. How would the failure modes shown in Figure 1-1 differ if an obsolete material such as vanadium steel were used to make the hip implant?

1-3. Discuss the feasibility and implications of replacing the entire arm.

1-4. Discuss the ethical problems associated with using fetal brain tissue for transplantation purposes to treat Parkinson's disease; or fetal bone marrow to treat leukemia.

SYMBOLS/DEFINITIONS

Latin Letters

f — Probability of failure.
r — Reliability or probability of success; $r = 1 - f$.

Terms

Biocompatibility: Acceptance of an artificial implant by the surrounding tissues and by the body as a whole.
Biological material: A material produced by a biological system.
Biomaterial: A synthetic material used to replace part of a living system or to function in intimate contact with living tissue.

BIBLIOGRAPHY

L. L. Hench and E. C. Ethridge, *Biomaterials: An Interfacial Approach*, Academic Press, New York, 1982.
W. Lynch, *Implants: Reconstructing the Human Body*, Van Nostrand-Reinhold Co., Princeton, N.J., 1982.
J. B. Park, *Biomaterials Science and Engineering*, Plenum Press, New York, 1984.
L. Stark and G. Agarwal (eds.), *Biomaterials*, Plenum Press, New York, 1969.
J. G. Webster (ed.), *Encyclopedia of Medical Devices and Instrumentation*, J. Wiley and Sons, New York, 1988.
D. F. Williams and R. Roaf, *Implants in Surgery*, W. B. Saunders Co., Philadelphia, 1973.

CHAPTER 2

THE STRUCTURE OF SOLIDS

The properties of a material are determined by its structure and chemical composition. Since chemical behavior depends ultimately upon the internal structural arrangement of the atoms, all material properties may be attributed to structure. Structure occurs on many levels of scale. These scales may be somewhat arbitrarily defined as the atomic or molecular (0.1–1 nm), nanoscale or ultrastructural (1 nm–1 μm), microstructural (1 μm–1 mm), and macrostructural (>1 mm). In pure elements, alloys, ceramics, and in polymers, the major structural features are on the atomic/molecular scale. Polycrystalline materials such as cast metals consist of grains, which may be quite large; however, the boundaries between the grains are atomic-scale features.

The first five sections of this chapter deal with the atomic/molecular structural aspects of materials. These sections constitute a review for those readers who have had some exposure to materials science. The final section deals with larger scale structure associated with composite and cellular materials.

This and the next two chapters give a brief review on the background of the materials science and engineering mostly pertinent to the subsequent chapters. Any reader who cannot comprehend Chapters 2–4 should review or study the basic materials science and engineering texts such as those given in the Bibliography.

2.1. ATOMIC BONDING

All solids are made up of atoms that are held together with the interaction of the outermost (*valence*) electrons. The valence electrons can move freely in the solid but can only exist in certain stable patterns within the confines of the solid. The nature of the patterns varies according to the *ionic, metallic,* or *covalent* bonding. In metallic bonds the electrons are *loosely* held to the ions, which makes the bond nondirectional. Therefore, in many metals it is easy for *plastic deformations* to occur (i.e., the ions can rearrange themselves permanently to the applied external forces). The ionic bonds are formed by exchanging electrons between metallic and nonmetallic atoms. The metallic atoms, such as Na, donate electrons and become positive ions (Na^+), while the nonmetallic atoms, such as Cl, receive electrons and become negative ions (Cl^-). The valence electrons are much more likely to be found in the space around the negative ions than the positive ions, thus making the bonds very directional. The structures of ionic solids are limited in their atomic arrangement due to the strong repulsive forces of like ions. Therefore, the positive ions are surrounded by the negative ions, and vice versa. The covalent bonds are formed when atoms share the valence electrons to satisfy their partially filled electronic orbitals. The greater the overlap of the valence orbitals or shells, the stronger the bonds become, but bond strength is limited by the strong repulsive forces between nuclei. Covalent bonds are also highly directional and strong as can be attested by diamond, which is the hardest material known.

In addition to the primary bonds, there are *secondary bonds*, which can be a major factor contributing to the material properties. Two major secondary bonds are the hydrogen and van der Waals bonds. The *hydrogen* bonds can arise when the hydrogen atom is covalently bonded to an electronegative atom so that it becomes a positive ion. The electrostatic force between them can be substantial since the hydrogen ion is quite small and can approach the negative ion very closely. The *van der Waals* forces arise when electrons are not distributed equally among ions which can form dipoles. The dipole–dipole interactions do not give rise to directional bonds and the effect is over a short distance. These bonds are much weaker than the hydrogen bonds as given in Table 2-1.

Although we have categorized the bonds as discussed above, the real materials may show some combinations of the bonding characteristics. For example, silicon atoms share electrons covalently but a fraction of the electrons can be freed and permit limited conductivity (semiconductivity). Thus, the silicon has covalent as well as some metallic bonding characteristics as shown in Figure 2-1.

Table 2-1. Strengths of Different Chemical Bonds as Reflected in Their Heat of Vaporization[a]

Bond type	Substance	Heat of vaporization (kJ/mol)
van der Waals	N_2	13
Hydrogen	Phenol	31
	HF	47
Metallic	Na	180
	Fe	652
Ionic	NaCl	1062
	MgO	1880
Covalent	Diamond	1180
	SiO_2	2810

[a] From B. Harris and A. R. Bunsell, *Structure and Properties of Engineering Materials*, Longmans, London, 1977, p. 21.

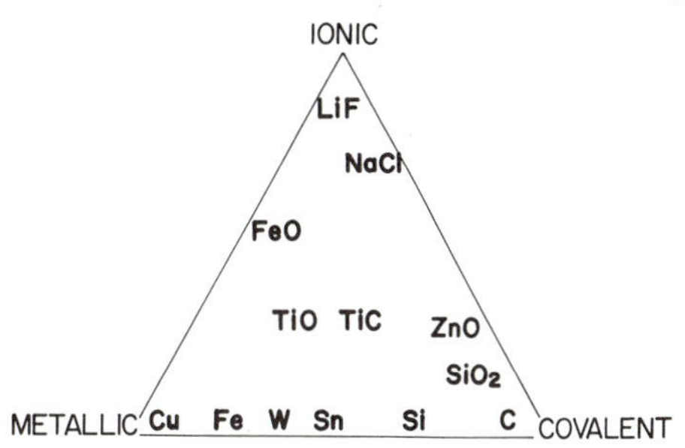

Figure 2-1. Most materials possess a combination of different bonds, making a generalization of bonding difficult.

2.2. CRYSTAL STRUCTURE

2.2.1. Atoms of the Same Size

The arrangement of atoms can be treated as an arrangement of hard spheres in view of their maintenance of characteristic equilibrium distances (bond length). The measurement of this distance is done by using X rays, which have short wavelengths, of the order of 1 angstrom ($\text{Å} = 10^{-10}$ m), approaching the atomic radius. When the atoms are arranged in a regular array, the structure can be represented by a *unit cell*, which has a characteristic dimension, the *lattice constant*, a, as shown in Figure 2-2. If this atomic structure is extended into three dimensions, the corresponding crystal structure will be cubic. This is a simple cubic space lattice, which is one of the three types of cubic crystals.

The *face-centered cubic* (fcc) structure is another cubic crystal as shown in Figure 2-3. This structure is called *close-packed* (actually it should be called the closest packed) in three dimensions. Because each atom touches 12 neighbors [hence the *coordination number* (CN) = 12] rather than 6 as in simple cubic, it results in a most efficiently packed structure. The *hexagonal close-packed* (hcp) structure is arranged by repeating layers of every other plane, that is, the atoms in the third layer occupy sites directly over the atoms in the first layer as shown in Figure 2-4. This can be represented as ABAB . . . while the fcc structure can be represented by three layers of planes ABCABC . . . . Both hcp and fcc have the same packing efficiency (74%); both structures have the most efficient packed planes of atoms with the same coordination number.

Another cubic structure is the *body-centered cubic* (bcc) in which an atom is located in the center of the cube as shown in Figure 2-5. This structure has a

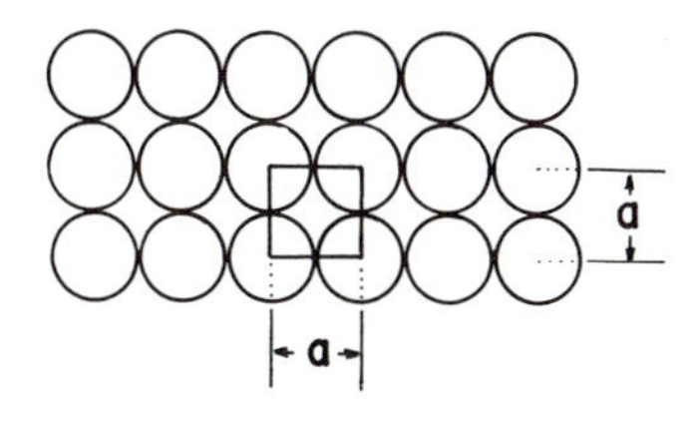

Figure 2-2. Stacking of hard balls (atoms) in simple cubic structure (a is the lattice spacing).

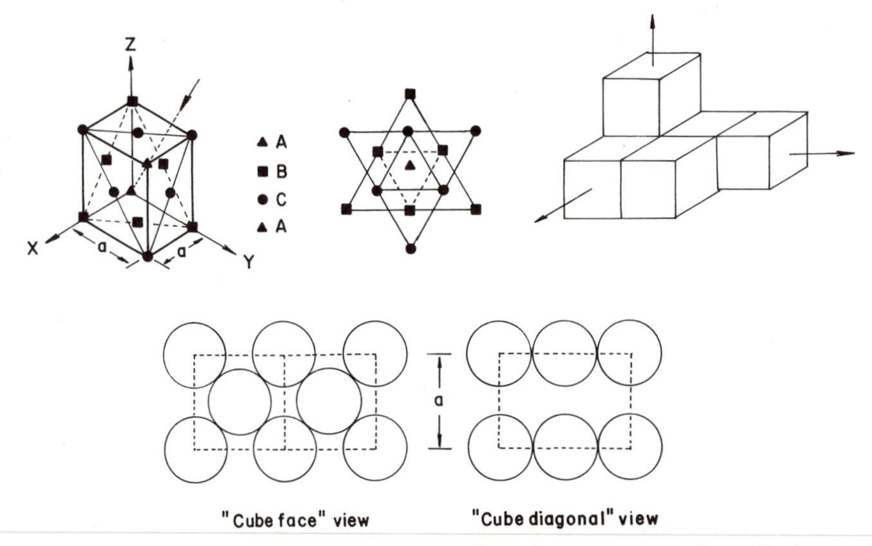

Figure 2-3. The face-centered cubic structure. Note the arrangement of atomic planes.

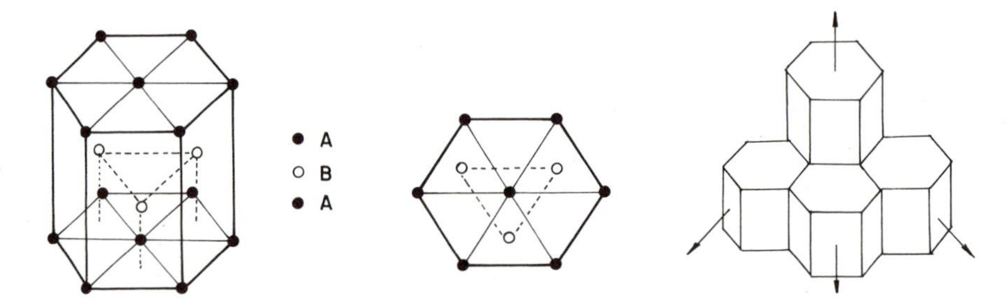

Figure 2-4. The hexagonal close-packed structure. Compare the arrangement of atomic planes with the fcc structure.

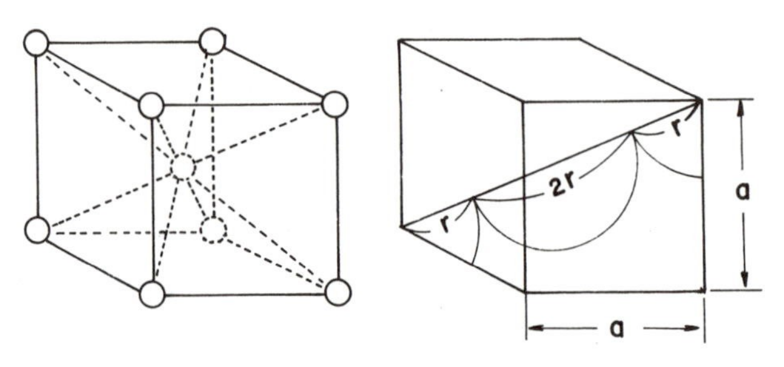

Figure 2-5. Body-centered cubic unit cell. Note that each unit cell has two atoms and $4r = a/\sqrt{3}$.

lower packing efficiency (68%) than the fcc structure. Other crystal structures include orthorhombic, in which the unit cell is a rectangular parallelepiped with unequal sides; hexagonal, with hexagonal prisms as unit cells; monoclinic, in which the unit cell is an oblique parallelepiped with one oblique angle and unequal sides; and triclinic, in which the unit cell has unequal sides and all oblique angles. Some examples of the crystal structures of real materials are given in Table 2-2.

Example 2-1

Iron (Fe) has a bcc structure at room temperature with atomic radius of 1.24 Å. Calculate its density (atomic weight of Fe is 55.85 g/mol).

Answer

From Figure 2-5, $a = 4r/\sqrt{3}$ and the density (ρ) is given by

$$r = \frac{\text{weight/unit cell}}{\text{volume/unit cell}}$$

$$= \frac{2 \text{ atoms/u.c.} \times 55.85 \text{ g/mol}}{(4 \times 1.24/\sqrt{3} \times 10^{-24} \text{ cm})^3/\text{u.c. } 6.02 \times 10^{23} \text{ atoms/mol}}$$

$$= \underline{7.87 \text{ g/cm}^3}$$

2.2.2. Atoms of Different Size

We seldom use pure materials for implants. Most of the materials used for implants are made of more than two elements. When two or more different sizes

Table 2-2. Examples of Crystal Structures

Material	Crystal structure
Cr	bcc
Co	hcp (below 417°C)
	fcc (above 417°C)
Fe	
Ferrite (α)	bcc (below 912°C)
Austenite (γ)	fcc (912–1394°C)
Delta iron (δ)	bcc (above 1394°C)
Mo	bcc
Ni	fcc
Ti	hcp (below 900°C)
	bcc (above 900°C)
Rock salt (NaCl)	fcc
Alumina (Al_2O_3)	hcp
Polyethylene	orthorhombic
Polyisoprene	orthorhombic

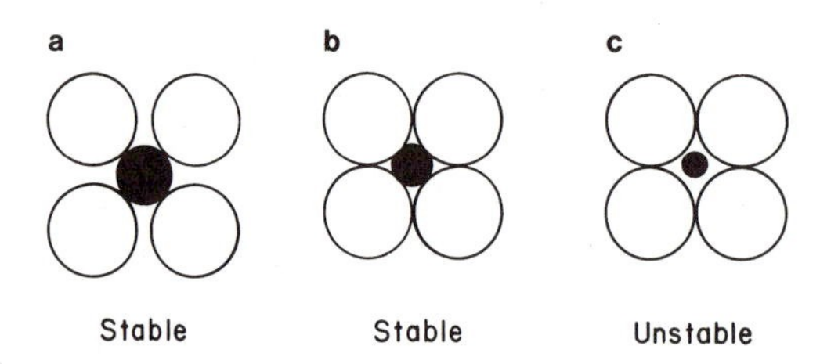

Figure 2-6. Possible arrangements of interstitial atoms. The critical radius ratio for (b) is given by $r + R = \sqrt{2}\ R$, hence $r/R = 0.414$ as given in Figure 2-7.

Structural Geometry	Radius Ratio	Coordination Number
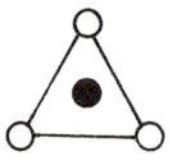	0.155	3
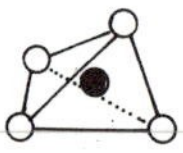	0.225	4
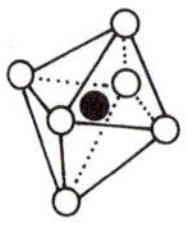	0.414	6
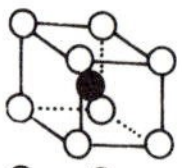	0.732	8
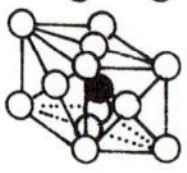	1.0	12

Figure 2-7. Minimum radius ratios and coordination numbers.

of atoms are mixed together in a solid, two factors must be considered: (1) the type of site and (2) the number of sites occupied. Consider the stability of the structure shown in Figure 2-6. In Figure 2-6a and b the interstitial atoms touch the larger atoms and hence they are stable, but they are not stable in c. At some critical value the interstitial atom will fit the space between six atoms (only four atoms are shown in two dimensions), which will give the maximum interaction between atoms and consequently the most stable structure results. Thus, at a certain radius ratio of the host and interstitial atoms the arrangement will be most stable. Figure 2-7 gives the minimum radius ratios for given coordination numbers. Note that these radius ratios are determined solely by geometric considerations.

Example 2-2

Calculate the minimum radius ratios for a coordination number of 6.

Answer

For coordination number 6, from Figure 2-7 looking down from the diagram below showing a two-dimensional representation of a structure with coordination number 6, we have

$$\cos 45^\circ = \frac{R}{R+r}, \qquad \frac{1}{\sqrt{2}} = \frac{R}{R+r}, \qquad \sqrt{2}R = R + r, \qquad \frac{r}{R} = \sqrt{2} - 1 = \underline{0.414}$$

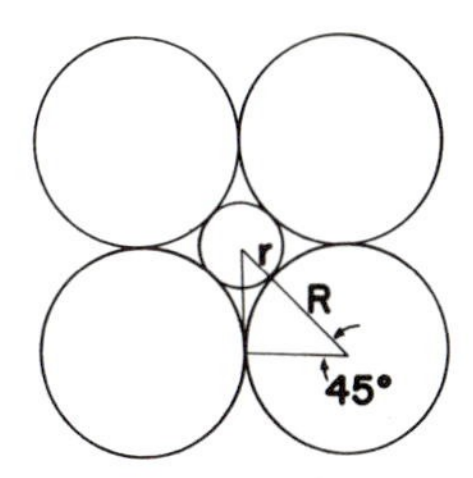

2.3. IMPERFECTIONS IN CRYSTALLINE STRUCTURES

The imperfections in crystalline solids are sometimes called *defects* and they play a major role in determining their physical properties. *Point defects* commonly appear as lattice vacancies, substitutional or interstitial atoms as shown in Figure 2-8. The interstitial or substitutional atoms are sometimes called alloying elements if put in intentionally, and impurities if they are unintentional.

The *line defects* are created when an extra plane of atoms is displaced or dislocated out of its regular lattice space registry (Figure 2-9). The line defects or dislocations will lower the strength of a solid crystal enormously since it takes much less energy to move or deform a whole plane of atoms one atomic distance at a time rather than all at once. This is analogous to moving a carpet on the floor or a heavy refrigerator. The carpet cannot be pulled easily if one tries to move an inch all at once but if one folds it and follows the fold until the fold reaches to the other end, it can be moved without too much force. Similarly, the refrigerator can be moved easily if one puts one or two logs under it. However, if one puts too many logs (analogous to dislocations) underneath, it becomes harder to move the refrigerator. Correspondingly, if a lot of dislocations are

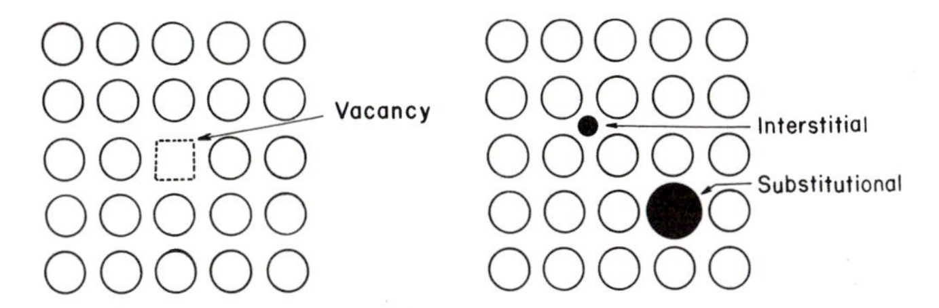

Figure 2-8. Point defects in the form of vacancies and interstitials.

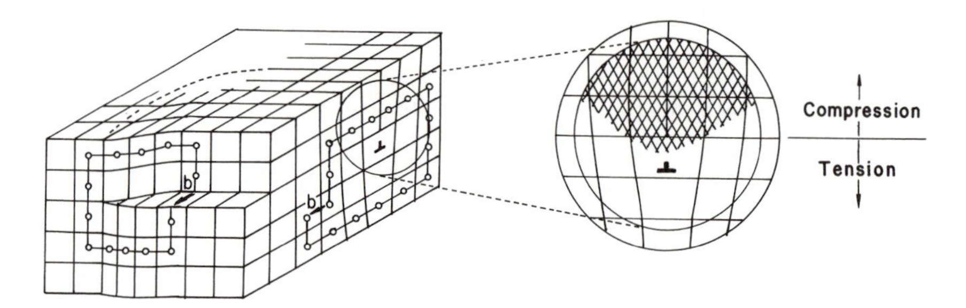

Figure 2-9. Line defects. The displacement is perpendicular to the edge dislocation but parallel to the screw dislocation (left-hand side). The unit length *b* has magnitude (one lattice space) and direction and is called a Burger's vector.

introduced in a solid, the strength increases considerably. The reason is that the dislocations become entangled with each other impeding their movement. The blacksmith practices this principle when he heats a horseshoe red-hot and hammers it. The hammering introduces dislocations. He has to repeat the heating and hammering process in order to increase the number of dislocations without breaking the horseshoe.

Planar defects exist at the grain boundaries. The *grain boundaries* are created when two or more crystals are mismatched at the boundaries; this occurs during crystallization. Within each grain all of the atoms are in a lattice of one specific orientation. Other grains have the same crystal lattice but different orientations

Figure 2-10. Midsection of a femoral component of a hip joint implant that shows grains (Co–Cr alloy). Notice the size distribution along the stem and from the core to the surface.

creating a region of mismatch. The grain boundary is less dense than the bulk, hence most diffusion of gas or liquid takes place along the grain boundaries. Grain boundaries can be seen by polishing and subsequent etching of a "polycrystalline" material. This is due to the fact that the grain boundary atoms possess higher energy than the bulk, resulting in a more chemically reactive site at the boundary. Figure 2-10 shows a polished surface of a metal implant. The size of the grains plays an important role in determining physical properties of a material. In general, a fine-grained structure is stronger than a coarse one for a given material at a low recrystallization temperature since the former contains more grain boundaries, which in turn interfere with the movement of atoms during deformation resulting in a stronger material.

2.4. LONG-CHAIN MOLECULAR COMPOUNDS (POLYMERS)

Polymers have very long-chain molecules that are formed by covalent bonding along the backbone chain. The long chains are held together either by secondary bonding forces such as van der Waals and hydrogen bonds or by primary covalent bonding forces through cross-links between chains. The long chains are very flexible and can be tangled easily. In addition, each chain can have side groups, branches, and copolymeric chains or blocks, which can also interfere with the long-range ordering of chains. For example, paraffin wax has the same chemical formula as polyethylene [$(CH_2CH_2)_n$], but will crystallize almost completely because of its much shorter chain lengths. However, when the chains become extremely long [from 40 to 50 *repeating units* ($-CH_2CH_2-$) to several thousands as in linear polyethylene], they cannot be crystallized completely (up to 80–90% crystallization is possible). Also, branched polyethylene in which side chains are attached to the main backbone chain at positions that a hydrogen atom normally occupies, will not crystallize easily due to the *steric hindrance* of side chains resulting in a more noncrystalline structure. The partially crystallized structure is called *semicrystalline*, which is the most commonly occurring structure for linear polymers. The semicrystalline structure is represented by disordered noncrystalline regions and ordered crystalline regions, which may contain folded chains as shown in Figure 2-11.

The degree of polymerization (DP) is one of the most important parameters in determining polymer properties. It is defined as average number of mers or

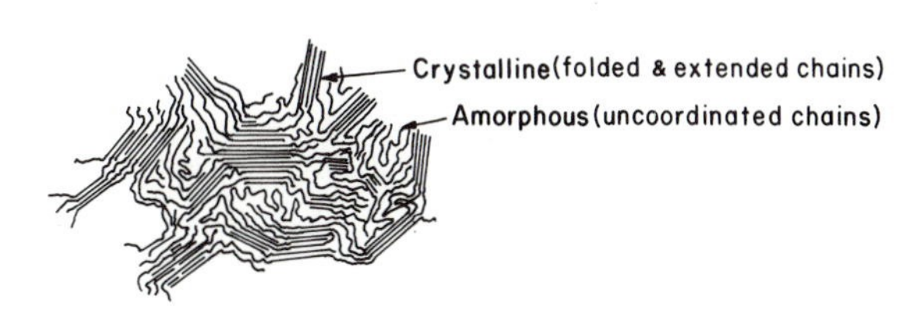

Figure 2-11. Fringed (micelle) model of linear polymer with semicrystalline structure.

repeating units per molecule, i.e., chain. Each chain may have a small or large number of mers depending on the condition of polymerization. Also the length of each chain may be different. Therefore, we deal with the *average* degree of polymerization or average molecular weight (M). The relationship between molecular weight and degree of polymerization can be expressed as

$$M = \text{DP} \times \text{molecular weight of mer (or repeating unit)} \tag{2-1}$$

The average molecular weight can be calculated according to the weight fraction (W_i) in each molecular weight fraction (MW_i):

$$M = \frac{\sum W_i \cdot MW_i}{\sum W_i} = \sum W_i MW_i \tag{2-2}$$

since $\sum W_i = 1$. This is *weight* average molecular weight.

As the molecular chains become longer by the progress of polymerization, their relative mobility decreases. The chain mobility is also related to the physical properties of the final polymer. Generally, the higher the molecular weight, the lesser the mobility of chains, which results in higher strength and greater thermal stability. The polymer chains can be arranged in three ways; linear, branched, and a cross-linked or three-dimensional network as shown in Figure 2-12. Linear polymers such as polyvinyls, polyamides, and polyesters are much easier to crystallize than the cross-linked or branched polymers. However, they cannot be crystallized 100% as with metals. Instead they become semicrystalline polymers.

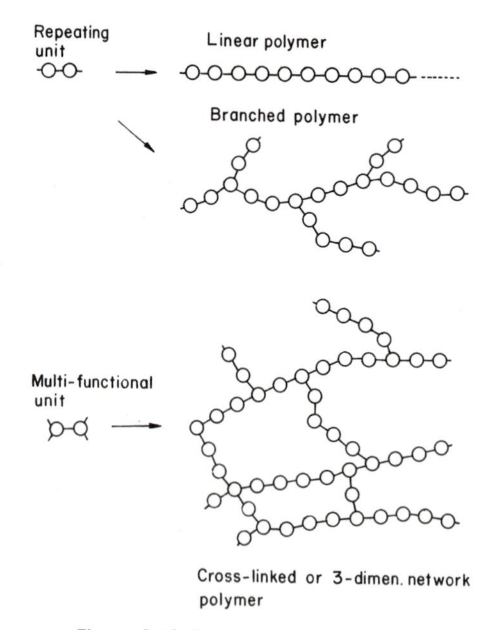

Figure 2-12. Types of polymer chains.

The arrangement of chains in crystalline regions is believed to be a combination of *folded* and *extended* chains. The chain folds, which are seemingly more difficult to form, are necessary to explain observed single-crystal structures in which the crystal thickness is too small to accommodate the length of the chain as determined by electron and X-ray diffraction studies. Figure 2-13 shows the two-dimensional representation of chain arrangements. The classical "fringed micelle" model (Figure 2-11) in which the amorphous and crystalline regions coexist has been modified to include chain folds in the crystalline regions.

The cross-linked or three-dimensional network polymers such as (poly)phenolformaldehyde cannot be crystallized at all and they become noncrystalline, amorphous polymers.

Vinyl polymers have a repeating unit $—CH_2—CHX—$ where X is some monovalent side group. There are three possible arrangements of side groups (X): (1) atactic, (2) isotactic, and (3) syndiotactic as shown in Figure 2-14. In *atactic* arrangements the side groups are randomly distributed while in *syndiotactic and isotactic* arrangements they are either in alternating positions or in one side of the main chain. If side groups are small like polyethylene (X = H) and the chains are linear, the polymer crystallizes easily. However, if the side groups are large as in polyvinyl chloride (X = Cl) and polystyrene ($X = C_6H_6$, benzene ring) and are randomly distributed along the chains (atactic), then a noncrystalline structure will be formed. The isotactic and syndiotactic polymers usually crystallize even when the side groups are large. Note that polyethylene does not have tacticity since it has symmetric side groups.

Copolymerization, in which two or more homopolymers (one type of repeating unit throughout its structure) are chemically combined, always disrupts the

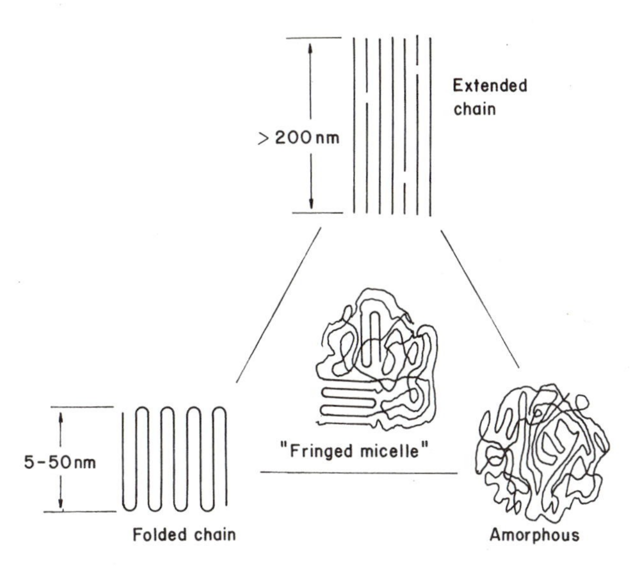

Figure 2-13. Two-dimensional representation of polymer solid.

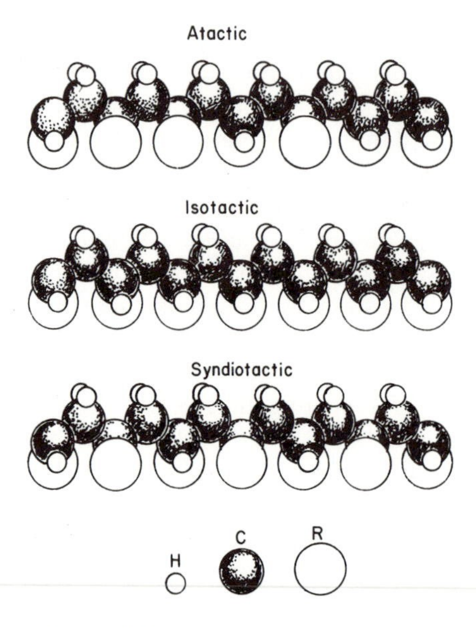

Figure 2-14. Tacticity of vinyl polymers.

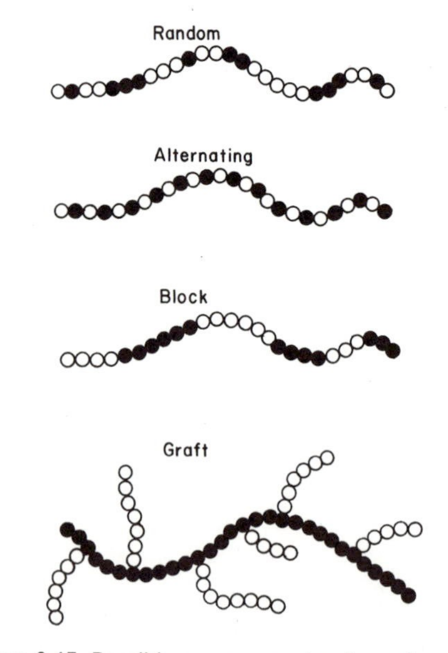

Figure 2-15. Possible arrangements of copolymers.

regularity of polymer chains, thus promoting the formation of noncrystalline structure (Figure 2-15). The addition of *plasticizers* to prevent crystallization by keeping the chains separated from one another will result in a noncrystalline version of a polymer that normally crystallizes. An example is celluloid, which is made of normally crystalline nitrocellulose plasticized with camphor. Plasticizers are also used to make rigid noncrystalline polymers like polyvinyl chloride (PVC) into a more flexible solid (a good example is Tygon® tubing).

Elastomers or rubbers are polymers that exhibit large stretchability at room temperature and can snap back to their original dimensions when the load is released. The elastomers are noncrystalline polymers that have an intermediate structure consisting of long-chain molecules in three-dimensional networks (see next section for more details). The chains also have "kinks" or "bends" in them that straighten when a load is applied. For example, the chains of *cis*-polyisoprene (natural rubber) are bent at the double bond due to the methyl group interfering with the neighboring hydrogen in the repeating unit $[-CH_2-C(CH_3)=CH-CH_2-]$. If the methyl group is on the opposite side of the hydrogen, then it becomes *trans*-polyisoprene, which will crystallize due to the absence of the steric hindrance present in the *cis* form. The resulting polymer is a very rigid solid called gutta-percha, which is not an elastomer.

Below the *glass transition temperature* (T_g; second-order transition temperature between viscous liquid and solid), natural rubber loses its compliance and becomes a glasslike material. Therefore, to be flexible, all elastomers should have T_g well below room temperature. What makes the elastomers not behave like liquids above T_g is in fact due to the *cross-links* between chains, which act as pinning points. Without cross-links the polymer would deform permanently. An example is latex, which behaves as a viscous liquid. Latex can be cross-linked with sulfur (*vulcanization*) by breaking double bonds (C=C) and forming C---S—S---C bonds between the chains. The more cross-links are introduced, the more rigid the structure becomes. If all of the chains are cross-linked together, the material will become a three-dimensional rigid polymer.

2.5. SUPERCOOLED AND NETWORK SOLIDS

Some solids such as window glass do not have a regular crystalline structure. Solids with such an atomic structure are called *amorphous* or noncrystalline materials. They are usually supercooled from the liquid state and thus retain a liquidlike molecular structure. Consequently, the density is always less than that of the crystalline state of the same material indicating inclusion of some voids (*free volume*, Figure 2-16). Due to the quasi-equilibrium state of the structure, the amorphous material tends to crystallize. It is also more brittle and less strong than the crystalline counterpart. It is very difficult to make metals amorphous since the metal atoms are extremely mobile. The ceramics and polymers can be made amorphous because of the relatively sluggish mobility of their molecules.

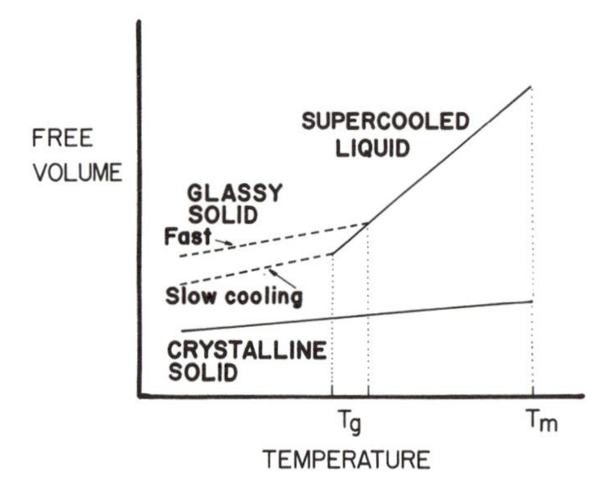

Figure 2-16. Change of volume versus temperature of a solid. The glass transition temperature (T_g) depends on the rate of cooling, and below T_g the material behaves as a solid like a window glass.

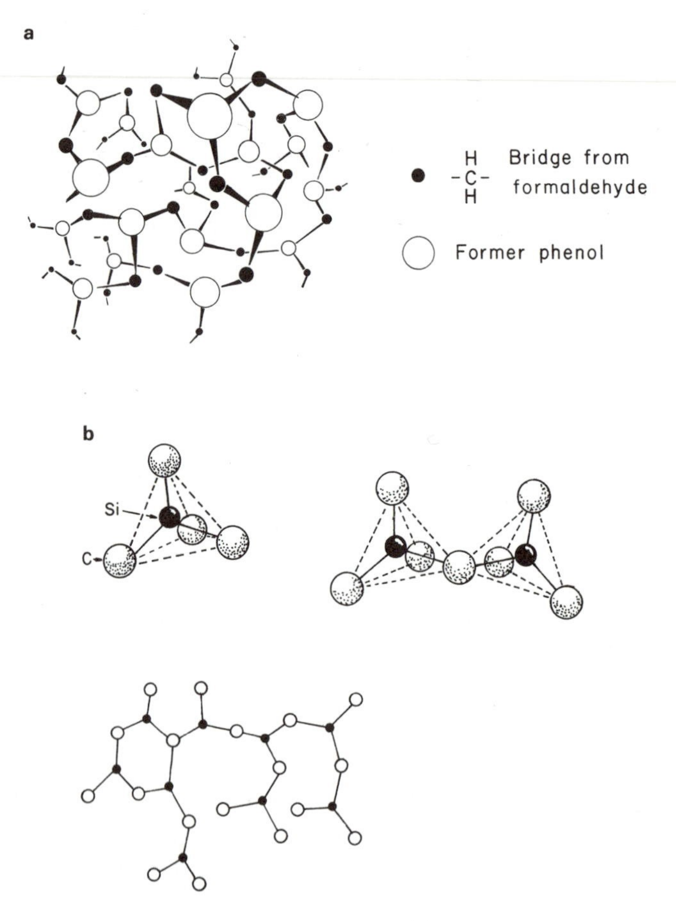

Figure 2-17. Network structure of noncrystalline solids: (a) phenolformaldehyde (Bakelite®); (b) silica glass. The subunit SiO_4 is a tetrahedron with a silicon atom in the center.

The *network structure* of a solid results in a three-dimensional, amorphous structure since the restrictions on the bonds and rigidity of subunits prevent them from crystallizing. Common network structure materials are phenolformaldehyde (Bakelite®) polymer and silica (SiO_2) glass as shown in Figure 2-17. Network structure of phenolformaldehyde is formed by *cross-linking* through the phenol rings while the Si—O tetrahedra are joined corner to corner via oxygen atoms. The three-dimensional network solids do not flow at high temperatures.

Example 2-3

Calculate the free volume of 100 g of iodine supercooled from the liquid state, which has a density of 4.8 g/cm^3. Assume the density of amorphous iodine is 4.3 g/cm^3 and crystalline density is 4.93 g/cm^3.

Answer

The fraction of supercooled iodine can be calculated by extrapolation. We consider the density to be proportional to crystallinity and calculate the slope of the line:

$$\frac{4.93 - 4.8}{4.93 - 4.3} = 0.21$$

The weight of the supercooled liquid is $0.21 \times 100\text{ g} = 21\text{ g}$; therefore, the total free volume is

$$\left(\frac{1}{4.3}\frac{\text{cm}^3}{\text{g}} - \frac{1}{4.93}\frac{\text{cm}^3}{\text{g}}\right) \times 21\text{ g} = \underline{0.65\text{ cm}^3}$$

Upon complete crystallization, the volume of the iodine will decrease by 0.65 cm^3.

2.6. COMPOSITE MATERIAL STRUCTURE

Composite materials are those that consist of two or more distinct parts. The term *composite* is usually reserved for those materials in which the distinct phases are separated on a scale larger than the atomic, and in which properties such as the elastic modulus are significantly altered in comparison with those of a homogeneous material. Accordingly, bone and fiberglass are viewed as composite materials, but alloys such as brass, or metals such as steel with carbide particles are not. Although many engineering materials, including biomaterials, are not composites, virtually all natural biological materials are composites.

The properties of a composite material depend upon the shape of the inhomogeneities (second phase material), upon the volume fraction occupied by them, and upon the stiffness and integrity of the interface between the constituents. The shape of the inhomogeneities in a composite material are classified as follows. Possible inclusion shapes are the particle, with no long dimension; the fiber, with

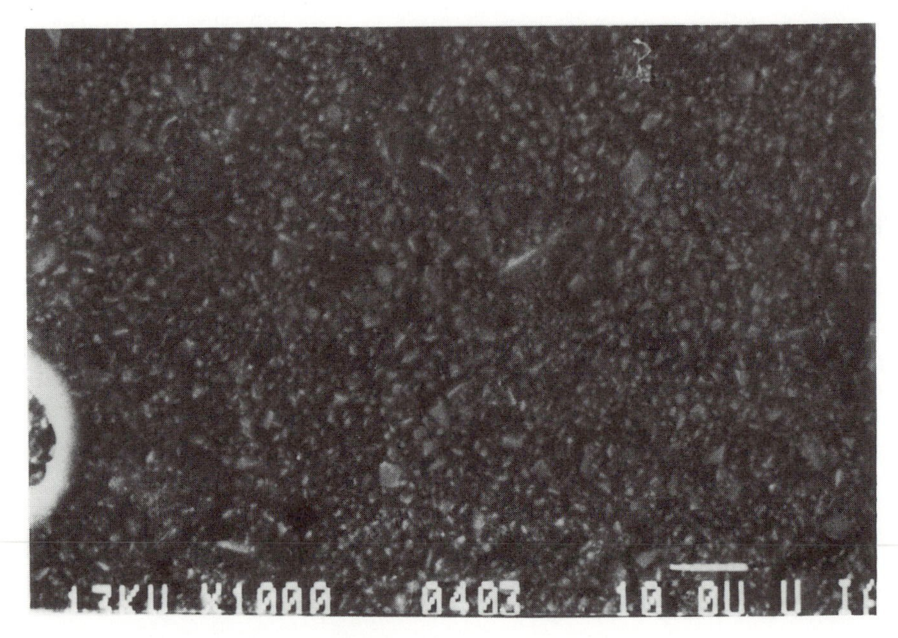

Figure 2-18. Dental composite. The particles are silica (SiO_2) and the matrix is polymeric. D. Boyer, unpublished data, University of Iowa, 1989.

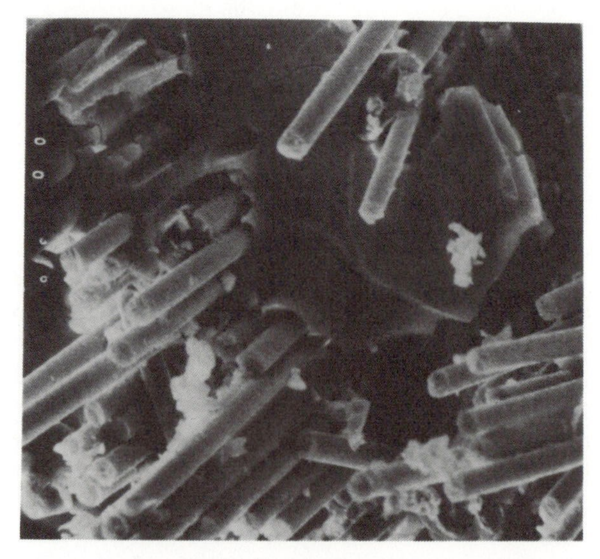

Figure 2-19. Glass-fiber-epoxy composite: fracture surface showing fiber pullout. From B. D. Agarwal and L. J. Broutman, *Analysis and Performance of Fiber Composites*, J. Wiley and Sons, New York, 1980.

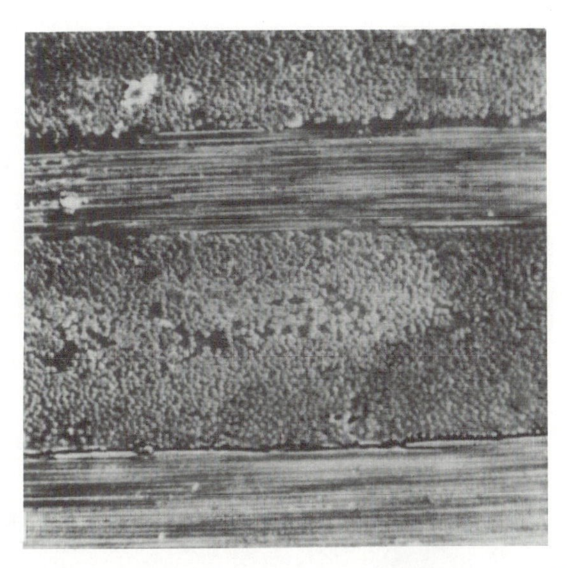

Figure 2-20. Structure of a cross-ply laminate. From B. D. Agarwal and L. J. Broutman, *Analysis and Performance of Fiber Composites*, J. Wiley and Sons, New York, 1980.

one long dimension; and the platelet or lamina, with two long dimensions. Cellular solids are those in which the "inclusions" are voids, filled with air or liquid. In the context of biomaterials, it is necessary to distinguish the above cells, which are structural, from biological cells, which occur only in living organisms.

Examples of composite material structures are as follows. The dental composite filling material shown in Figure 2-18 has a particulate structure. This material is packed into the tooth cavity while still soft, and the resin is polymerized *in situ.* The silica particles serve to provide hardness and wear resistance superior to that of the resin alone. A typical fibrous solid is shown in Figure 2-19. The fibers serve to stiffen and strengthen the polymeric matrix. In this example, pullout of fibers during fracture absorbs mechanical energy, conferring toughness on the material. Fibers have been added to the polymeric parts of total joint replacement prostheses, in an attempt to improve the mechanical properties. Figure 2-20 shows a laminated structure in which each lamina is fibrous while Figure 2-21 shows representative synthetic cellular materials.

The properties of composite materials depend very much upon structure, as they do in homogeneous materials. Composites differ in that the engineer has considerable control over the larger scale structure, and hence over the desired properties. The relevant structure–property relations will be developed in Chapter 8.

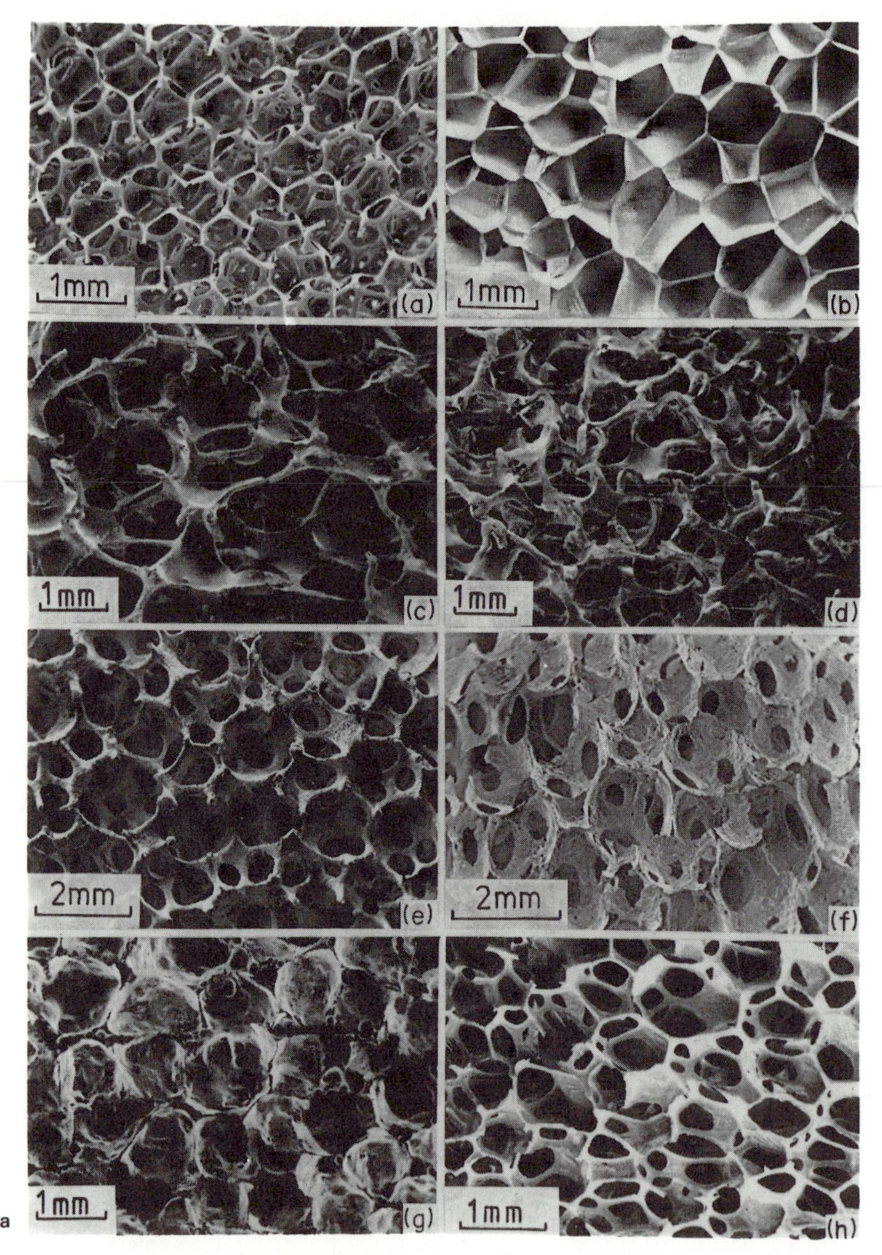

Figure 2-21. Synthetic (a) and natural (b) cellular materials. From L. J. Gibson and M. F. Ashby, *Cellular Solids*, Pergamon Press, Elmsford, N.Y., p. 16.

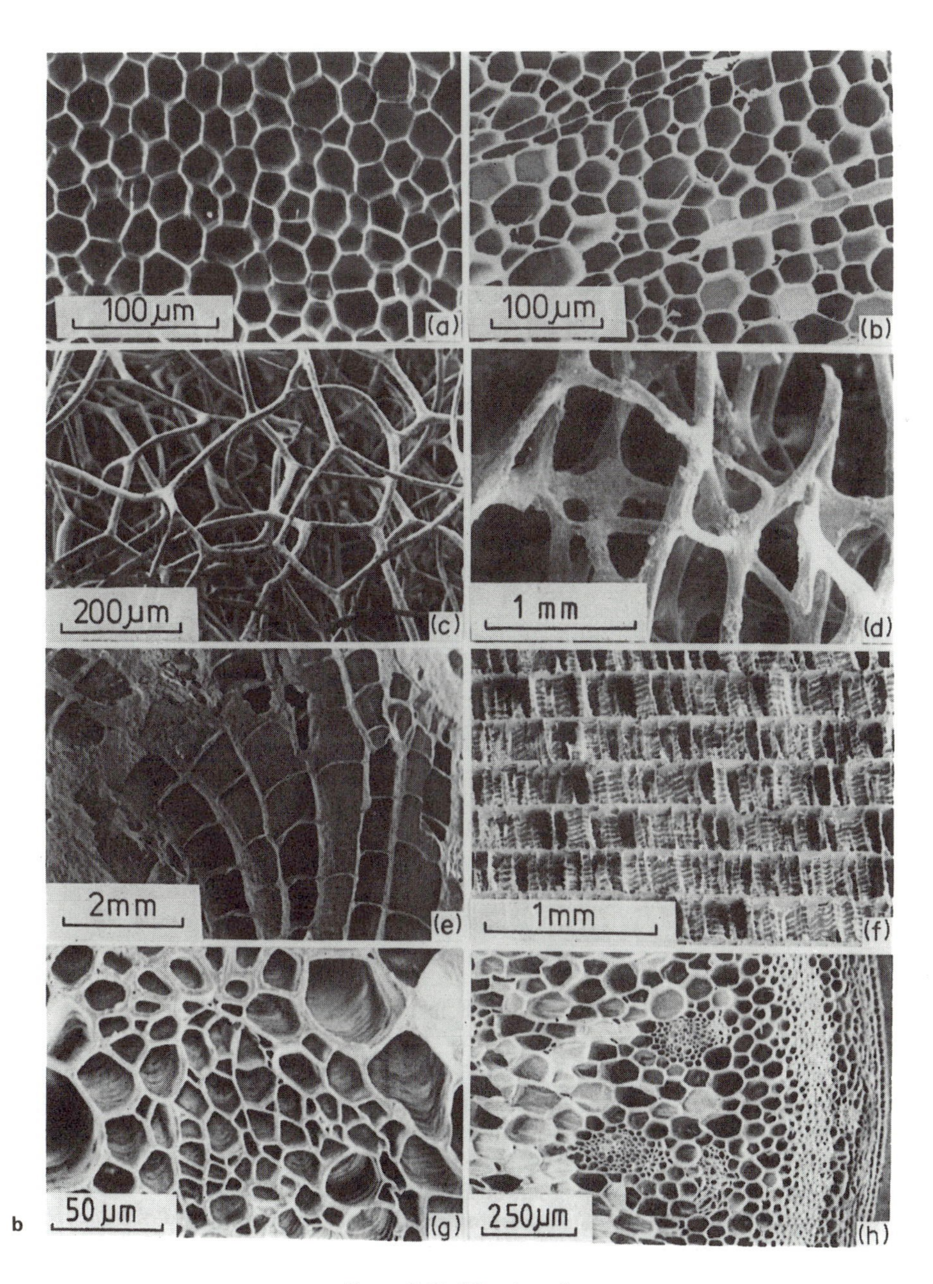

Figure 2-21. (*Continued*).

PROBLEMS

2-1. Identify the type of bond (ionic or covalent) in the following compounds: (a) ammonia (NH_3); (b) salt (NaCl); (c) carbon tetrachloride (CCl_4); (d) hydrogen peroxide (H_2O_2); (e) ozone (O_3); (f) ethylene ($CH_3{=}CH_3$); (g) water (H_2O); (h) magnesium oxide (MgO); (i) diamond (C).

2-2. Calculate the number of atoms present per cm^3 for alumina (Al_2O_3), which has a density of $3.8\ g/cm^3$.

2-3. Calculate the diameters of the smallest cations that have a 6-fold and 8-fold coordination with O^{2-} ions (see Table 6-1 and Figure 2-7).

2-4. Steel contains carbon which forms iron carbide (Fe_3C). Determine the weight percentage of carbon in Fe_3C.

2-5. A metal alloy of 92.5 wt% Ag and 7.5 wt% Cu is called sterling silver. Determine the atomic percentage of silver.

2-6. Titanium (Ti) is bcc above 882°C. The atomic radius increases 2% when the bcc structure changes to hcp during cooling. Determine the percentage volume change. Hint: there will be a change in the atomic packing factor.

2-7. The molecular weight of polymers can be either number or weight average (M_n or M_w), which are defined as

$$M_n = \frac{\Sigma (N_i M_i)}{\Sigma N_i} \quad \text{and} \quad M_w = \frac{\Sigma (W_i M_i)}{\Sigma W_i}$$

where N_i is the number of molecules with a molecular weight of M_i, M_i is the molecular weight of the ith species, and W_i is the weight fraction of molecules of the ith species. Show that

$$M_n = \frac{1}{\Sigma (W_i / M_i)}$$

2-8. Calculate the weight of an iron atom. The density of iron is $7.87\ g/cm^3$. Avogadro's number is 6.02×10^{23}. How many iron atoms are contained in a cubic centimeter?

2-9. A polyethylene is made of the following fractional distribution. Calculate M_n, M_w, and M_w/M_n (polydispersity).

W_i	0.10	0.20	0.30	0.30	0.10
M_i (kg/mol)	10	20	30	40	50

2-10. A fibrous material is made with all of the fibers parallel. What is the maximum volume fraction of fibers if the fibers have a circular cross section? Can you suggest any way to further increase the volume fraction of fibers?

SYMBOLS/DEFINITIONS

Greek Letter

ρ Density, mass per unit volume.

Latin Letters

a	Lattice constant, which is the spacing of atoms in a crystal lattice. For example, in cubic crystal systems all of the lattice constants are the same, i.e., $a = b = c$ so that the atomic spacing is the same in each principal direction.
bcc	Body-centered cubic lattice; one atom is positioned in the center of the cubic unit cell.
CN	Coordination number; number of atoms touching an atom in the middle.
fcc	Face-centered cubic lattice; one atom is positioned in each face of the cubic unit cell.
hcp	Hexagonal close-packed lattice; 12 atoms surround and touch a central atom of the same species, resulting in the hexagonal prism symmetry.
M_n	Number average molecular weight of a polymer.
M_w	Weight average molecular weight of a polymer.
r, R	Radius of an atom.
T_g	Glass transition temperature at which solidification without crystallization takes place from viscous liquid.

Terms

Composites	Materials obtained by combining two or more materials at a macroscale taking advantage of salient features of each material. An example is (high strength) fiber-reinforced epoxy resin.
Copolymers	Polymers made from two or more homopolymers; can be obtained by grafting, block, alternating, or random attachment of the other polymer segment.
Covalent bonding	Bonding of atoms or molecules by sharing valence electrons.
Elastomers	Rubbery materials. The restoring force comes from uncoiling or unkinking of coiled or kinked molecular chains. They can be highly stretched.
Free volume	The difference in volume occupied by the crystalline state (minimum) and noncrystalline state of a material for a given temperature and pressure.
Glass transition temperature	See T_g above.
Hydrogen bonding	A secondary bonding through dipole interactions in which a hydrogen ion is one of the dipoles.
Imperfections	Defects created in a perfect (crystalline) structure by vacancy, interstitial and substitutional atoms by the introduction of an extra plane of atoms (dislocations), or by mismatching at the crystals during solidification (grain boundaries).
Ionic bonding	Bonding of atoms or molecules through electrostatic interaction of positive and negative ions.
Lattice constant	The spacing of atoms in a crystal lattice. For example, in cubic crystal systems all of the lattice constants are the same, i.e., $a = b = c$, so that the atomic spacing is the same in each principal direction.
Metallic bonding	Bonding of atoms or molecules through loosely bound valence electrons around positive metal ions.
Minimum radius ratio (r/R)	The ratio between the radius of a smaller atom to be fitted into the space among the larger atoms' radius based on geometric consideration.

Packing efficiency	The (atomic) volume per unit volume (or space).
Plasticizer	Substance made of small molecules, mixed with (amorphous) polymers to make the chains slide more easily past each other making the polymer less rigid.
Polycrystalline	Structure of a material that is an aggregate of single crystals (grains).
Repeating unit	The smallest unit representing a polymer molecular chain.
Semiconductivity	Electrical conductivity of a semiconductor lies between that of a conductor and an insulator. There is an energy band gap on the order of 0.1 eV (Sn) to 6 eV (C, diamond), which governs the conductivity and its dependence on temperature.
Semicrystalline solid	Solid that contains both crystalline and noncrystalline regions and usually occurs in polymers due to their long-chain molecules.
Steric hindrance	Geometrical interference that restrains movements of molecular groups such as side chains and main chains of a polymer.
Tacticity	Arrangement of asymmetrical side groups along the backbone chain of polymers; groups can be distributed at random (atactic), one side (isotactic), or alternating (syndiotactic).
Unit cell	The smallest repeating unit of a space lattice representing the whole crystal structure.
Valence electrons	The outermost (shell) electrons of an atom.
van der Waals bonding	A secondary bonding arising through the fluctuating dipole–dipole interactions.
Vinyl polymers	Thermoplastic linear polymers synthesized by free radical polymerization of vinyl monomers having a common structure of $CH_2{=}CHR$.
Vulcanization	Cross-linking of rubbers by sulfur.

BIBLIOGRAPHY

F. W. Billmeyer, Jr., *Textbook of Polymer Science*, 3rd ed., Interscience Publ., J. Wiley and Sons, New York, 1984.

D. Flanagan (ed.), *Materials*, W. H. Freeman and Co., San Francisco, 1967.

R. H. Krock and M. L. Ebner, *Ceramics, Plastics and Metals*, Chapter 3, D. C. Heath and Co., Lexington, Mass., 1965.

W. G. Moffatt, G. W. Pearsall, and J. Wulff, *The Structure and Properties of Materials*, Vol. I, *Structures*, Chapters 1–3, J. Wiley and Sons, New York, 1964.

L. Pauling, *The Nature of the Chemical Bonding*, Chapter 2, Cornell University Press, Ithaca, N.Y., 1960.

M. J. Starfield and M. A. Shrager, *Introductory Materials Science*, McGraw–Hill Book Co., New York, 1972.

L. H. Van Vlack, *Elements of Materials Science and Engineering*, 6th ed., Chapters 1–3, Addison–Wesley Publ. Co., Reading, Mass., 1989.

CHAPTER 3

CHARACTERIZATION OF MATERIALS I

The characterization of materials is an important step to be taken before utilizing the materials for any purpose. Depending on the purpose, one can subject the material to mechanical, thermal, chemical, optical, electrical, and other characterizations to make sure that the material under consideration can function without failure for the life of the final product. We will consider only mechanical, thermal, and surface properties in this chapter while in the next chapter we will study electrical, optical, and diffusional properties.

3.1. MECHANICAL PROPERTIES

Among the most important properties for the application of materials in medicine and dentistry are the mechanical properties. We will study the fundamental mechanical properties that will be used in the later chapters.

3.1.1. Stress–Strain Behavior

For a material that undergoes a mechanical deformation, the *stress* is defined as a force per unit area, which is usually expressed in newtons per square meter (pascal, Pa) or pounds force per square inch (psi).

$$\text{Stress } (\sigma) = \frac{\text{force}}{\text{cross-sectional area}} \left[\frac{\text{N}}{\text{m}^2}\right] \text{ or } \left[\frac{\text{lbf}}{\text{in}^2}\right] \tag{3-1}$$

A load (or force) can be applied upon a material in *tension*, *compression*, and *shear* or any combination of these *forces* (or stresses). Tensile stresses are generated in response to loads (forces) that pull an object apart (Figure 3-1a), while compressive stresses tend to squeeze it together (Figure 3-1b). Shear stresses

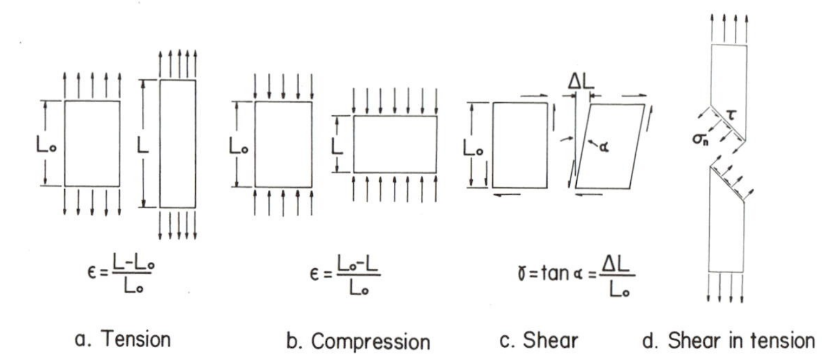

Figure 3-1. Three different modes of deformation. The shear stresses can be produced by tension or compression as in *d*.

resist loads that deform or separate by sliding layers of molecules past each other on one or more planes (Figure 3-1c). The shear stresses can also be found in uniaxial tension or compression since the applied stress produces the maximum shear stress on planes at 45° to the direction of loading (Figure 3-1d).

The deformation of an object in response to an applied load is called *strain*:

$$\text{Strain } (\varepsilon) = \frac{\text{deformed length} - \text{original length}}{\text{original length}} \left[\frac{\text{m}}{\text{m}}\right] \text{ or } \left[\frac{\text{in}}{\text{in}}\right] \tag{3-2}$$

It is also possible to denote strain by the stretch ratio, i.e., deformed length/original length. The deformations associated with different types of stresses are called tensile, compressive, and shear strain (see Figure 3-1).

If the stress-strain behavior is plotted on a graph, a curve that represents a continuous response of the material toward the imposed force can be obtained as shown in Figure 3-2. The stress-strain curve of a solid sometimes can be demarcated by the yield point (YS) into elastic and plastic regions. In the elastic region, the strain ε increases in direct proportion to the applied stress σ (Hooke's law):

$$\sigma = E\varepsilon : \text{stress} = (\text{initial slope})(\text{strain}) \tag{3-3}$$

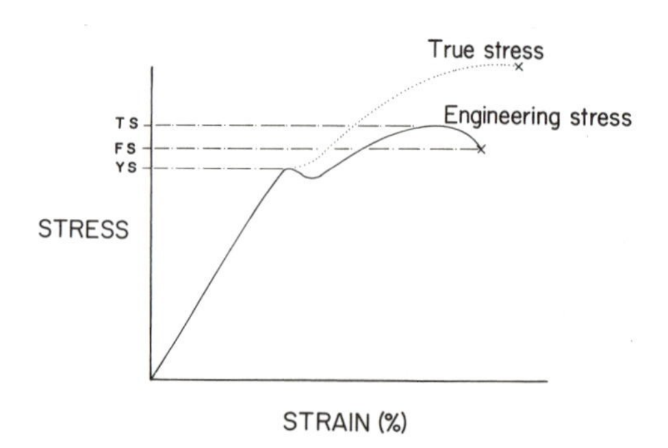

Figure 3-2. Stress–strain behavior of an idealized material.

The slope (E) or proportionality constant of the tensile/compressive stress–strain curve is called *Young's modulus* or the *modulus of elasticity.* It is the value of the increment of stress over the increment of strain. The stiffer a material is, the higher the value of E and the more difficult it is to deform. Similar analysis can be performed for deformation by shear, in which the shear modulus G is defined as the initial slope of the curve of shear stress versus shear strain. The unit for the modulus is the same as that of stress since strain is dimensionless. The shear modulus of an isotropic material is related to its Young's modulus by

$$E = 2G(1 + \nu) \tag{3-4}$$

in which ν is the *Poisson's ratio* of the material. Poisson's ratio is defined as the negative ratio of the transverse strain to the longitudinal strain for tensile or compressive loading of a bar. Poisson's ratio is close to 1/3 for common stiff materials, and is slightly less than 1/2 for rubbery materials and for soft biological tissues.

In the *plastic region,* strain changes are no longer proportional to the applied stress. Further, when the applied stress is removed, the material will not return to its original shape but will be permanently deformed, which is called a plastic deformation. Figure 3-3 is a schematic illustration of what will happen on the atomic scale when a material is deformed. Note that the individual atoms are distorted and stretched while part of the strain is accounted for by a limited movement of atoms past one another. When the load is released before atoms can slide over other atoms, the atoms will go back to their original positions, making it an elastic deformation. When a material is deformed plastically, the atoms are moved past each other in such a way they will have new neighbors and when the load is released they can no longer go back to their original positions.

Referring back to Figure 3-2, a peak stress can be seen which is often followed by an apparent decrease until a point is reached where the material ruptures. The peak stress is known as the *tensile* or *ultimate tensile strength* (TS); the stress where failure occurs is called the *failure* or *fracture strength* (FS).

In many materials such as stainless steels, definite yield points occur. This point is characterized by temporarily increasing strain without further increase

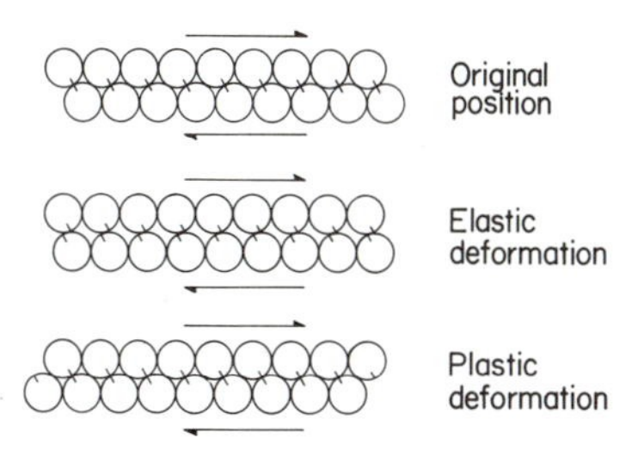

Figure 3-3. Schematic diagrams of a two-dimensional atomic model showing elastic and plastic deformation.

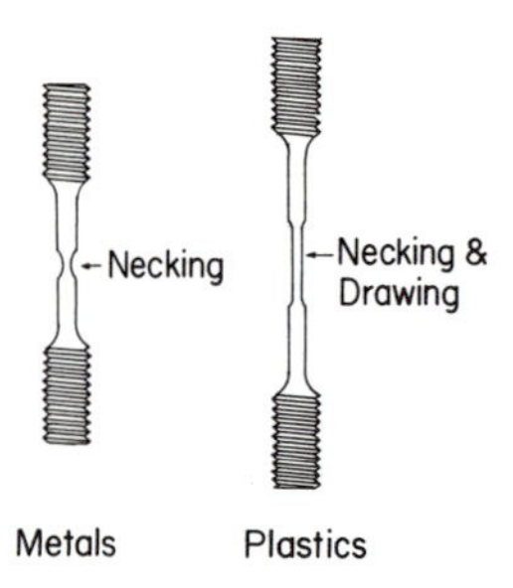

Figure 3-4. Deformation characteristics of metals and plastics under stress. Note that metals rupture without further elongation after necking occurs; by contrast, in plastics the necked region undergoes further deformation called drawing.

in stress. Sometimes, it is difficult to decipher the yield point; therefore, an *offset* (usually 0.2%) yield point is used in lieu of the original yield point.

Thus far we have been examining the *engineering stress–strain* curves, which differ from the *true stress–strain* curves since the former curve is obtained by assuming a constant cross-sectional area over which the load is acting from the initial loading until final rupture. This assumption is not correct, which accounts for the peak seen at the ultimate tensile stress. For example, as a specimen is loaded in tension, sometimes *necking* (Figure 3-4) occurs which reduces the area over which the load is acting. If additional measurements are made of the changes in cross-sectional area that occur, and the true area is used in the calculations, then a dotted curve like that in Figure 3-2 is obtained.

Example 3-1

The following data were obtained for an aluminum alloy. The figure below shows stress versus strain for an aluminum alloy. A standard tensile test specimen with a 2-inch gage length and 0.505-inch diameter was used.

(a) Plot the engineering stress-strain curve.
(b) Determine the Young's modulus, yield strength (0.2% offset), and tensile strength.
(c) Determine the engineering and true fracture strength. The diameter of the broken pieces was 0.4 in.

Load (kilo-lbf)	Gage length (in)
2	2.000
4	2.004
8	2.008
10	2.010
12	2.011
13	2.014
14	2.020
16	2.050
16 (maximum)	2.099
15.6 (fracture)	2.134

Answer

(a) See the plot from the calculations based on cross-sectional area = $\pi(\text{diameter})^2/4 = 0.2\ \text{in}^2$.

Stress (ksi)	Strain (%)
20	0.2
40	0.4
50	0.5
60	0.6
65	0.7
70	1.0
80	2.5
80	5.0
78	6.7

(b) Young's modulus = 40 ksi/0.004 = 1×10^7 psi (69 MPa)
0.2% yield stress: from graph; 80,000 psi (560 MPa)
Tensile strength: 80,000 psi (560 MPa)

(c) Engineering fracture strength
$=15{,}600\ \text{lb}/0.2\ \text{in}^2 = 78{,}000$ psi (538 MPa)
True fracture strength
$=15{,}600\ \text{lb}/\pi(0.2\ \text{in})^2 = 124{,}140$ psi (856 MPa)

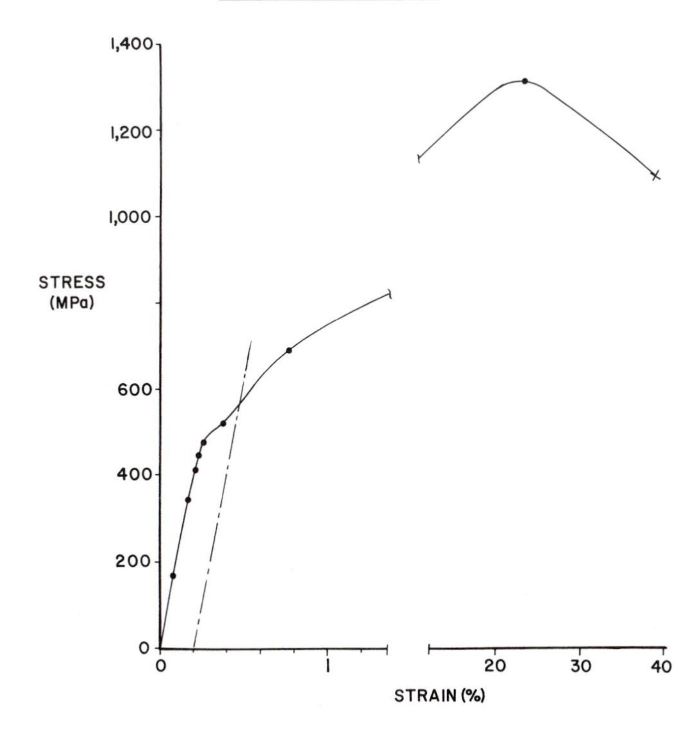

3.1.2. Mechanical Failure

Mechanical failure usually occurs by fracture. The fracture of a material can be characterized by the amount of energy per unit volume required to produce the failure. The quantity is called *toughness* and can be expressed in terms of stress and strain:

$$\text{Toughness} = \int_{\varepsilon_0}^{\varepsilon_f} \sigma \, \mathrm{d}\varepsilon = \int_{l_0}^{l_f} \sigma \frac{\mathrm{d}l}{l} \tag{3-5}$$

Expressed another way, toughness is the summation of stress times the normalized distance over which it acts (strain) taken in small increments. The area under the stress–strain curve provides a simple method of estimating toughness as shown in Figure 3-5.

A material that can withstand high stresses and can undergo considerable plastic deformation (*ductile–tough* material) is tougher than one that resists high stresses but has no capacity for deformation (*hard–brittle* material) or one that has a high capacity for deformation but can only withstand relatively low stresses (*ductile–soft* or plastic material).

Brittle materials exhibit fracture strengths far below the theoretical strength predicted based on known atomic bond strengths. Moreover, there is much variation in strength from specimen to specimen, so that the practical strength is difficult to predict. These facts, along with the comparative weakness of ceramics in tension, are the major reasons why ceramic and glassy materials are not used extensively for implantation despite their excellent compatibility with tissues.

The comparative weakness of brittle materials is explained as follows. The stress on a brittle material is not uniformly distributed over the entire cross-sectional area if a crack or flaw is present as shown in Figure 3-6. If the crack

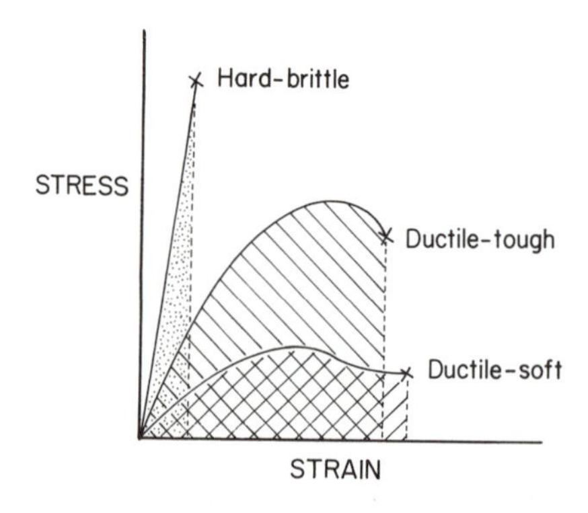

Figure 3-5. Stress–strain curves of different types of materials. The areas underneath the curves are the measure of toughness.

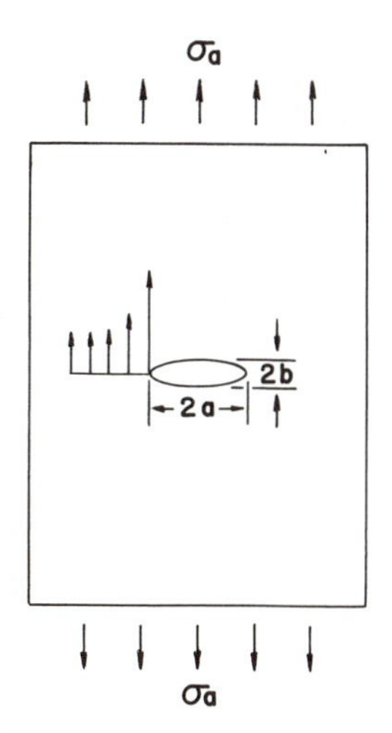

Figure 3-6. An elliptic microcrack inclusion in a brittle material.

is a narrow elliptic hole in a specimen subjected to a tensile stress, the maximum stress (σ_{max}) acting at the ends of the hole is given by

$$\sigma_{max} = \sigma_{app}\left(1 + \frac{2a}{b}\right) \tag{3-6}$$

where σ_{app} is the applied (or nominal) tensile stress experienced away from the crack. One can rearrange Eq. (3-6):

$$\frac{\sigma_{max}}{\sigma_{app}} = \left(1 + \frac{2a}{b}\right) \tag{3-7}$$

where the ratio $\sigma_{max}/\sigma_{app}$ is called the *stress concentration factor* (scf), which can be substantial if the a/b ratio is high, i.e., a sharp crack. If the crack tip has a radius of curvature r $(=b^2/a)$, then Eq. (3-6) can be rewritten as

$$\sigma_{max} = \sigma_{app}\left[1 + 2\left(\frac{a}{r}\right)^{1/2}\right] \tag{3-8}$$

Since $a \gg r$ for a crack,

$$\sigma_{max} \cong 2\sigma_{app}\left(\frac{a}{r}\right)^{1/2} \tag{3-9}$$

Equation (3-9) indicates that the stress concentration becomes very large for a sharp crack tip as well as for long cracks. Thus, the propagation of a sharp crack can be blunted if one increases the crack tip radius. (Some may have noticed that when a crack develops on a large display window a drill hole is made on the crack tip in order to arrest the crack.)

Griffith proposed an energy approach to fracture. The elastic energy stored in a test specimen of unit thickness is

$$\sigma \times \varepsilon = \pi a^2 \sigma \left(\frac{\sigma}{E} \right) = \frac{\pi (a\sigma)^2}{E} \tag{3-10}$$

Observe that the elastic energy for a brittle material is twice the area under the stress–strain curve. The elastic energy is used to create two new surfaces as the crack propagates. The surface energy $4\gamma a$ (γ is the surface energy) should be smaller than the elastic energy for the crack to grow. Thus, the incremental changes of both energies for the crack to grow can be written as

$$\frac{\mathrm{d}}{\mathrm{d}a} \left(\frac{\pi (a\sigma)^2}{E} \right) = \frac{\mathrm{d}}{\mathrm{d}a} (4\gamma a) \tag{3-11}$$

Hence,

$$\sigma = \sigma_f = \sqrt{\frac{2\gamma E}{\pi a}} \tag{3-12}$$

Since for a given material E and γ are constants,

$$\sigma_f = \frac{K}{\sqrt{\pi a}} \tag{3-13}$$

In this case K has the units of psi $\sqrt{\text{in}}$ or MPa $\sqrt{\text{m}}$ and is proportional to the energy required for fracture. K is also called *fracture toughness.* Stress concentrations also occur in ductile materials, but their effect is usually not as serious as in brittle ones since local yielding that occurs in the region of peak stress will effectively blunt the crack and alleviate the stress concentration.

Example 3-2

Estimate the size of the surface flaw in a glass whose modulus of elasticity and surface energy are 70 GPa and 800 erg/cm^2 respectively. Assume that the glass breaks at a tensile stress of 100 MPa.

Answer

From Eq. (3-12), and keeping in mind the transformation from cgs to SI units,

$$a = \frac{2\gamma E}{\pi \sigma_f^2}$$

$$= \frac{2 \times 800 \text{ dyne/cm} \times 70 \text{ GPa}}{\pi (100 \text{ MPa})^2}$$

$$= \underline{3.565\ \mu\text{m}}$$

(Note that if the crack is on the surface, its length is a; if it is inside the specimen, it is $2a$. Remember 1 erg = 1 dyne cm.)

When a material is subjected to a constant or a repeated load below the fracture stress, it can fail after some time. This is called static or dynamic (cyclic) fatigue, respectively. The effect of cyclic stresses (Figure 3-7) is to initiate microcracks at centers of stress concentration within the material or on the surface resulting in the growth and propagation of cracks leading to failure. The rate of

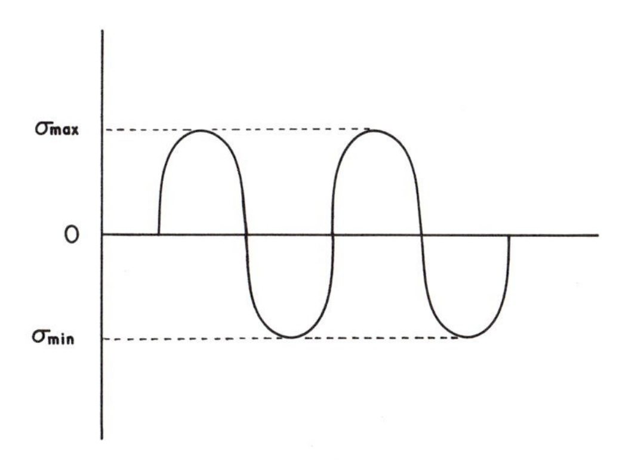

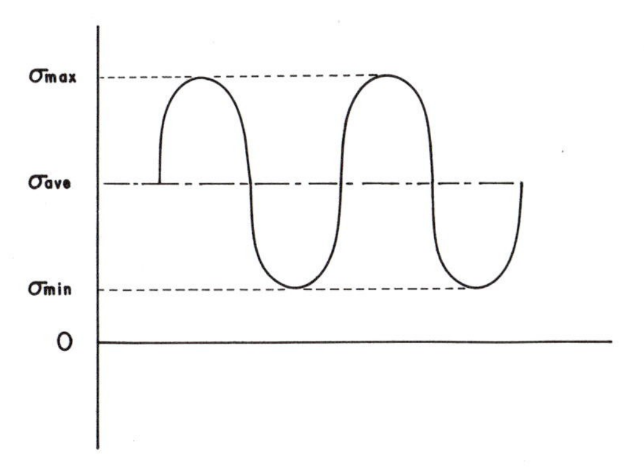

Figure 3-7. Cyclic stresses. σ_{min} and σ_{max} are the maximum and minimum values of the cyclic stresses. The range of stresses $\Delta\sigma = \sigma_{\text{min}} - \sigma_{\text{max}}$ and average stress $\sigma_{\text{ave}} = (\sigma_{\text{max}} + \sigma_{\text{min}})/2$. The top curve is for fluctuating and the bottom curve is for reversed cyclic loading.

crack growth can be plotted in a log–log scale versus time as shown in Figure 3-8. The most significant portion of the curve is the crack propagation stage, which can be estimated as follows:

$$\frac{da}{dN} = A(\Delta K)^m \tag{3-14}$$

where a, N, and ΔK are the crack length, number of cycles, and the range of the stress intensity factor; from Eq. (3-13), $\Delta K = \Delta\sigma\sqrt{\pi a}$. The A and m are the intercept and the slope of the linear portion of the curve.

Another method of testing the fatigue properties is to monitor the number of cycles to failure at various stress levels as shown in Figure 3-9. This test requires a large number of specimens compared with the crack propagation test. The *endurance limit* is the stress below which the material will not fail in fatigue no matter how many cycles are applied. Not all materials exhibit an endurance limit.

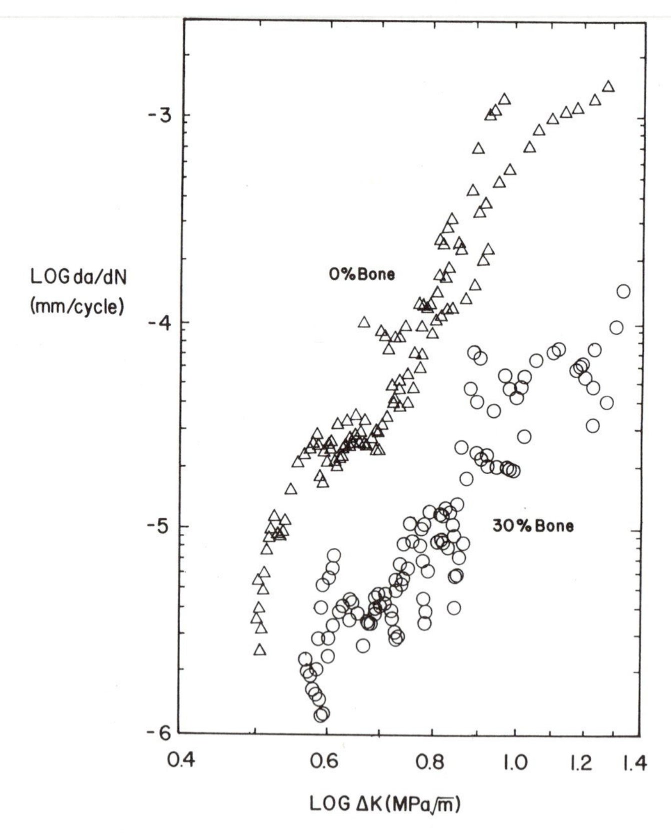

Figure 3-8. Log da/dN versus log ΔK for polymethyl methacrylate bone cement. From Y. K. Liu, J. B. Park, G. O. Njus, and D. Stienstra, "Bone Particle Impregnated Bone Cement. I. *In Vitro* Studies," *J. Biomed. Mater. Res., 21,* 247–261, 1987.

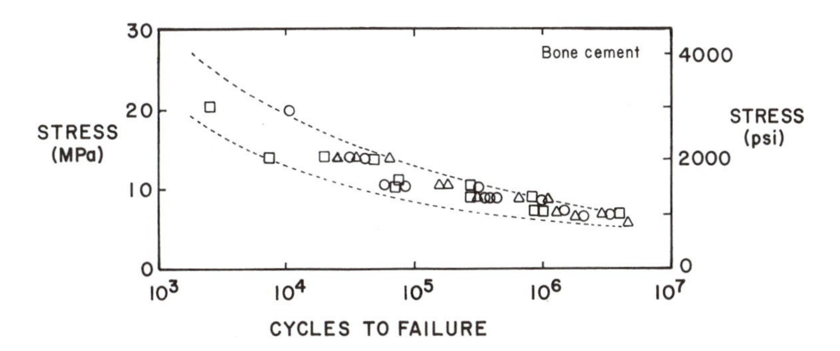

Figure 3-9. Stress versus log *N* (number of cycles) of PMMA bone cement. The Simplex P bone cement test specimens were fabricated at pressures between 5 and 50 psi and tested in air at 22°C. From T. A. Freitag and S. L. Cannon, "Fracture Characteristics of Acrylic Bone Cement. II. Fatigue," *J. Biomed. Mater. Res.*, *11*, 609–624, 1977.

Since implants are often flexed many times during a patient's life, the fatigue properties of materials are very important in implant design.

3.1.3. Viscoelasticity

3.1.3.1. Viscoelastic Material Behavior. Viscoelastic materials are those for which the relationship between stress and strain depends on time. In such materials the stiffness will depend on the rate of application of the load. In addition, mechanical energy is dissipated by conversion to heat in the deformation of viscoelastic materials. All materials exhibit some viscoelastic response. In metals such as steel or aluminum at room temperature, as well as in quartz, the response at small deformation is almost purely elastic. Metals can behave plastically at large deformation, but ideally plastic deformation is independent of time. By contrast, materials such as synthetic polymers, wood, and human tissue display significant viscoelastic effects.

3.1.3.2. Characterization of Viscoelastic Materials. *Creep* is a slow, progressive deformation of a material under constant stress. Suppose the history of stress σ as it depends on time t to be a step function beginning at time zero:

$$\sigma(t) = \sigma_0 H(t) \tag{3-15}$$

$H(t)$ is the unit Heaviside step function defined as zero for t less than zero, and one for t equal to or greater than zero. The strain $\varepsilon(t)$ will increase as shown in Figure 3-10. The ratio

$$J(t) = \varepsilon(t)/\sigma_0 \tag{3-16}$$

is called the creep compliance. In linear materials, it is independent of stress level. If the load is released at a later time t_s, the strain will exhibit recovery, as shown in Figure 3-10. For linear materials, we may invoke the Boltzmann

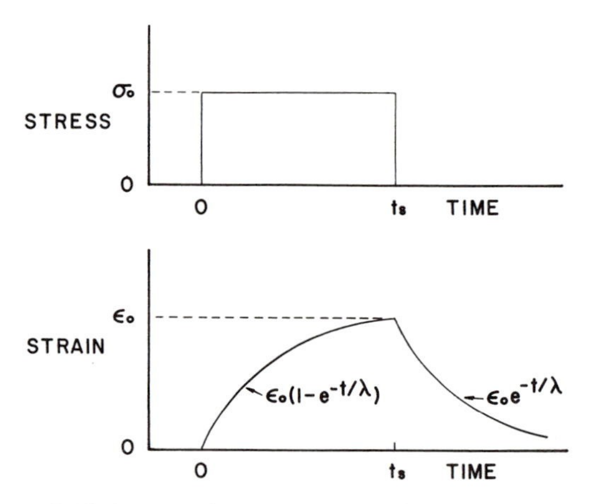

Figure 3-10. Creep and creep recovery of viscoelastic material.

superposition principle, which states that the effect of a compound cause is the sum of the effects of the individual causes. The stress may then be written as a superposition of a step up followed by a step down:

$$\sigma(t) = \sigma_0[H(t) - H(t - t_1)] \tag{3-17}$$

so the strain is

$$\varepsilon(t) = \sigma_0[J(t) - J(t - t_1)] \tag{3-18}$$

The strain may or may not recover back to zero, depending on the material.

Stress relaxation is the gradual decrease of stress when the material is held at constant extension. If we suppose the strain history to be a step function beginning at time zero: $\varepsilon(t) = \varepsilon_0 H(t)$, the stress $\sigma(t)$ will decrease as shown in Figure 3-11. The ratio

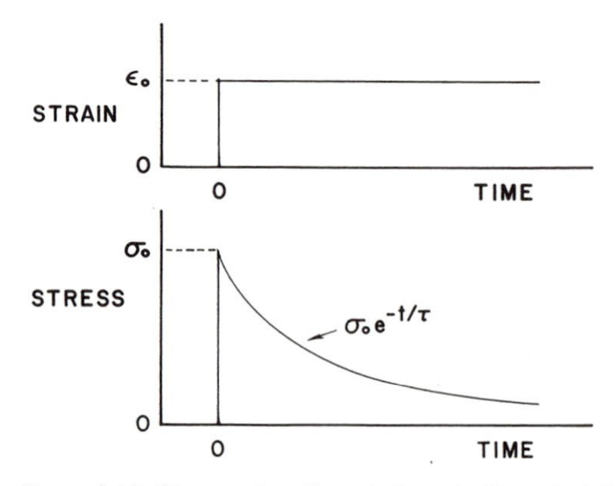

Figure 3-11. Stress relaxation of viscoelastic material.

$$E(t) = \sigma(t)/\varepsilon_0 \tag{3-19}$$

is called the relaxation modulus. In linear materials, it is independent of strain level.

If a sinusoidally varying stress,

$$\sigma(t) = \sigma_0 \sin(2\pi f t) \tag{3-20}$$

of frequency f, is applied to a linearly viscoelastic material, the strain

$$\varepsilon(t) = \varepsilon_0 \sin(2\pi f t - \delta) \tag{3-21}$$

will also be sinusoidal in time but will lag the stress by a phase angle δ, as shown in Figure 3-12. The loss angle δ is a measure of the viscoelastic damping of the material. Both the loss angle and the dynamic stiffness $E = \sigma_0/\varepsilon_0$ depend on frequency. The tangent of the loss angle is referred to as the *loss tangent*: tan δ.

For a given material, the above properties are not independent. For example, the creep and relaxation functions are related by a convolution; the dynamic properties are related to the creep or relaxation behavior by Fourier transformation; and the loss angle and dynamic stiffness are related by the Kramers Kronig relations. We remark that the larger the value of the loss tangent, the more rapidly the dynamic stiffness changes with frequency, and the more rapidly the relaxation modulus and creep compliance change with time. An approximate correspondence between the frequency scale for dynamic behavior and the time scale for creep and relaxation is the relation

$$t = 1/2\pi f \tag{3-22}$$

Furthermore, the viscoeleastic properties of a material in tension need not follow the same time or frequency dependence as those in shear.

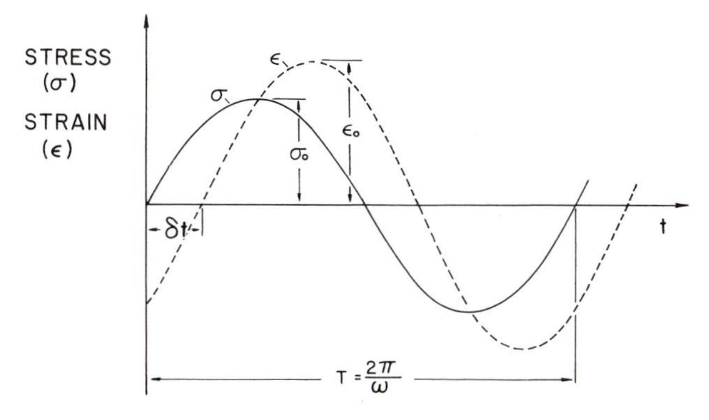

Figure 3-12. Loss angle δ for oscillatory loading of viscoelastic material. $\delta = 2\pi\delta t/T$.

3.1.3.3. Prediction of the Response. Viscoeleastic materials may be used under conditions of complex loading. It is possible to predict the response of such materials based on the material properties discussed above. The Boltzmann superposition principle for linear materials is applied to decompose the load history into a series of differential creep and recovery episodes. Summing the effects of these, one obtains the Boltzmann superposition integral:

$$\varepsilon(t) = \int_0^{\tau} J(t-\tau)\frac{d\sigma}{d\tau}\,d\tau \tag{3-23}$$

or conversely,

$$\sigma(t) = \int_0^{\tau} E(t-\tau)\frac{d\varepsilon}{d\tau}\,d\tau \tag{3-24}$$

in which τ is a time variable of integration. Consequently, if the response of a material to step stress or strain has been determined experimentally, the response to *any* load history can be found for the purpose of analysis or design.

3.1.3.4. Mechanical Models. Simple mechanical models may be considered as an aid for visualizing viscoelastic response. The models consist of springs, which are purely elastic, and dashpots, which are purely viscous. In a spring, the stress is proportional to the strain;

$$\sigma = k\varepsilon \tag{3-25}$$

[see Eq. (3-3)], while in a dashpot, the stress is proportional to the time derivative of the strain (the strain rate):

$$\sigma = \eta\, d\varepsilon/dt \tag{3-26}$$

For example, the Kelvin model consists of a spring in parallel with a dashpot. The strain is the same in both elements, but the stress in the Kelvin model is the sum of the spring and dashpot stresses. After some reductions, one may obtain a differential form of the stress–strain relation:

$$\eta\, d\varepsilon/dt + k\varepsilon = \sigma \tag{3-27}$$

The solution of this differential equation for a step stress input gives the creep compliance

$$J(t) = [1 - e^{-kt/\eta}]/k \tag{3-28}$$

Real materials, both synthetic and natural, exhibit behavior that cannot be described by the simple mechanical models. In a two- or three-element model, the creep or relaxation is completed within about one logarithmic decade (a factor of ten in time) while in real materials, the creep or relaxation is distributed over many decades.

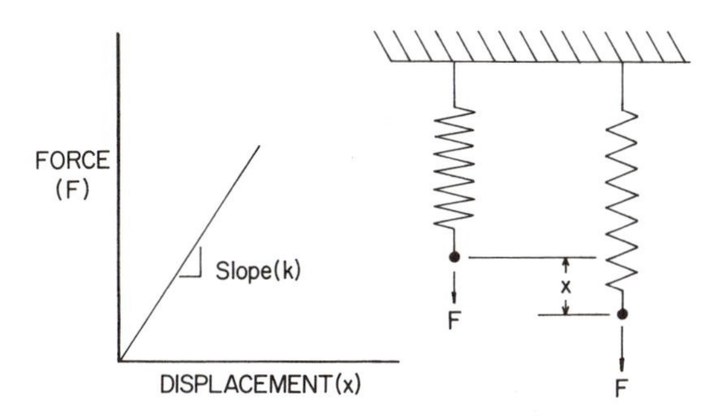

Figure 3-13. Force versus displacement of a spring.

Although the simple equation (3-3) can describe the elastic behavior of many materials at low strains as represented by a spring in Figure 3-13, it cannot be used to characterize the polymers and tissues that are some of the major concerns of this book. The fluidlike behavior of a material (such as water and oil) can be described in terms of (shear) stress and (shear) strain as in the elastic solids, but the proportionality constant (viscosity, η) is derived from the relationship given in Eq. (3-26). It is noted that the stress and strain are shear in this case rather than tensile or compressive although the same symbols are used to avoid complications.

A mechanical analog (dashpot) can be used to model the viscous behavior of Eq. (3-26) as shown in Figure 3-14. An automobile shock-absorbing cylinder that contains oil as the damping fluid has a similar construction. By examining Eq. (3-26) one can see that the stress is *time* dependent; i.e., if the deformation is accomplished in very short time ($dt \to 0$), then the stress becomes infinite. On the other hand, if the deformation is achieved slowly ($dt \to \infty$), the stress approaches zero regardless of the viscosity value.

Real materials that have both elastic and viscous aspects to their behavior are known as *viscoelastic* materials. The simple equations (3-25) and (3-26), when combined together as if the material is made of springs and dashpots, can describe,

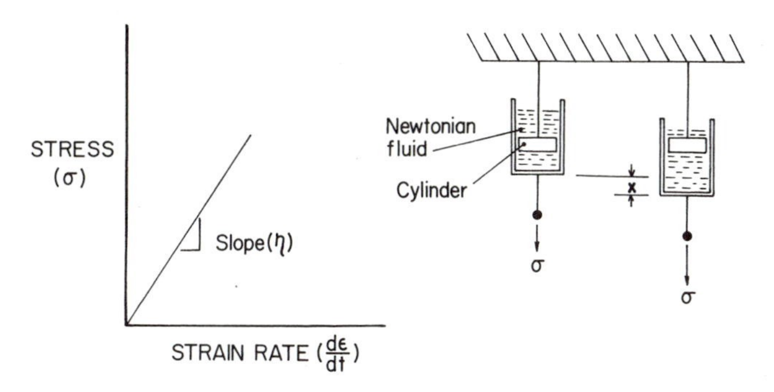

Figure 3-14. Stress versus strain rate of a dashpot.

in principle, the viscoelastic behavior of a material. The stress–time behavior of a spring and dashpot combination can be represented as shown in Figure 3-16. If the spring and dashpot are arranged in series and parallel they are called Maxwell and Voigt (or Kelvin) models, respectively. Remember that Eq. (3-25) does not involve time, implying the spring acts instantaneously when stressed. Hence, if the Maxwell model is stressed suddenly, the spring reacts instantaneously while the dashpot cannot react since the piston of the dashpot cannot move due to the infinite stress required by the surrounding fluid. However, if we hold the Maxwell model after instantaneous deformation, the dashpot will react due to the retraction of the spring and this will take time (dt = finite). The foregoing description can be expressed concisely by a simple mathematical formulation. In general, the response to stress by the Maxwell model will result on the strain as a summation,

$$\varepsilon_{\text{total}} = \varepsilon_{\text{spring}} + \varepsilon_{\text{dashpot}} \tag{3-29}$$

Differentiating both sides and using Eqs. (3-25) and (3-26) will result in Eq. (3-27). Replacing the spring constant k with Young's modulus E in view of the linear behavior, then

$$\frac{d\varepsilon}{dt} = \frac{1}{E}\frac{d\sigma}{dt} + \frac{\sigma}{\eta} \tag{3-30}$$

Equation (3-30) can be applied easily to a simple mechanical test such as stress relaxation in which the specimen is strained instantaneously and the relaxation of the load is monitored while the specimen is held at a constant length as shown in Figure 3-16. Thus, the strain rate becomes zero ($d\varepsilon/dt = 0$) and at $t = 0$ and $\sigma = \sigma_0$, in σ_0 = constant, hence

$$\frac{\sigma}{\sigma_0} = \exp\left(-\frac{E}{\eta}t\right) \tag{3-31}$$

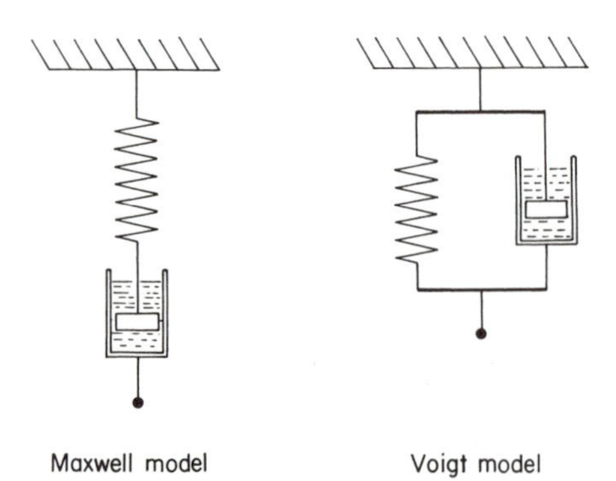

Figure 3-15. Two-element viscoelastic models.

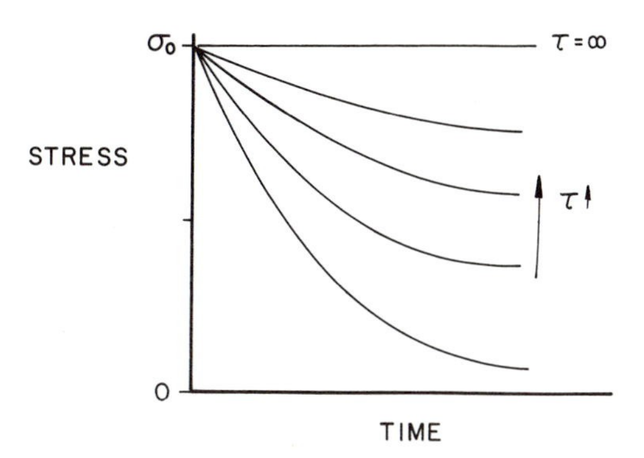

Figure 3-16. Relaxation curve for Maxwell model with different values of relaxation time (τ).

The constant η/E has dimensions of time and is defined as the *relaxation time* τ and Eq. (3-31) becomes

$$\sigma = \sigma_0 \exp(-t/\tau) = \frac{\sigma_0}{\exp(t/\tau)} \tag{3-32}$$

Examining Eq. (3-32), one can see that if the relaxation time is short compared with the present time t, then the stress σ at a given time becomes small. On the other hand, if the relaxation time is long, then the stress σ is nearly the same as the original stress σ_0 as shown in Figure 3-16.

Example 3-3

A stress of 1 MPa is required to stretch a 2-cm aorta strip to 2.3 cm. After 1 hour in the same stretched position, the strip exerted a stress of 0.75 MPa. Assume the property of the aorta does not vary appreciably during the experiment.

(a) What is the relaxation time, assuming a simple exponential decay model?
(b) What stress would be exerted by the aorta strip in the same stretched position after 5 hours?

Answer

(a) From Eq. (3-32)

$$\sigma = \sigma_0 \exp(-t/\tau)$$

$$\frac{0.75}{1} = \exp\left(-\frac{1}{\tau}\right)$$

Therefore, $\tau = \underline{3.48 \text{ hour}}$

(b) Substituting the relaxation time

$$\sigma = 1 \exp\left(-\frac{5}{3.48}\right) = \underline{0.24 \text{ MPa}}$$

In comparison, window glass has a very large relaxation time of many years (minimal stress relaxation over short times) while the water and oil have a short relaxation time. Thus, when stressed, their shape changes immediately to relieve the applied stress (instantaneous stress relaxation).

Similar analysis can be made with the Voigt model. In this case the strain of spring and dashpot represents the total strain, i.e.,

$$\varepsilon_{\text{total}} = \varepsilon_{\text{spring}} = \varepsilon_{\text{dashpot}} = \varepsilon \tag{3-33}$$

The total stress is the summation of that in the spring and dashpot

$$\sigma_{\text{total}} = \sigma_{\text{spring}} + \sigma_{\text{dashpot}} = \sigma \tag{3-34}$$

Substituting Eqs. (3-25) and (3-26) into (3-34), with E again substituted for k,

$$\sigma = E\varepsilon + \eta \frac{\mathrm{d}\varepsilon}{\mathrm{d}t} \tag{3-35}$$

If a stress is applied and the stress is removed after a certain time, then the strain recovers with time in a way that can be derived from Eq. (3-35):

$$\varepsilon(t) = \varepsilon_0 \exp\left(-\frac{E}{\eta} t\right) \tag{3-36}$$

where ε_0 is the strain at the time of stress removal. The constant η/E is termed the *retardation time* λ for this *creep recovery* process. Since the strain is being recovered from the original strain ε_0, Eq. (3-36) can be rewritten:

$$\varepsilon_{\text{recovery}} = \varepsilon_0 - \varepsilon_0 \exp(-t/\lambda) = \varepsilon_0[1 - \exp(-t/\lambda)] \tag{3-37}$$

Figure 3-17 shows the creep response of a Voigt model with varying retardation times.

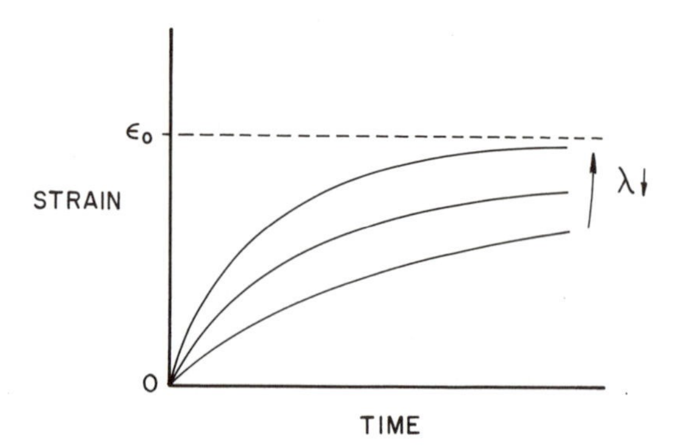

Figure 3-17. Retarded elastic deformation (creep and recovery) curves for Voigt model with various retardation times (λ).

Example 3-4

A piece of polyethylene is stretched 20% of its length. When the stress was released, it recovered 50% of its strain after 1 hour at room temperature.
(a) What is the retardation time assuming a single-exponential model?
(b) What is the amount of strain recovered after 5 hours at room temperature?

Answer

(a) From Eq. (3-37)

$$\varepsilon = \varepsilon_0[1 - \exp(-t/\lambda)]$$

$$\frac{\varepsilon}{\varepsilon_0} = 0.5 = 1 - \exp\left(-\frac{1}{\lambda}\right)$$

Therefore, $\lambda = \underline{1.443 \text{ hour}}$

(b)

$$\varepsilon = 0.2[1 - \exp(-5/1.443)]$$

$$= \underline{0.194}$$

(or 19.4% which is 96.9% recovery of the strain since the original strain is 20%).

3.1.3.5. Behavior of Viscoelastic Materials. The viscoelastic behavior of various materials is shown in Figure 3-18. The dynamic stiffness (Figure 3-18a) and loss tangent (Figure 3-18b) are shown since they are the properties most easy to conceptualize. Observe that the scales are logarithmic. The very low frequencies correspond to very long times in creep: $t = 1/2\pi f$. The behavior of a three-element spring-dashpot model is also shown in Figure 3-18b for comparison. Observe that the glass-to-rubber transition in polymers is associated with a large peak in the loss tangent in the frequency domain. Cross-linked polymers exhibit a nonzero limit to the stiffness at low frequency; the stiffness of un-cross-linked polymers tends to become zero at sufficiently low frequency or sufficiently long time. Such materials are viscoelastic liquids. Some such materials may superficially appear solid; the excess creep becomes apparent only after months or years. Blood is also a viscoelastic liquid.

3.1.3.6. Applications. There are a variety of consequences of viscoelastic behavior that influence the application of viscoelastic materials. For example, in those applications for which a steady-state stress is applied, the creep behavior is of the greatest importance. The expected service life of implant materials may be very long; consequently, attention to the long-term creep behavior is in order. Blood vessels experience a steady-state internal pressure that gives rise to circumferential stress, so creep in blood vessel materials is important. Creep also occurs in the polyethylene socket component of the total hip replacement, but it is not a major failure mechanism. Stress relaxation is relevant to such situations as the loosening of screws that have been tightened to a specified extension. Bone screws are an example. In applications involving vibration, the width of the vibration amplitude versus frequency curve is proportional to the loss tangent. The

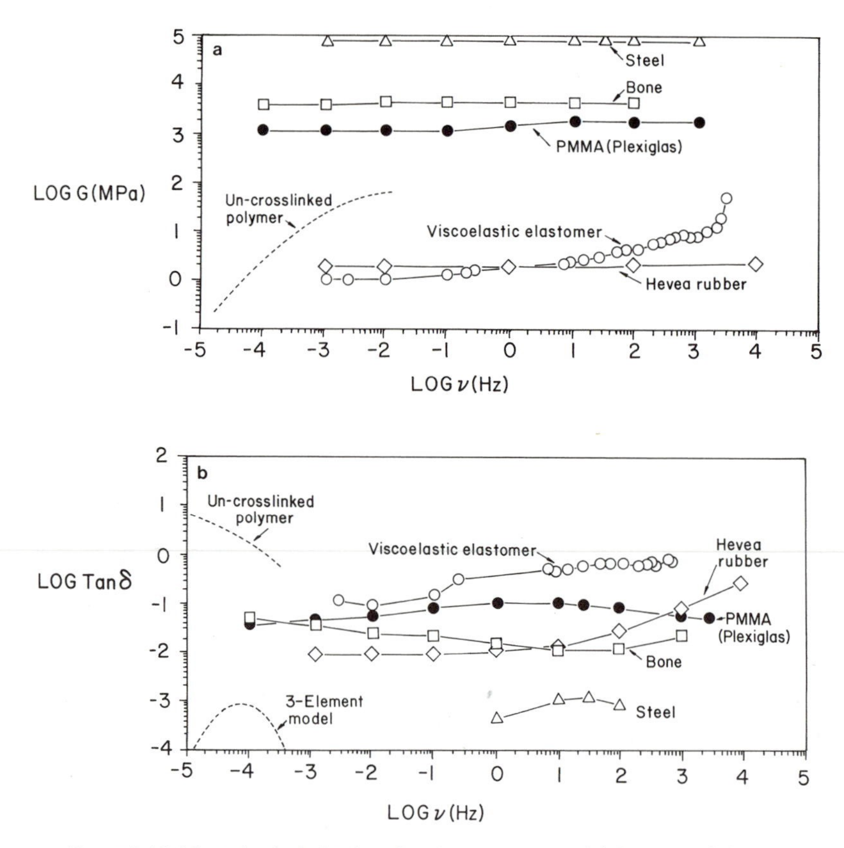

Figure 3-18. Viscoelastic behavior of various materials. (a) Stiffness; (b) loss.

maximum vibration amplitude at resonance is inversely proportional to the loss tangent. Vibrations damp out more rapidly for higher values of loss tangent. Viscoeleastic materials may therefore be used to reduce vibration in machinery or to protect the body from vibration or from mechanical shock. At higher frequency, sound or ultrasound waves are attenuated in proportion to the loss tangent. Diagnostic ultrasound is therefore absorbed in tissues, and will not penetrate far into a body composed of highly viscoelastic materials.

3.2. THERMAL PROPERTIES

The most familiar thermal properties are the melting and freezing (solidification) temperatures. These are phase transformations that occur at specific temperatures. These transformation temperatures depend on the bond energy; e.g., the higher the bonding strength, the higher the melting temperature. If the material is made of different elements or compounds, then it may have a range of melting or solidification temperatures, that is, the liquid coexists with solid over a range of temperatures unlike a pure material.

The thermal energy spent on converting 1 gram of a material from solid to liquid is called heat of fusion. The unit is joules per gram where 1 joule is equivalent to 1 newton meter. The heat of fusion is closely related to the melting temperature (T_m); i.e., the higher T_m, the higher the heat of fusion although there are many exceptions (Table 3-1).

The thermal energy spent on changing the temperature of a material by 1°C per unit mass is called *specific heat*. Traditionally, water is usually chosen as a standard substance and 1 calorie is the heat required to raise 1 gram of water from 15°C to 16°C but now the standard unit of heat is the joule; thus, the specific heat is in units of J/g (1 calorie is equivalent to 4.187 J).

The change in length Δl for a unit length l_0 per unit temperature is called the *linear coefficient of expansion* (α), which can be expressed as

$$\alpha = \frac{\Delta l}{l_0 \Delta T} \tag{3-38}$$

The thermal expansion may depend on direction in a single crystal or composite, and it may depend on temperature. If the material is homogeneous and isotropic, then the *volumetric* thermal expansion coefficient (V) can be approximated as

$$V \approx 3\alpha \tag{3-39}$$

Another important thermal property is the *thermal conductivity*, which is defined as the amount of heat passed for a given time, thickness and area of the

Table 3-1. Thermal Properties of Materials

Substance	Melting temp. (°C)	Specific heat (J/g)	Heat of fusion (J/g)	Thermal conductivity (W/mK)	Linear thermal expansion coeff. ($\times 10^{-6}$/°C)
Mercury	−38.87	0.138	12.7	68	60.6
Gold	1063	0.13	67	297	14.4
Silver	960.5	0.2345	108.9	421	19.2
Copper	1083	0.385	205.2	384	16.8
Platinum	1773	0.134	113	70	—
Enamel	—	0.75	—	0.82	11.4
Dentine	—	1.17	—	0.59	8.3
Acrylic	70[a]	1.465	—	0.2	81.0
Water	0	4.187	334.9 (ice)	—	—
Paraffin	52	2.889	146.5	—	—
Beeswax	62	—	175.8	0.4	350
Alcohol	−117	—	2.29	104.7	—
Glycerin	18	2.428	75.4	—	—
Amalgam	480	—	—	23	22.1–28
Porcelain	—	1.09	—	1	4.1

[a] Softening temperature (T_g).

material. The unit is watt/mK where 1 watt is equivalent to 1 J per second. Generally, thermal conductivity of metals is much higher than ceramics and polymers due to the free electrons in metals, which act as energy conductors.

Example 3-5

In order to fill a cavity, a cylindrical hole of 2-mm diameter is made in a molar tooth. The length of the hole is 4 mm. Consider that it is filled with amalgam, then consider it to be filled with acrylic resin. Assume the temperature variation is 50°C. The modulus of elasticity of amalgam and resin are 20 and 2.5 GPa, respectively. Moreover, $E_{\text{dentin}} = 18$ GPa, $\alpha_{\text{amalgam}} = 25 \times 10^{-6}/°\text{C}$, $\alpha_{\text{resin}} = 81 \times 10^{-6}/°\text{C}$, $\alpha_{\text{enamel}} = 11 \times 10^{-6}/°\text{C}$, $\nu = 1/3$.

(a) Calculate the volume changes for the fillings.
(b) Calculate, based upon an elementary one-dimensional model and neglecting any remaining enamel, the force developed between the dentine and fillers.
(c) Calculate the force between the filling and the remaining tooth structure, considered as a cylindrical shell in a more realistic two-dimensional model. Also determine the stress at the interface and the stress in the enamel and discuss the results.

Answer

(a) Since the volume expansion coefficient can be defined as in Eq. (3-39)

$$\frac{\Delta V}{V_0 \Delta T} = 3\alpha, \text{ therefore } \Delta V = V_0 \times 3\alpha \times \Delta T$$

The net volume change after the filling will be

$$\Delta V_{\text{amalgam}} = \pi \ (1 \text{ mm})^2 \times 4 \text{ mm} \times 3(25 - 8.3) \times 10^{-6} \times 50 = \underline{0.03 \text{ mm}^3}$$

$$\Delta V_{\text{resin}} = \pi \ (1 \text{ mm})^2 \times 4 \text{ mm} \times 3(81 - 8.3) \times 10^{-6} \times 50 = \underline{0.14 \text{ mm}^3}$$

(b) Since $F = \sigma \times A$, $A = \pi Dh = \pi \times 2 \times 4 \times 10^{-6} \text{ m}^2 = 25.13 \times 10^{-6} \text{ m}^2$, and $\sigma = E \times \Delta\varepsilon$, $\Delta\varepsilon = \Delta T[\alpha_{\text{amalgam or resin}} - \alpha_{\text{dentin}}]$

$$F_{\text{amalgam}} = 25.13 \times 10^{-6} \times 20 \times 10^9 \times 16.7 \times 10^{-6} \times 50 = \underline{420 \text{ N}}$$

$$F_{\text{resin}} = 25.13 \times 10^{-6} \times 2.5 \times 10^9 \times 72.7 \times 10^{-6} \times 50 = \underline{228 \text{ N}}$$

Note that although the resin expands more than 4 times in volume than does the amalgam, the force exerted on the tooth is one-half that of amalgam. This is due to the difference in stiffness. The actual force exerted by the fillings will be smaller than the calculated values since they can expand freely toward the top of the holes. The force would be felt faster by having amalgam as a filler than for the resin since amalgam can conduct heat much faster than the resin.

(c) Consider the remaining tooth structure to be a cylindrical shell of thickness t, length L, and radius R. Calculate the strain in the filling ε_f, in terms of the stresses σ in different directions, Poisson's ratio ν, the coefficient of thermal expansion α, and the temperature change ΔT. First, write the three-dimensional constitutive equation relating stress, strain, and temperature change:

$$\varepsilon_{fx} = [\sigma_x - \nu(\sigma_y + \sigma_z)]/E + \alpha\Delta T$$

Similarly, in the y direction,

$$\varepsilon_{fy} = [\sigma_y - \nu(\sigma_x + \sigma_z)]/E + \alpha\Delta T$$

Let us neglect stress in the z direction. Moreover, the stresses should be equal in the x and y directions in view of the cylindrical geometry assumed. So the radial and tangential strain in the filling (subscript f) are given by

$$\varepsilon_f = \sigma(1 - \nu_f)/E_f + \alpha_f\Delta T$$

Since Poisson's ratio is about 1/3, neglect of the elementary three-dimensional aspect of this problem would generate an error of about 30%. This would perhaps be acceptable in view of the other simplifications involved in this problem.

The surrounding tooth (represented by subscript t) may be thought of as a thin-walled pressure vessel in this approximation, so the tangential stress is much greater than the radial stress. The tangential strain is given by

$$\varepsilon_t = \sigma_t/E_t + \alpha_t\Delta T$$

If no gap opens up between filling and tooth (since the filling expands more than the tooth), the tangential strains will be equal, so

$$\sigma_f(1 - \nu_f)/E_f + \alpha_f\Delta T = \sigma_t/E_t + \alpha_f\Delta T$$

Now consider the radial force F. Stress is force per area, so for the filling, we consider equal radial and tangential stresses; thus,

$$\sigma_f = F/2\pi RL$$

For the tooth, represented as a thin cylindrical shell with pressure P from within (radial stress), the tangential stress is given by

$$\sigma_t = PR/t = [R/t][F/2\pi RL] = F/2\pi Lt$$

Substituting the stresses above and simplifying,

$$F = [2\pi(\alpha_f - \alpha_t)\Delta TL]/[(1/E_t t) + ((1 - \nu_f)/E_f R)]$$

Observe that the force between filling and tooth is a result of the *difference* in thermal expansion. Of course the actual geometry of a filled tooth is more complex than we have assumed here. A more accurate result would be obtained by finite element analysis involving the true geometry.

We obtain, after substituting appropriate values, forces of 574 N for a resin filling and 528 N for an amalgam filling. Moreover, we calculate the tangential tensile stress in the enamel with an amalgam filling to be 21 MPa. This is in the neighborhood of the fracture stress of enamel. Ordinarily, the dentist would avoid having such a thin layer of tooth structure remaining. Moreover, the filling is placed at a temperature between room temperature (20°C) and body temperature (37°C). Even though a range of 50°C can occur in the mouth, a portion of that range is below ambient, as in the eating of ice cream. Cooling of the filling can result in a gap forming between the filling and tooth, and leakage can occur in that gap.

3.3. PHASE DIAGRAMS

When two or more metallic elements are melted and cooled, they form an intermetallic compound, a *solid solution*, or more commonly a mixture thereof. Such combinations are called alloys. The alloys can exist as either single or multiple phase depending on temperature and composition. A *phase* is defined as a physically homogeneous part of a material system. Thus, liquid and gas are each single phases, but there can be more than one phase for a solid such as fcc iron and bcc iron depending on pressure and temperature. Among multiphase metals, steels are iron-based alloys containing various amounts of a carbide (usually Fe_3C) phase. In this case the carbon atoms occupy the interstitial sites of the iron atoms (see Figure 2-8); this is called an *interstitial* solid solution. Most metal atoms are too large to exist in the interstitial sites. If the two metal atoms are roughly the same size, have the same bonding tendencies, and tend to crystallize in the same types of crystal structure, then a *substitutional* solid solution may form. This structure is composed of a random mixture of two different atoms as shown in Figure 3-19. Unless the elements are very similar in properties, such a solution will exhibit a limited solubility; i.e., as more substitutional atoms are added in the matrix, the lattice will be more and more distorted until phase separation occurs at the solubility limit. In some systems, such as the Cu–Ni system, as shown in Figure 3-19, complete solid solubility exists.

The phase diagram is constructed first by preparing known compositions of Cu–Ni; melting and cooling them under thermal equilibrium. During the cooling

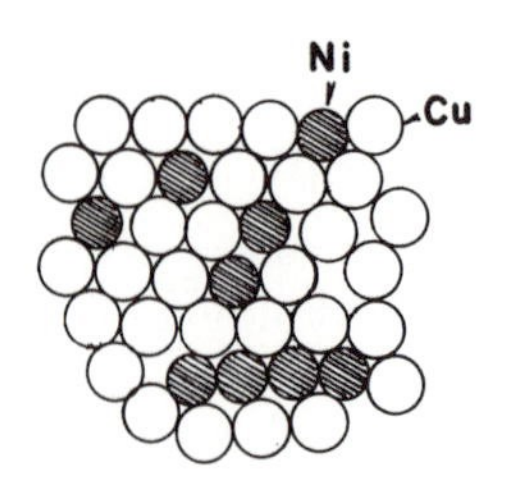

Figure 3-19. Substitutional solid solution of Cu–Ni system.

cycle one has to determine at what temperatures the first solid phase (α) appears and all of the liquid disappears. These points will determine the *liquidus* and *solidus* line in the phase diagram. From this phase diagram one can determine the types of phase and amount of each element present for a given composition and temperature. Thus, if we cool 40 wt% Ni–60 wt% Cu liquid solution, from Figure 3-20:

Temp. (°C)	Phase (relative amount)	Composition of each phase
Above 1270	Liquid (all)	40 Ni–60 Cu
1250	Liquid (63%)	33 Ni–67 Cu
	α(37%)	52 Ni–48 Cu
1220	Liquid (5%)	26 Ni–74 Cu
	α (95%)	43 Ni–57 Cu
Below 1210	α (all)	40 Ni–60 Cu

The relative amount of each phase present at a given temperature and composition is determined by the *lever rule* after making a horizontal (*tie*) line at the temperature of interest. Let C_A and C_B be the composition of element A (NI) and B (Cu) in the two-phase region met by the tie line (say 1240°C) with the same composition given above (40 wt% Ni = C_A), then the amount of liquid (L) phase can be calculated as follows:

$$\frac{\mathrm{L}}{\alpha + \mathrm{L}} = \frac{C_\mathrm{I} - C_\mathrm{A}}{C_\mathrm{I} - C_\alpha} = \frac{52 - 40}{52 - 33} = \frac{12}{19} = 0.63 \tag{3-40}$$

where C_A is the original composition of element A. This principle can be applied in more complicated systems such as Ag-Cu (eutectic) or Fe–C (eutectic + eutectoid) systems as shown in Figures 3-21 and 3-22. The eutectic and eutectoid

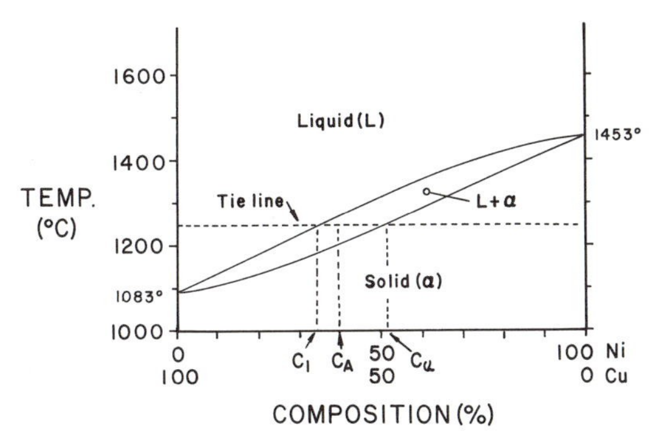

Figure 3-20. The Cu–Ni phase diagram: an example of complete solid solubility. See text for explanation of the tie line.

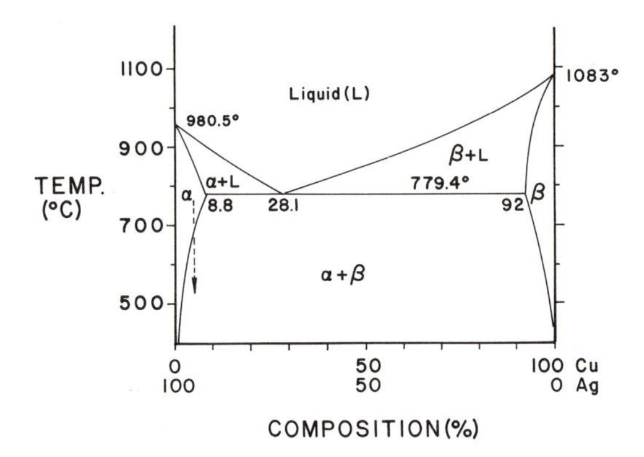

Figure 3-21. The Cu–Ag phase diagram: the dashed line indicates the precipitation hardening by quenching the α phase.

reactions are defined as

$$
\begin{aligned}
&L_2 \Leftrightarrow S_1 + S_3 \text{ (eutectic)} \\
&S_2 \Leftrightarrow S_1 + S_3 \text{ (eutectoid)}
\end{aligned}
\tag{3-41}
$$

where L refers to liquid, S refers to solid phases, and the numbers indicate the phases. There is a relatively larger amount of one of the components at those temperatures. For example, the copper content increases from 8.8% (S_1 or α), 28.1% (L), and 92% (S_3 or β) for the Cu–Ag alloy at 779.4°C as can be deduced from Figure 3-21. Note that the last liquid will disappear at the eutectic temperature and composition (see Figure 3-21).

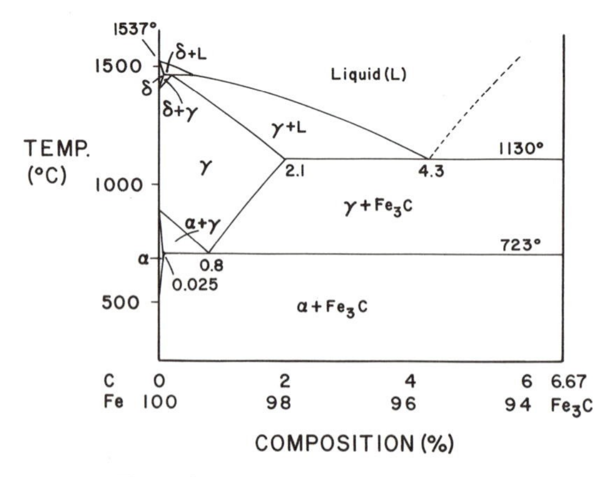

Figure 3-22. The Fe–C phase diagram.

Example 3-6

Copper and silver metal are mixed thoroughly in powder form, in proportions 80 wt% Cu and 20 wt% Ag and heated to well above the melting temperature of the alloy. The liquid metal is then cooled and allowed to reach thermodynamic equilibrium. Give the composition of each phase and the relative amount of each phase.

(a) At 1000°C.
(b) At 780°C.
(c) At 700°C.

Answer

From Figure 3-21:

(a) At 1000°C, all liquid (80 wt% Cu + 20 wt% Ag)
(b) At 780°C, β (92 wt% Cu + 8 wt% Ag) 81 wt %
L (28.1 wt % Cu + 71.9 wt% Ag) 19 wt%

$$\frac{\beta}{\beta + \mathrm{L}} = \frac{28.1 - 80}{28.1 - 92} = \underline{0.81}$$

(c) At 700°C, α (6 wt% Cu + 94 wt% Ag) 15 wt%
β (93 wt% Cu + 7 wt% Ag) 85 wt%

$$\frac{\alpha}{\alpha + \beta} = \frac{93 - 80}{93 - 6} = \frac{13}{87} = \underline{0.15}$$

3.4. STRENGTHENING BY HEAT TREATMENTS

3.4.1. Metals

One of the strengthening processes is the precipitation (or age)-hardening of alloys by heat treatments. This is accomplished by rapidly cooling (quenching) a solid solution of decreasing solubility as shown in Figure 3-21 along the dashed vertical line. If the quenching is done properly, there will not be enough time for the second phase (β) to form. Hence, a quasi-thermal equilibrium exists, but depending on the amount of thermal energy (hence temperature) and time the second phase (β) will form (precipitation). If the β-phase particles are small and uniformly dispersed throughout the matrix, their presence can increase the strength greatly. It is important that they be dispersed within a grain as well as at grain boundaries so that the dislocations can be impeded during the deformation process as in the case of cold-working.

In relation to the precipitation process, the diffusionless *martensitic* transformation process is another mechanism of strengthening steel and other alloys. When fcc iron or steel is quenched from the austenitic temperature range (γ phase in Figure 3-22), there is no time for carbon and other alloying elements to form $\alpha + \underline{\mathrm{C}}$ phases. Almost all of the carbon atoms must diffuse to form iron carbide ($\underline{\mathrm{C}}$) as well as the carbide formers (Cr, Mo, and V), which should concentrate in the carbide, whereas ferrite formers (Ni and Si) must diffuse into the ferrite (α). These reactions take time at low temperature (below 400°C). Since the fcc structure of austenite is not in equilibrium, a driving force develops and at low enough temperatures this driving force becomes sufficient to force transformation by shear. The resulting structure is a *tetragonal martensite* instead of the body-centered cubic ferrite. The martensite is extremely hard because it is noncubic (has fewer slip systems than the bcc structure) and the interstitially entrapped carbon prevents slip. Martensite is the hardest iron-rich phase material but is extremely brittle. Hence, tempering by heating (600°C) and slow cooling is necessary to make the material tough and strong:

$$\text{Martensite} \xrightarrow{\text{tempering}} \alpha + \text{Carbide } (\underline{\text{C}}; \text{Fe}_3\text{C}) \tag{3-42}$$

3.4.2. Ceramics and Glasses

As mentioned earlier, ceramics and glasses are hard and brittle due to their *nonyielding* character during deformation, which in turn is due to their bonding characteristics. Because of this brittleness, they are subject to the effect of stress concentration at the microcracks present in the material when in the tensile deformation mode (see Section 3.1.2). Therefore, if we want to increase their strength, we can employ means to overcome these problems by (1) introducing surface compression that has to be overcome before the net stress becomes tensile, and (2) reducing stress concentration by minimizing sharp cracks on the surface and in the bulk.

We can accomplish the surface compression by thermal treatment (quenching from high to low temperature and surface crystallization) or by chemical treatment (ion exchange). In the latter process, a small ion such as Na^+ is exchanged with a larger one such as K^+. Reduction of the crack radius for glasses can be accomplished by a simple fire-polishing process, which will also concomitantly introduce surface compression making the glass very strong (this can increase its strength up to 200 times the original strength).

3.4.3. Polymers and Elastomers

Polymers and elastomers cannot be heat treated to increase their strength in general. The strengths of these materials are sensitive to the chemical composition, side group, branching, molecular weight, polydispersity, and degree of crosslinking. We will consider this topic in Chapter 8.

3.5. SURFACE PROPERTIES AND ADHESION

Surface properties are important in many material-related problems. The surface property is directly related to the bulk property since the surface is the discontinuous boundary between different phases. If ice is being melted, then there are two surfaces created between three phases, i.e., liquid (water), gas (air and water vapor), and solid (ice).

Table 3-2. Surface Tension of Materials

Substance	Temperature (°C)	Surface tension (N/m)
Mercury	20	0.465
Lead	327	0.452
Zinc	419	0.785
Copper	1131	1.103
Gold	1120	1.128

[a] $1\ N/m = 10^3\ ergs/cm^2 = 10^3\ dynes/cm$.

The *surface tension* develops near the phase boundaries since the equilibrium bonding arrangements are disrupted leading to an excess energy which will minimize the surface area. Other means of minimizing the surface energy are to attract foreign materials (*adsorption*) or bonding with the adsorbent (*chemisorption*).

The conventional units used to describe surfaces are dynes per cm or ergs per cm^2 for surface energy (or tension) but these units are exactly the same since 1 erg is 1 dyne cm. The SI unit is N/m as given in Table 3-2.

If a liquid is dropped on a solid surface, then the liquid droplet will spread or make a spherical globule as shown in Figure 3-23. At equilibrium the sum of surface tensions (γ_{GS}, γ_{LS}, and γ_{GL}) among the three phases (gas, liquid, and solid) in the solid plane should be zero since the liquid is free to move until force equilibrium is established. Therefore,

$$\gamma_{GS} - \gamma_{LS} - \gamma_{GL} \cos\theta = 0$$

$$\cos\theta = \frac{\gamma_{GS} - \gamma_{LS}}{\gamma_{GL}} \tag{3-43}$$

where θ is called the contact angle. The wetting characteristic can be described:

$$\begin{aligned} &\theta = 0 && \text{complete wetting} \\ &0 < \theta < 90^\circ && \text{partial wetting} \\ &\theta > 90^\circ && \text{nonwetting} \end{aligned} \tag{3-44}$$

Note that Eq. (3-44) gives only ratios rather than absolute values of surface tension. Some values of contact angle are given in Table 3-3.

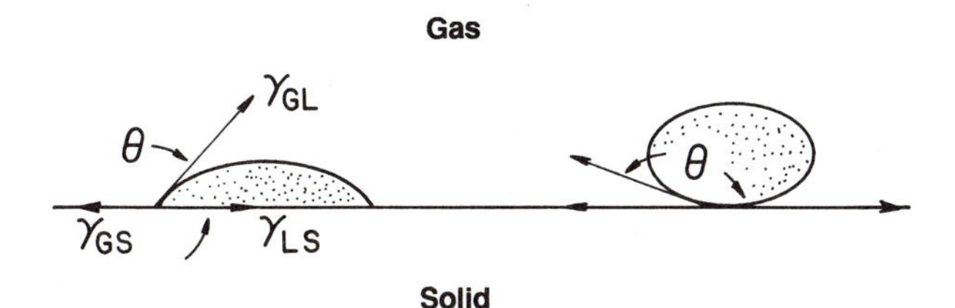

Figure 3-23. Wetting and nonwetting of a liquid on the flat surface of a solid. Note the contact angle.

Table 3-3. Contact Angle Values

Liquid	Substrate	Contact angle
Methylene iodine	Soda-lime glass	29°
(CH_2I_2)	Fused quartz	33°
Water	Paraffin wax	107°
Mercury	Soda-lime glass	140°

Table 3-4. Critical Surface Tension of Polymers

Polymer	γ_c (dynes/cm)
Polyhexamethylene adipamide, nylon 66	46
Polyethylene terephthalate	43
Poly(6-amino caproic acid), nylon 6	42
Polyvinyl chloride	39
Polyvinyl alcohol	37
Polymethyl methacrylate	33–44
Polyethylene	31
Polystyrene	30–35
Polydimethyl siloxane	24
Polytetrafluoroethylene	18.5

The lowest surface tension of a liquid (γ_{GL}) in contact with a solid surface with a contact angle (θ) greater than zero is termed the *critical surface tension* (γ_c). The critical surface tension can be obtained by measuring contact angles of a series of homologous liquids. Surface tensions of some polymers are given in Table 3-4.

When two surfaces are bonded together, it is called adhesion if the two materials are different and cohesion if the same. All cemented surfaces with a cementing agent are bonded by adhesion; hence, the cementing agent is an adhesive. For the maximum interfacial strength, the thickness of the adhesive layer must be optimal as shown in Figure 3-24.

In dental and medical applications, the adhesives should be considered a temporary remedy since the tissues are living, replacing the old cells with new ones, thus destroying the initial bonding. This problem led to the development of porous implants, which allow tissues to grow into the interstices (pores), making a viable, interlocking system between implants and tissues.

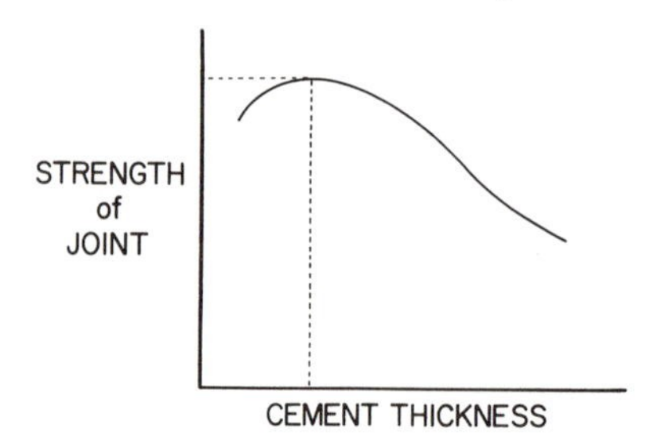

Figure 3-24. Variation of the strength of a joint versus thickness of cement between the adherend.

PROBLEMS

3-1. Which of the following materials will be best described by the three stress–strain curves of Figure 3-5?

(a) Ceramics and glasses.
(b) Plastics (polymers) such as polyethylene.
(c) Glassy polymers such as Plexiglas® (polymethyl methacrylate).
(d) Soft tissues such as skin, blood vessel walls, etc.
(e) Hard mineral tissues of bone and teeth.
(f) Steels.
(g) Rubber bands.

3-2. Poisson's ratio (ν) is defined by the following expression:

$$\nu = -\varepsilon_x / \varepsilon_z$$

where ε is strain; the load is applied in the z direction in simple tension or compression.

Silicone rubber has a Poisson's ratio of 0.4999. Consider a silicone rubber drain hose for surgical procedure; the hose has an outer diameter of 4 mm and a wall thickness of 0.5 mm. How much does the lumen (interior space) of the catheter constrict if the hose is stretched by 20%?

3-3. The following data were obtained using a stainless steel tensile specimen:

Stress (MPa)	Strain (%)	Stress (MPa)	Strain (%)
98	0.06	700	0.50
160	0.10	770	0.60
280	0.16	830	0.70
350	0.20	870	0.80
500	0.30	920	0.90
620	0.40	930	1.00

(a) Plot the stress–strain curve.
(b) Determine the modulus of elasticity, 0.2% offset yield strength, fracture strength, and toughness.

3-4. An applied strain of 0.5 produces an immediate stress of 2 MPa in a piece of rubber, but after 10 days, the stress is only 1 MPa. For the following, assume the material obeys a single-exponential relaxation process.

(a) What is the relaxation time?
(b) What is the stress after 100 days?

3-5. The following data were obtained using unknown liquids and a polyethylene sheet:

Liquid	γ (ergs/cm^2)	Contact angle (°)
A	75	96
B	40	63
C	30	15
Polyethylene	29	—

(a) Determine the interfacial surface tension for each liquid and polyethylene.
(b) What conclusions can you draw regarding γ_{GL} and γ_{SL}?

3-6. The surface properties change after a material is implanted inside the body. Explain how they will be changed and what methods should be used to understand the interaction between the tissue and implant. Can you use the data obtained from an *in vitro* surface experiment *in vivo*?

3-7. What is the endurance limit for a solid polymer (polymethyl methacrylate) in Figure 3-9? If a tooth implant is made of this material with a 4-mm diameter in cylindrical shape, how long will it last? Assume the maximum force of chewing is 100 N in compression.

3-8. Plot temperature (near T_m) versus surface tension given in Table 3-2. What conclusions can you draw?

3-9. Show that the maximum (resolved) shear stress operates on a 45° angle with respect to the force (stress) being applied. Hint: the shear stress $\tau = F_s/A_s$ can be resolved into a component upon a slip direction that has an angle of λ and a force normal to the slip plane. The slip plane has an angle of ϕ, then $\tau = \sigma \cos \lambda \cos \phi$, where σ is the applied stress on the cross-sectional area A.

SYMBOLS/DEFINITIONS

Greek Letters

α: Linear thermal expansion coefficient, amount of length change per unit length per unit temperature.

δ: Loss angle, the phase angle between stress and strain in oscillatory loading of viscoelastic materials; it is zero in an elastic material.

ε: Strain, change in length per unit length.

γ: Surface energy. See below.

γ_c: Critical surface tension, the lowest surface energy a solid can exhibit.

η: Viscosity, measure of the flow characteristic of a material (Pa s or poise).

λ: Retardation time at which the strain level is reduced to its original value by $1/e$ amount at a constant stress.

ν: Poisson's ratio, the ratio of lateral contraction to longitudinal extension in simple tension.

θ: Contact angle that is formed between liquid and solid substrate due to the partial or nonwetting nature of the surface.

σ: Stress (tensile, compressive, or shear), force per unit cross-sectional area.

τ: Relaxation time at which the stress level is reduced to its original value by $1/e$ amount at a constant strain.

Latin Letters

a: Surface crack length of an elliptic crack ($2a$ if the crack is in the bulk).

b: One-half width of an elliptic crack.

E: Young's modulus or modulus of elasticity, slope of the stress–strain curve in the elastic portion.

FS: Failure or fracture strength of a material.

$H(t)$: Heaviside step function; 0 for $t < 0$, 1 for $t > 0$.

$J(t)$: Creep compliance, ratio of time-varying strain to constant stress, $\varepsilon(t)/\sigma_0$.

k: Spring constant.

K: Fracture toughness derived from the micromechanics of crack propagation. Units are $\text{MPa}\sqrt{\text{m}}$ or $\text{psi}\sqrt{\text{in.}}$

r: Crack tip radius.

scf: Stress concentration factor, the ratio of the stress at the tip of a crack or hole to nominal stress away from the tip.

TS: Maximum or ultimate tensile strength of a material.

YS: Yield point or stress beyond which the material will be permanently deformed.

Terms

Adhesion:	Joining of two different materials.
Adsorption:	Physical attachment of foreign material (usually gas) on a surface.
Boltzmann superposition principle:	The effect of a compound cause is the sum of the effects of the individual causes. This is the statement of linearity for viscoelastic materials.
Chemisorption:	Chemical attachment of foreign material (usually gas) on a surface.
Cohesion:	Joining of identical materials.
Convolution:	A type of integral equation used in the analysis of viscoelastic materials.
Creep:	Increase of strain with time in viscoelastic materials under constant stress.
Endurance limit:	Stress level below which the material will not fail by cyclic fatigue loading no matter how many cycles (a practical limit is often chosen as 10^7 cycles).
Engineering stress:	Stress calculated based on its original cross-sectional area.
Fourier transform:	An integral transform equation involving sinusoids. In the context of viscoelastic materials, Fourier transforms relate the time and frequency domains.
Heaviside step function:	The unit step function; it has a value of zero for arguments less than zero, and one for arguments greater than zero. It is used in the analysis of viscoelastic materials.
Hooke's law:	Stress is linearly proportional to the strain. Most materials follow this law at low strains.
Isotropic:	The properties of the material are the same in every direction. Materials such as steel and glassy polymers are usually isotropic, but composite materials and biological materials are not.
Kramers Kronig relations:	Relationships between the compliance and loss functions of frequency in viscoelastic materials.
Martensite:	Iron carbon alloy (steel) obtained by quenching from austenite (γ), it has a body-centered tetragonal crystal structure.
Maxwell model:	A mechanical analog model consisting of a spring and a dashpot in series for describing viscoelastic material properties.
Necking:	Unstable irreversible flow of material locally during tensile deformation, resulting in a necklike shape.
Relaxation modulus:	Stress relaxation of viscoelastic materials is the decrease in stress that occurs under constraint strain. The relaxation modulus is the ratio of stress to strain during stress relaxation.
Surface tension (surface energy):	Amount of free energy exhibited at the surface of a material.
Tempering:	Toughening of martensite by heat treatment, the structure becomes more stable by converting to ferrite (α) and carbide microstructure.
Thermal conductivity:	Amount of heat (thermal energy) passed for a given thickness, time, and cross-sectional area of a material.
Toughness:	Amount of energy expended before its fracture or failure.
True stress:	Stress calculated based on a specimen's true cross-sectional area.
Voigt or Kelvin model:	A mechanical analog model of describing material properties by arranging a spring and dashpot in parallel.
Yield point:	The point of the stress–strain curve where transition occurs from elastic to plastic deformation, i.e., the curve deviates from the initial linear portion.

BIBLIOGRAPHY

M. F. Ashby and D. R. H. Jones, *Engineering Materials 1*, Pergamon Press, Elmsford, N.Y., 1980.

W. D. Callister, Jr., *Materials Science and Engineering: An Introduction* (2nd Ed.), Chapters 6 and 9, J. Wiley and Sons, New York, 1991.

A. H. Cottrell, *The Mechanical Properties of Matter*, Chapters 4 and 8, J. Wiley and Sons, New York, 1964.

H. W. Hayden, W. G. Moffatt, and J. Wulff, *The Structure and Properties of Materials*, Vol. III, Chapters 1 and 2, J. Wiley and Sons, New York, 1965.

F. A. McClintock and A. S. Argon (eds.), *Mechanical Behavior of Materials*, Chapters 1 and 2, Addison-Wesley Publ. Co., Reading, Mass., 1966.

L. H. Van Vlack, *Elements of Materials Science and Engineering* (6th Ed.), Chapters 4 and 5, Addison-Wesley Publ. Co., Reading, Mass., 1970.

L. H. Van Vlack, *A Textbook of Materials Technology*, Chapters 2 and 9, Addison-Wesley Publ. Co., Reading, Mass., 1973.

CHAPTER 4

CHARACTERIZATION OF MATERIALS II

In addition to the mechanical and thermal properties of materials, other physical properties could be important in particular applications of biomaterials. Properties considered in this chapter include electrical, optical, absorption of X rays, density, porosity, acoustic, ultrasonic, and diffusion.

4.1. ELECTRICAL PROPERTIES

Electrical properties of materials are important in such applications as pacemakers and stimulators, as well as in piezoelectric implants to stimulate bone growth.

Electrical resistance R is defined as the ratio between the potential difference (voltage) V applied to the object and the current i that flows through:

$$R = \frac{V}{i} \tag{4-1}$$

If the potential difference V is measured in volts (V) and the current i in amperes (A), the resistance R is in ohms, denoted by a capital Greek omega Ω. *Ohm's law* states that voltage is proportional to current in a conductor, so that the resistance R is independent of voltage. Metals obey Ohm's law if the temperature does not change much, but semiconductors do not. The resistance of an object depends upon both the material of which it is made, and the shape. The characteristic of *resistivity*, by contrast, is associated with the material itself. Resistivity ρ_e is defined as the ratio of electric field E to current density J, which is current per cross-sectional area (the electric field is the gradient in electric potential):

$$\rho_e = \frac{E}{J} \tag{4-2}$$

The units of resistivity are ohm-meter (Ω-m).

Example 4-1

Consider a pacemaker wire of circular cylindrical shape, $d = 0.1$ mm in diameter and $L = 100$ mm long, made of silver. Determine the electrical resistance.

Answer

In view of the uniform cross section, the electric field and the current density will be uniform over the length of the wire (L). The electric field is the gradient in potential, so

$$E = V/L$$

The current density is the current per unit area, or

$$J = i/A = 4i/\pi d^2$$

The resistance is therefore

$$R = V/i = 4EL/J\pi d^2 = 4L\rho_e/\pi d^2$$

For silver, $\rho_e = 1.6 \times 10^{-8}$ ohm-m, so $\underline{R = 0.2 \text{ ohm}}$. This is much smaller than the resistance of the surrounding tissue, so it is not likely to be a problem. Single-strand pacemaker wire would be vulnerable to mechanical fatigue arising from flexure from the beating heart, so multistrand wires are used.

The electrical resistivity of materials varies over many orders of magnitude. Insulators, or materials with very high resistivity, are used to isolate electrical equipment, including implantable devices such as pacemakers and other stimulators, from the body tissues. Polymers and ceramics tend to be good insulators. Electrical resistivities of representative materials are given in Table 4-1.

Table 4-1. Electrical Resistivity of Various Materials

Material	Resistivity (Ω − m)
PMMA	10^{14}
Al_2O_3	10^{12}
SiO_2	10^{12}
Bone (wet, longitudinal)	46
Muscle (wet, longitudinal)	2
Physiological saline	0.7
Stainless steel	7.3×10^{-7}
Platinum	10^{-7}
Silver	1.6×10^{-8}
Copper	1.7×10^{-8}

Piezoelectricity is a coupling between mechanical deformation and electrical polarization of a material. Specifically, mechanical stress results in electric polarization, the direct effect; and an applied electric field causes strain, the converse effect. The piezoelectric constitutive equations are as follows:

$$D_j = \sum_{jk} d_{ijk}\sigma_{jk} + \sum_{j} K_{ij}E_j + p_j\Delta T \qquad (4\text{-}3)$$

$$\varepsilon_{ij} = \sum_{kl} S_{ijkl}\sigma_{kl} + \sum_{k} d_{kij}E_k + \alpha_{ij}\Delta T \qquad (4\text{-}4)$$

in which D is the electric displacement, σ the stress, d the piezoelectric sensitivity tensor, K the dielectric permittivity, E the electric field, p the pyroelectric coefficient, T the temperature, ε the strain, S the elastic compliance, and α the thermal expansion coefficient. Only materials with sufficient asymmetry exhibit piezoelectricity or pyroelectricity and consequently have d and p coefficients different from zero. The physical origin of piezoelectricity lies in the presence of asymmetric charged groups in the material as shown in Figure 4-1. As the material is deformed, the charges move with respect to each other so that change in dipole moment occurs. Figure 4-1 also shows the polarization that results from stress via several d coefficients. As for the elastic compliance S, the first term in Eq. (4-4) represents Hooke's law as discussed in the previous chapter. All piezoelectric materials are anisotropic, and the S coefficients represent different compliances for different directions of loading. Young's modulus E is the inverse of the compliance S in the direction considered.

Fukada and Yasuda first demonstrated that dry bone is piezoelectric in the classic sense. The piezoelectric properties of bone are of interest in view of their hypothesized role in bone remodeling. Wet collagen, however, does not exhibit

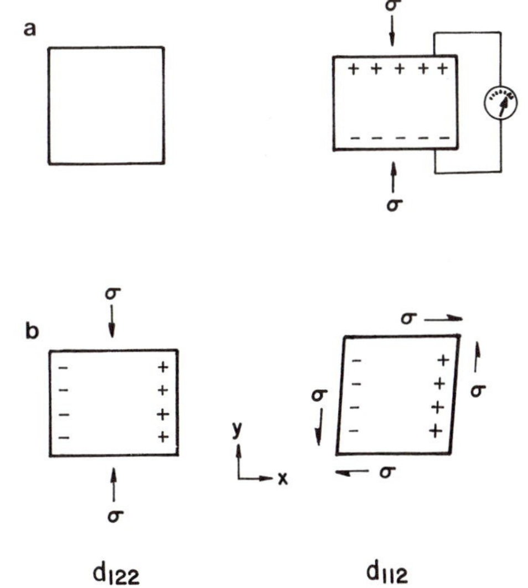

Figure 4-1. Piezoelectric materials. (a) Physical origin of piezoelectricity: Charge separation in asymmetric unit cell under deformation. (b) Piezoelectric d coefficients from Eqs. (4-3) and (4-4). Left: Polarization in $x(1)$ direction due to compressive stress in $y(22)$ direction; right: polarization in x direction due to shear stress in $xy(12)$ direction.

piezoelectric response. Studies of the dielectric and piezoelectric properties of fully hydrated bone raise some doubt as to whether wet bone is piezoelectric at all at physiological frequencies. Piezoelectric effects occur in the kilohertz range, well above the range of physiologically significant frequencies. Both the dielectric and piezoelectric properties of bone depend strongly upon frequency. The magnitude of the piezoelectric sensitivity coefficients of bone depends on frequency, on direction of load, and on relative humidity. Values up to 0.7 pC/N ($\cong 10^{-12}$ C/N) have been observed in bone, to be compared with 0.7 and 2.3 pC/N for different directions in quartz, and 600 pC/N in some piezoelectric ceramics. It is, however, uncertain whether bone is piezoelectric in the classic sense at the relatively low frequencies that dominate in the normal loading of bone. *Streaming potentials* can result in stress-generated potentials at relatively low frequencies even in the presence of dielectric relaxation but this process is as yet poorly understood. The electrical potentials observed in transient deformation of wet bone *in vivo* may be mostly due to streaming potentials.

Compact bone also exhibits a permanent electric polarization as well as *pyroelectricity*, which is a change of polarization with temperature. These phenomena are attributed to the polar structure of the collagen molecule; these molecules are oriented in bone. The orientation of permanent polarization has been mapped in various bones and has been correlated with developmental events.

Electrical properties of bone are relevant not only as a hypothesized feedback mechanism for bone remodeling, but also in the context of external electrical stimulation of bone to aid its healing and repair.

Example 4-2

Suggest an application of artificial piezoelectric materials in the body.

Answer

In some cases natural growth or repair of bone may be inadequate. Inclusion of an active piezoelectric element in a bone plate or joint replacement will generate electrical signals *in vivo*, which will stimulate the growth of bone. This approach has been used on a research basis.

Example 4-3

A piezoelectric stimulator 1 cm^2 in cross-sectional area and 1 mm thick is incorporated in a composite bone plate. It experiences 1% of the stress seen in a healthy leg bone during walking (8 MPa). The material is a lead titanate zirconate ceramic for which the relevant piezoelectric coefficient is 100 pC/N and the dielectric constant is 1000. Determine the peak voltage produced by the device. For the purpose of calculation, neglect the leakage of charge through the conductive pathways in bone.

Answer

The charge density may be calculated as follows. The charge density q/A is the piezoelectric coefficient times the stress; the stress is 0.01×8 MPa as given above.

$$q/A = 100\ \text{pC/N}\,[0.01][8 \times 10^{6}\ \text{N/m}^2] = [10^{-10}\ \text{C/N}][8 \times 10^{4}\ \text{N/m}^2] = 8\ \mu\text{C/m}^2$$

Under the assumptions given, the implant behaves as a capacitor of capacitance C, for which the charge q is $q = CV$, in which V is the voltage. $V = q/[k\varepsilon_0 A/d]$, with k the dielectric constant, ε_0 the permittivity of space, A the cross-sectional area, and d the thickness. Using the charge density from above, $V = [8\ \mu C/m^2][1\ mm]/[1000 \times 8.85 \times 10^{-12}\ F/m^2] = \underline{0.9\ volt}$. We have neglected the parallel capacitance and the leakage conductance of the surrounding tissue. As for the parallel capacitance, the dielectric constant of muscle, for example, is about 10^5 at 100 Hz and 10^8 at 0.01 Hz. The actual stimulating voltage will therefore be considerably less than determined above and will moreover decay rapidly with time, following each step in walking as a result of the conductivity of the surrounding bone and muscle.

4.2. OPTICAL PROPERTIES

Optical properties of materials are relevant to their performance when used in the eye, as well as to cosmetic aspects of dental materials. A ray of light incident upon a transparent material will be partly reflected and partly transmitted. The transmitted ray is bent or *refracted* by the material. It is observed experimentally (and can be deduced from Maxwell's equations) that the incident ray, normal to the surface, and refracted ray all lie in the same plane, and the angle of incidence equals the angle of reflection. The angle of the refracted ray depends upon a material property known as the *refractive index*, usually denoted by n. The refractive index is defined as the ratio of the speed of light in vacuum to the speed of light in the medium. The relation between the angles of incidence and of refraction is given by *Snell's law* as follows:

$$n_1 \sin\theta_1 = n_2 \sin\theta_2 \tag{4-5}$$

in which θ_1 is the angle of the incident ray with respect to the normal to the surface, n_1 is the refractive index of the medium containing the incident ray, and θ_2 is the angle of the refracted ray to the normal to the surface of the material, which has refractive index n_2 as shown in Figure 4-2. Some representative indices of refraction are given in Table 4-2, for yellow-orange light at a wavelength of 589 nm.

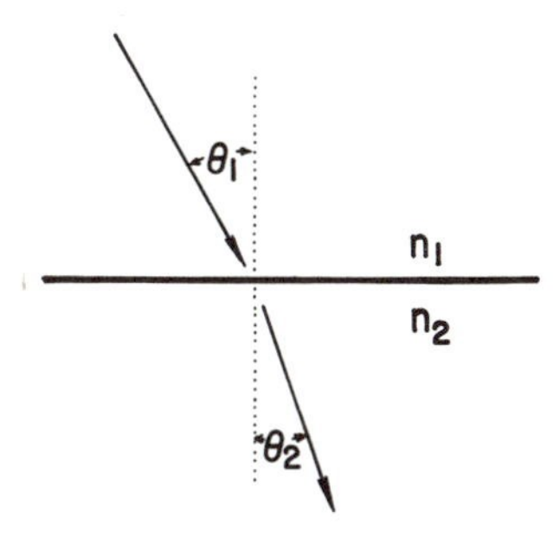

Figure 4-2. Snell's law for refracted light.

Table 4-2. Refractive Index of Some Materials

Material	Refractive index
Vacuum	1.0
Air	1.0003
Water	1.33
Human aqueous humor	1.336
Human vitreous humor	1.338
Human cornea	1.376
Human lens	1.42
HEMA hydrogel, wet	1.44
PMMA	1.49
Polyethylene (film)	1.5
Crown glass	1.52
Flint glass	1.66

In ophthalmological biomaterials, transparent materials find application in lenses. Refraction of light by a convex lens is shown in Figure 4-3. The focal length of such a lens is defined as the distance from the lens to the image plane when parallel rays of light (from far away) impinge upon the lens. The focal length f of a simple, thin lens (in air or vacuum) depends on its refractive index n and the surface radii of curvature r_1 and r_2 as follows:

$$\frac{1}{f} = (n - 1)\left(\frac{1}{r_1} - \frac{1}{r_2}\right) \tag{4-6}$$

This is the "lens makers equation" and is derivable from Snell's law. The sign convention for the radii of curvature is that they are considered positive if the center of curvature is on the right side of the lens. Consequently, both surfaces of a double convex lens contribute positive power. In biomedical applications, the lens is likely to have one or more interfaces with tissue fluid or the tissues

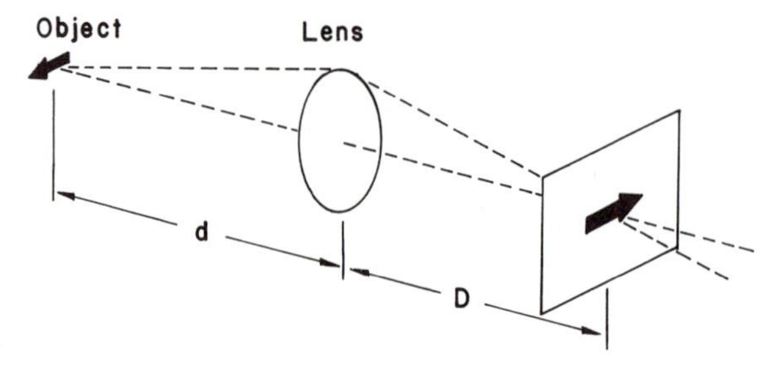

Figure 4-3. Refraction of light by a convex lens.

of the eye. The optical power of lenses is usually expressed in diopters (D) in the ophthalmic setting:

$$P(\mathrm{D}) = \frac{1}{f(\text{meters})} \tag{4-7}$$

so that a converging lens with a focal length of 100 mm has a power of +10 D. Diverging lenses have negative focal lengths, hence negative optical power. A typical, normal human eye has an optical power of about 60 D of which about 43 D is associated with the curvature of the cornea. The eye measures about 22 mm from cornea to retina.

Example 4-4

Consider a PMMA intraocular lens of diameter $d = 8.3$ mm, thickness $t = 2.4$ mm, anterior radius of curvature 17.8 mm, and posterior radius of curvature 10.7 mm. Determine the optical power and focal length of this lens in air and in water.

Answer

Use the thin lens equation (4-6). In air, assume $n = 1$ for air and take $n = 1.49$ for PMMA from Table 4-2. $P = (1.49 - 1)(1/0.0178\text{ m} - 1/-0.0107\text{ m}) = 73\text{ m}^{-1} = \underline{73\text{ D}}$, so $f = 1/P = \underline{13.6\text{ mm}}$ in air. Water has an index of refraction $n = 1.33$, so $P = (1.49 - 1.33)(1/0.0178\text{ m} - 1/-0.0107\text{ m}) = 24\text{ m}^{-1} = \underline{24\text{ D}}$, so $f = 1/P = \underline{42\text{ mm}}$ in water. We remark that the normal lens of the eye has an optical power of about 19 D.

Transparent biomaterials are used to make contact lenses and intraocular lenses. PMMA is the material of choice for intraocular lenses and for hard contact lenses. Its main disadvantage as a contact lens material is that its permeability to oxygen is low, so that the cornea, which receives its oxygen by diffusion from the air, suffers hypoxia. Other materials are used in ocular applications. Representative material properties are given in Table 4-3. Polycarbonate is an amorphous thermoplastic that has been used in the manufacture of contact lenses, particularly those of strong power. The comparatively high refractive index of polycarbonate permits a thinner, lighter lens to be made. Soft contact lenses are commonly made of the water-absorbing hydrogel material poly-HEMA (hydroxyethyl methacrylate), which is highly permeable to oxygen.

Table 4-3. Physical Properties of Some Transparent Materials

Material	Density (g/cm^3)	Refractive index	Young's modulus (MPa)	Tensile strength (MPa)
PMMA	1.19	1.49	2800	55
Silicone rubber	0.99–1.5	1.43	6	2.4–6.9
Silicone rubber, contact lens	1.09	1.43	6	1.4
Polycarbonate	1.2	1.59	2200	60

4.3. X-RAY ABSORPTION

The ability of materials to absorb X rays is of interest in the context of the visibility of an implanted object in diagnostic radiographs. X rays are electromagnetic waves similar to light except that their wavelength is much shorter and their energy is much higher. The index of refraction for X rays in matter of any kind is very nearly unity. Consequently, X rays are neither bent nor reflected to any appreciable extent as they interact with matter. The contrast that gives rise to an image in an X-ray film comes about from differences in materials' ability to absorb the X rays. The absorption follows *Beer's law*:

$$I = I_0 \, \mathrm{e}^{-\alpha x} \tag{4-8}$$

in which I is the intensity at a depth x and α is the absorption coefficient. Absorption of X rays in matter is governed by the *photoelectric effect*, in which the incident X-ray photon is absorbed by one of the bound electrons in an atom of material, and the electron is ejected; and by the *Compton effect*, in which the X-ray photon is scattered from a free or weakly bound electron in the material. The absorption due to the photoelectric effect process is proportional to the fifth power of the atomic number N, and increases with the wavelength (λ) (hence decreases with the energy) in the X-ray energy range 100 to 350 keV:

$$\alpha = N^5 \lambda^{7/2} \tag{4-9}$$

At energies below 20 keV, a resonant effect occurs in which absorption becomes very strong when the X-ray energy is equal to the binding energy of the electrons in the inner "K" shell of the atoms of the absorber material. Clinical X-ray diagnostic equipment operates at tube voltages of from 20 to 200 kV. The emitted X rays are at energies (in electron volts) equal to or less than the tube voltage. Most radiological techniques involve tube voltages between 60 and 100 kV, for which absorption by the photoelectric effect and the Compton effect are comparably important. Since the X-ray energy

$$E = h\nu = hc/\lambda \tag{4-10}$$

in which h is Planck's constant and c is the speed of light, the X rays have wavelengths from 100 pm (0.1 nm) at 10 keV to 5 pm at 200 keV. These wavelengths are much smaller than those of visible light, 400 to 700 nm.

It is clear that the heavier elements absorb X rays strongly as given in Table 4-4. Heavy metals such as lead are commonly used to shield X-ray equipment. Human soft tissue contains a great deal of the lighter elements hydrogen, carbon, and oxygen and is consequently relatively transparent to X rays. Bone, by virtue of its calcium and phosphorus content, absorbs more strongly and

Table 4-4. Mass Absorption Coefficient of Various Materials

Material	Atomic No.	Density ρ (g/cm^3)	Specific absorption coefficient μ/ρ (cm^2/g)
Al	13	2.70	48.7
P	15	1.82	73
Ca	20	1.55	172
Cr	24	7.19	259
Fe	26	7.87	324
Co	27	8.9	354
Pb	82	11.34	241

For Cu$K\alpha$ X rays, wavelength $\lambda = 1.54$ Å or 0.154 nm.

therefore shows up well in X-ray images. Metallic implants absorb strongly and also are highly visible in X-ray images. Polymers, by contrast, are relatively transparent to X rays. Barium sulfate is incorporated in bone cement to make it visible in diagnostic X-ray images.

4.4. DENSITY AND POROSITY

The density ρ of a material is defined as the ratio of mass to volume for a sample of the material,

$$\rho = \frac{m}{V} \tag{4-11}$$

A biomaterial that replaces an equivalent volume of tissue may have a different weight, as a result of the differences in density. In some applications this can be problematical. Densities of representative materials are given in Table 4-5.

Porous materials are used in a variety of biomedical applications including implants and filters for extracorporeal devices such as heart–lung machines. In other applications such as bone plates, porosity may be an undesirable characteristic since pores concentrate stress and decrease mechanical strength. Perhaps the most important physical quantity associated with porous materials is the *solid volume fraction* V_s. The *porosity*, often expressed as a percent figure, is given by

$$\text{Porosity} = 1 - V_s \tag{4-12}$$

It is also noted that there are three measurements of volume, i.e., true, apparent, and total (bulk) volume,

Table 4-5. Density of Some Materials

Material	Density (g/cm^3)
Air (at STP)	0.0013
Fat	0.94
Polyethylene, UHMW	0.94
Water	1.0
Soft tissue	1.01–1.06
Rubber	1.1–1.2
Silicone rubber	0.99–1.50
PMMA	1.19
Compact bone	1.8–2.1
Glass	2.4–2.8
Aluminum	2.8
Titanium	4.5
Stainless steel	7.93
Wrought CoCr	9.2
Gold	19.3

$$\text{True volume} = \text{total (bulk) volume} - \text{total pore volume} \tag{4-13}$$

$$\text{Apparent volume} = \text{total (bulk) volume} - \text{open pore volume} \tag{4-14}$$

$$\text{Total pore volume} = \text{open pore volume} + \text{closed pore volume} \tag{4-15}$$

We can also have three types of densities corresponding to the three definitions of the volume.

The *pore size* is important in situations in which tissue ingrowth is to be encouraged, or if the permeability of the porous material is of interest. Porous materials may be characterized by a single pore size, or may exhibit a distribution of pore sizes. Porosity and pore size can be measured in a variety of ways. If the density of the parent solid is known, a measurement of apparent density of a block of material suffices to determine the porosity. Mercury intrusion porosimetry is a more precise method that measures the porosity and the pore size distribution. In this method, mercury is forced into the pores under a known pressure and the relationship between pressure and mercury volume is determined. Since the mercury has a high surface tension and does not wet most materials, higher pressures are required to force the mercury into progressively smaller pores. If a single pore size predominates, it can be measured by optical or electron microscopy.

Example 4-5

A polyurethane open cell foam is to be used for a lining for an artificial leg. The solid volume fraction is 4%, the pores are 0.5 mm in diameter, and the solid polyurethane has a density of 0.9 g/cm^3. Determine the porosity and apparent density of the material, and the weight of a sheet 200 mm by 200 mm by 1 cm thick.

Answer

Porosity $P = 1 - V_s$, so $P = 1 - 0.04 = 0.96$, or 96% porosity
Apparent density $= \rho_{app} = V_s \rho_{true} = 0.04(0.9 \text{ g/cm}^3) = 0.036 \text{ g/cm}^3$
Mass $= V\rho_{app} = 20 \times 20 \times 1 \text{ (cm}^3\text{) } 0.036 \text{ g/cm}^3 = 14.1 \text{ g}$, so weight $= 0.032$ lb

The weight of the foam lining is much less than that of an artificial leg. In some cases, the effect of implant density is important. For example, intraocular lenses made of PMMA are denser than the soft tissues of the eye, as shown in Table 4-5. During rapid eye movements, the lens is accelerated by the structures to which it is attached, e.g., the iris. By contrast the natural eye lens is essentially neutrally buoyant. Damage to these eye structures by intraocular lenses has been observed clinically. Lenses of a silicone rubber composition of lower density are currently under investigation; these would offer the added benefit of being softer, hence easier to insert through a small incision. Another example of the density of implants is augmentation mammoplasty, in which the added weight of the implant can be problematical.

4.5. ACOUSTIC AND ULTRASONIC PROPERTIES

Acoustic and ultrasonic properties of biomaterials are relevant in the context of their importance in diagnostic ultrasound images. Important properties are the acoustic velocity v, the acoustic attenuation α, and the material density ρ. The relation for the attenuation of ultrasound is identical to that for X rays: Eq. (4-8).

Signals for ultrasonic imaging devices are generated by *reflection* of the waves from interfaces in the body, in contrast to diagnostic X rays in which differences in attenuation are exploited to make an image. The amplitude reflection coefficient associated with an interface between material 1, containing the incident wave, and material 2 is given by

$$R_A = \frac{Z_2 - Z_1}{Z_2 + Z_1} \tag{4-16}$$

The acoustic impedance Z is defined as

$$Z = \rho v \tag{4-17}$$

in which the acoustic velocity v is proportional to the square root of the bulk stiffness divided by the material mass density. By contrast to X rays, the atomic number has no bearing on ultrasound. The reflection of electromagnetic waves such as light is similarly determined from the index of refraction, which is also defined in terms of wave velocity.

Comparative acoustic properties of some relevant materials are given in Table 4-6. Attenuations are given for a frequency of 1 MHz, at the lower end of the range used clinically. The ultrasonic attenuation α (per unit length) is directly

0353'''02871

Table 4-6. Acoustic Properties of Some Materials

Material	Velocity v (m/sec)	Impedance Z (kRayl)[a]	Attenuation coeff. α (dB/cm)
Air (at STP)	330	0.04	12
Water	1480	148	0.002
Fat	1450	138	0.63
Blood	1570	161	0.18
Kidney	1560	162	1.0
Soft tissues (av.)	1540	163	0.7
Liver	1550	165	0.94
Muscle	1580	170	1.3–3.3
Bone	4080	780	15
PMMA	2670	320	
UHMWPE	2000	194	
Ti6Al4V	4955	2225	
Stainless steel	5800	4576	
Barium titanate	4460	2408	

[a] kRayl = 10^4 kg/m^2/sec.

related to the viscoelastic loss tangent tan δ discussed in Chapter 3; the attenuation therefore depends on frequency. The relation is

$$\alpha = (2\pi\nu/v)\tan\delta/2 \tag{4-18}$$

in which ν is the frequency of the wave and v is its velocity. As for biomaterials, we observe that polymeric implants, which do not show up well in X-ray images, will usually be highly visible in a diagnostic ultrasound scan as a result of the differences in acoustic properties in comparison with the natural tissues.

4.6. DIFFUSION PROPERTIES

Diffusion properties of materials are important in applications in which transport of biologically significant constituents is required. Examples include the transport of oxygen and carbon dioxide from the atmosphere to the blood in the artificial lung component of the heart–lung machine and the transport of oxygen to the cornea through contact lenses. The diffusion equation, which governs the motion of dissolved materials under a gradient of concentration C, is given by

$$\partial C/\partial t = D\nabla^2 C \tag{4-19}$$

in which D is the diffusion coefficient, and ∇^2 is the Laplacian, which in one dimension reduces to $\partial^2 C/\partial x^2$. The driving force for material transport may be a pressure gradient rather than a concentration gradient. Moreover, the geometry

Table 4-7. Gas Permeability of Various Materials

Material	Permeability to O_2	Applications
Silicone rubber	50	Contact lens, lung
Polyalkylsulfone	6	Lung
Polyethylenecellulose-perfluorobutyrate	5	Lung
Teflon film	1.1	Lung
Poly-HEMA	0.69	Contact lens
PMMA	0.0077	Contact lens

Units: $(cm^3/sec)[(cm\ thick/cm^2)(cm\ Hg \times 10^{-9})]$.

in many biomedical applications may be approximated by a thin film. In that case, the volumetric flux F (in units of volume per unit time) across a layer of area A is given by

$$F = KA\Delta P \tag{4-20}$$

in which ΔP is the pressure difference across the layer, and K is the permeability coefficient. Representative permeabilities for oxygen transport are given in Table 4-7. Permeabilities for other gases are in general different. Carbon dioxide, for example, diffuses through these materials from two to five times more rapidly than oxygen.

When high oxygen transport is desired, a material with a large permeability coefficient should be chosen, if all other aspects of the materials under question are comparable. In the case of contact lenses, poly-HEMA lenses are commonly used for soft lenses, even though the permeability is lower than that of silicone rubber. The permeability of poly-HEMA is adequate for the oxygenation of the cornea, and it is chosen for other reasons, such as ease of manufacture. As for membrane materials for oxygenators in heart–lung machines, the oxygen transport depends on the membrane thickness as well as the permeability. Minimum thickness is dictated by the membrane strength, so a strong material with a lower permeability may result in the highest oxygen flux.

PROBLEMS

4-1. The crystalline lens of a normal human eye has radii of curvature 10.2 mm and 6 mm, is convex, and has a thickness of 2.4 mm. The refractive index varies with position but may be taken as 1.386. Determine the optical power and focal length, recognizing that *in vivo* the lens is immersed in ocular fluid. We remark that "crystalline" in this context is an anatomical term that means transparent; it does not mean the lens has a regular atomic arrangement. The lens is actually fibrous.

4-2. Barium compounds and iodine compounds are used to enhance contrast in diagnostic radiology. Why?

4-3. Show that a contact lens that fits the cornea exactly will provide the same optical correction whether the lens is in fact on the cornea or separated by an infinitesimal air gap. *Hint*: Consider two lenses of different indices of refraction and one common curvature. Determine the focal length for two cases: lenses in contact and lenses separated by a layer of air.

4-4. Consider a hemispherical silicone augmentation mammoplasty 130 mm in diameter. Determine the weight of two such mammoplasties. Compare with a corresponding volume of natural tissue. Discuss the implications.

4-5. Determine the reflection coefficient for ultrasonic waves crossing a muscle to bone interface. Some of the needed material properties may be found elsewhere in this book. Discuss the implications for ultrasonic imaging through bone. Discuss the use of other kinds of waves for imaging through bone.

4-6. Calculate the voltage generated in bone by mechanical deformation associated with walking. Use the given piezoelectric coefficient for bone, and find any other needed material properties elsewhere in this book.

4-7. It is often difficult to remove the bone cement from leg bones from which a hip joint prosthesis must be removed for revision arthroplasty. An experimental technique for this purpose is extracorporeal shock wave lithotripsy (ESWL), a method that was originally developed to shatter kidney stones by intense sonic shock pulses without surgery. Calculate the relative sound intensity of a wave after passing through 5 cm of muscle and 2 cm of bone, as a percent of the initial intensity.

4-8. Discuss the advantages and disadvantages of the ESWL method described in Problem 4-7.

SYMBOLS/DEFINITIONS

Greek Letters

α	Attenuation coefficient for ultrasound or X rays.
λ	Wavelength of light, X rays, or ultrasound.
ν	Frequency of a wave.
ρ	Density.
ρ_e	Electrical resistivity.
∇^2	Laplacian operator, which in one dimension reduces to $\partial^2/\partial x^2$.

Latin Letters

c	Speed of light.
C	Concentration.
dB	Decibel. A logarithmic unit of ratios, often applied to sound pressure levels.
D	Diopter, see below.
E	Young's modulus, the ratio of stress to strain for simple tension or compression.
f	Focal length of a lens.
F	Volumetric flux, volume per unit time.
h	Planck's constant, associated with the quantum nature of light.
K	Permeability coefficient.
m	Mass.
n	Refractive index.
N	Atomic number.
R	Electrical resistance. Also, amplitude reflection coefficient.
S	Elastic compliance. For anisotropic materials the compliance tensor describes the strain-to-stress ratio for different directions. For example, $S_{1111} = 1/E_1$.
v	Ultrasonic wave speed.
V	Voltage.
V_s	Solid volume fraction of porous material.
Z	Acoustic impedance, equal to density times sound velocity.

Terms

Attenuation: Absorption of sound or ultrasound waves.

Beer's law: Absorption of X rays or light occurs such that the transmitted intensity decreases exponentially with distance.

Compton effect: A mechanism for the scattering of X rays be free or weakly bound electrons in matter.

Diopter: Optical power of lens, or degree of ray divergence of a bundle of light rays. The power in diopters is the inverse of the focal length measured in meters.

Maxwell's equations: Set of four field equations that govern electricity, magnetism, radio wave propagation, and the behavior of light.

Permeability: The ability of a material to pass a gaseous or ionic species, under a pressure gradient or a concentration gradient.

Photoelectric effect: A mechanism for the extinction of X rays by the X-ray photon knocking bound electrons out of atoms.

Piezoelectricity: Electrical polarization of a material in response to mechanical stress.

Pyroelectricity: Electrical polarization of a material in response to temperature change.

Refractive index: Ratio of speed of light in vacuum to speed of light in a material. It is a measure of the ability of a material to refract (bend) a beam of light.

Snell's law: Governing equation for the angle of incident and refracted light rays.

Streaming potential: Electric potential developed when charged particles such as ions flow through a tube or porous medium such as bone; potential may be attributed to the imbalance created by washing away the electric double layer.

Tensor: A mathematical expression with well-defined transformation properties under changes in coordinates. A vector is a tensor of rank one; a scalar is a tensor of rank zero. Stress and strain are examples of tensors of rank two.

BIBLIOGRAPHY

A. P. Arya, *Fundamentals of Nuclear Physics*, Allyn and Bacon, Boston, 1966.

N. Baier and G. Lowther, *Contact Lens Correction*, Butterworths, London, 1977.

S. Balter, "X-ray Equipment Design," in *Encyclopedia of Medical Devices and Instrumentation*, J. G. Webster (ed.), J. Wiley and Sons, New York, 1988.

D. O. Cooney, *Biomedical Engineering Principles*, Marcel Dekker, New York, 1976.

G. W. R. Davidson, III, "Biomaterials, Testing Structural Properties of," in *Encyclopedia of Medical Devices and Instrumentation*, J. G. Webster (ed.), J. Wiley and Sons, New York, 1988.

E. Fukada and I. Yasuda, "On the Piezoelectric Effect of Bone," *J. Physiol. Soc. Jpn.*, *12*, 1158, 1957.

A. Goldstein, "Ultrasonic Imaging," in *Encyclopedia of Medical Devices and Instrumentation*, J. G. Webster (ed.), J. Wiley and Sons, New York, 1988.

R. L. Kronenthal, *Polymers in Medicine and Surgery*, Plenum Press, New York, 1975.

M. R. O'Neal, "Contact Lenses," in *Encyclopedia of Medical Devices and Instrumentation*, J. G. Webster (ed.), J. Wiley and Sons, New York, 1988.

M. Spector, M. Miller, and N. Bealso, "Porous Materials," in *Encyclopedia of Medical Devices and Instrumentation*, J. G. Webster (ed.), J. Wiley and Sons, New York, 1988, p. 2335.

P. N. T. Wells (ed.), *Scientific Basis of Medical Imaging*, Churchill Livingstone, Edinburgh, 1982.

CHAPTER 5

METALLIC IMPLANT MATERIALS

Metals have been used in various forms as implants. The first metal developed specifically for human use was the "Sherman Vanadium Steel," which was used to manufacture bone fracture plates and screws. Most metals such as Fe, Cr, Co, Ni, Ti, Ta, Mo, and W used for manufacturing implants can be tolerated by the body in minute amounts. Sometimes those metallic elements, in naturally occurring forms, are essential in cell functions (Fe) or synthesis of a vitamin B_{12} (Co), but cannot be tolerated in large amounts in the body. The biocompatibility of the implant metals is of considerable concern because they can corrode in the hostile body environment. The consequence of corrosion is loss of material, which will weaken the implant, and probably more important, the corrosion products escape into the tissue resulting in undesirable effects. In this chapter we study the composition–structure–property relationship of metals and alloys used for implant fabrications.

5.1. STAINLESS STEELS

The first stainless steel used for implant materials was the 18-8 (type 302 in modern classification), which is stronger than the vanadium steel and more resistant to corrosion. Vanadium steel is no longer used in implants since its corrosion resistance is inadequate as discussed in Section 5.6. Later, 18-8sMo stainless steel was introduced, which contains molybdenum to improve the corrosion resistance in salt water. This alloy became known as type 316 stainless steel. In the 1950s the carbon content of 316 stainless steel was reduced from 0.08 wt% to 0.03 wt% maximum for better corrosion resistance in chloride solution; this became known as 316L.

5.1.1. Types and Compositions of Stainless Steels

Chromium is a major component of corrosion-resistant stainless steel. The minimum effective concentration of chromium is 11 wt%. The chromium is a reactive element but it and its alloys can be passivated to give an excellent corrosion resistance.

The *austenitic stainless steels*, especially types 316 and 316L, are most widely used for implants. These are *not* hardenable by heat treatment but can be hardened by cold-working. This group of stainless steels is nonmagnetic and possesses better corrosion resistance than any others. The inclusion of molybdenum enhances resistance to pitting corrosion in salt water. The ASTM (American Society for Testing and Materials) recommends type 316L rather than 316 for implant fabrication. The specifications for 316 and 316L stainless steels are given in Table 5-1.

The nickel serves to stabilize the austenitic phase at room temperature and, in addition, to enhance corrosion resistance. The austenitic phase stability can be influenced by both the Ni and Cr contents as shown in Figure 5-1 for 0.10 wt% carbon stainless steels.

5.1.2. Properties of Stainless Steel

Table 5-2 gives the mechanical properties of 316 and 316L stainless steels. As can be noted, a wide range of properties can be obtained depending on the heat treatment (to obtain softer materials) or cold-working (for greater strength and hardness). The designer must be careful when selecting materials of this type. Even the type 316L stainless steels may corrode inside the body under certain circumstances such as in a highly stressed and oxygen-depleted region.

Table 5-1. Compositions of 316 and 316L Stainless Steels[a]

	Composition (wt%)	
Element	Grade 1	Grade 2
Carbon	0.08 max	0.03 max
Manganese	2.00 max	2.00 max
Phosphorus[b]	0.03 max	0.03 max
Sulfur	0.03 max	0.03 max
Silicon	0.75 max	0.75 max
Chromium	17.00–20.00	17.00–20.00
Nickel	12.00–14.00	12.00–14.00
Molybdenum	2.00–4.00	2.00–4.00

[a] From *Annual Book of ASTM Standards*, Part 46, American Society for Testing and Materials, Philadelphia, 1980, p. 578.
[b] Slight variations are given (0.025 max) for special quality stainless steels (F318 and F139 of ASTM).

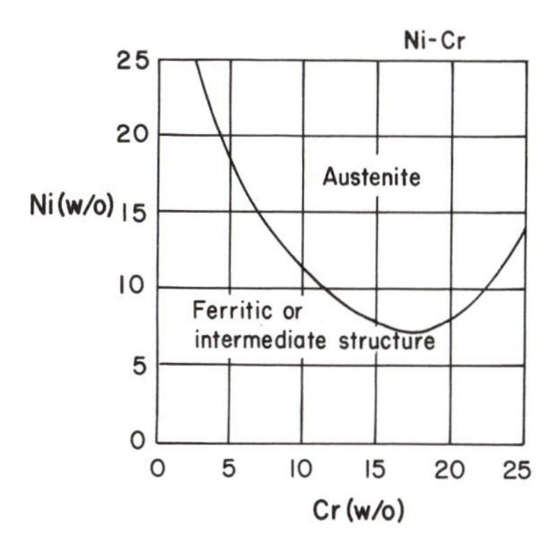

Figure 5-1. The effect of Ni and Cr contents on the austenitic phase of stainless steels containing 0.1 wt% C. From F. H. Keating, *Chromium–Nickel Austenitic Steels*, Butterworths, London, 1956.

They are, however, suitable to use in temporary devices such as fracture plates, screws, and hip nails.

5.1.3. Manufacturing of Implants Using Stainless Steel

The austenitic stainless steels work-harden very rapidly as shown in Figure 5-2 and therefore cannot be cold-worked without intermediate heat treatments. The heat treatments should not induce, however, the formation of chromium carbide (CCr_4) in the grain boundaries; that may cause corrosion. For the same reason, the austenitic stainless steel implants are not usually welded.

The distortion of components by the heat treatments can occur but this problem can be solved easily by controlling the uniformity of heating. Another undesirable effect of the heat treatment is the formation of surface oxide scales, which have to be removed either chemically (acid) or mechanically (sandblasting).

Table 5-2. Mechanical Properties of Stainless Steel Surgical Implants[a]

Condition	Ultimate tensile strength, min, psi (MPa)	Yield strength (0.2% offset), min, psi (MPa)	Elongation 2 in. (50.8 mm), min, %	Rockwell hardness, max
		Grade 1 (type 316)		
Annealed	75,000 (515)	30,000 (205)	40	95 HRB
Cold-finished	90,000 (620)	45,000 (310)	35	—
Cold-worked	125,000 (860)	100,000 (690)	12	300–350
		Grade 2 (type 316L)		
Annealed	73,000 (505)	28,000 (195)	40	95 HRB
Cold-finished	88,000 (605)	43,000 (295)	35	—
Cold-worked	125,000 (860)	100,000 (690)	12	—

[a] From *Annual Book of ASTM Standards*, Part 46, American Society for Testing and Materials, Philadelphia, 1980, p. 579.

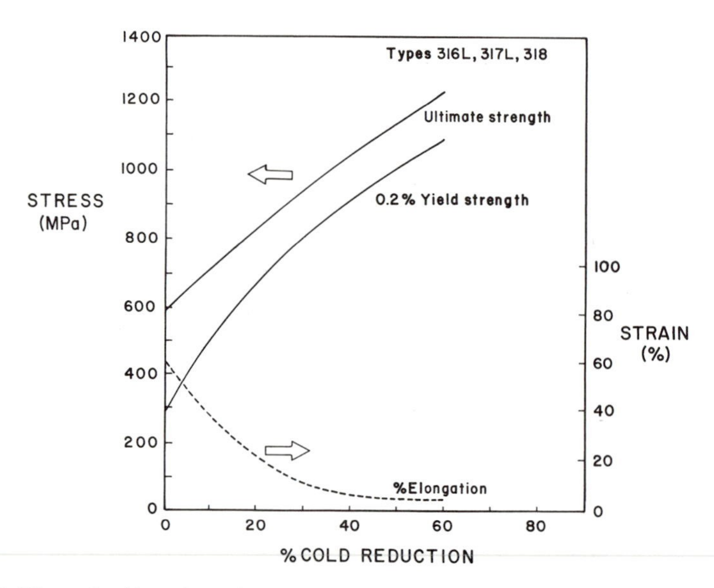

Figure 5-2. Effect of cold-work on the yield and ultimate tensile strength of 18-8 stainless steel. From *Source Book on Industrial Alloy and Engineering Data*, American Society for Metals, Metal Park, Ohio, 1978, p. 223.

After the scales are removed, the surface of the component is polished to a mirror or mat finish. The surface is then cleaned, degreased, and passivated in nitric acid (ASTM Standard F86). The component is washed and cleaned again before packaging and sterilizing.

Example 5-1

Calculate the amount of the volume change when iron is oxidized to FeO ($\rho = 5.95\ \mathrm{g/cm^3}$). The density of Fe is $7.87\ \mathrm{g/cm^3}$.

Answer

Since the molecular weight of Fe is 55.85 g/mol

$$\frac{55.87\ \mathrm{g/mol}}{7.87\ \mathrm{g/cm^3}} = 7.1\ \mathrm{cm^3/mol}$$

The molecular weight of FeO is 71.85 g/mol, hence

$$\frac{71.85\ \mathrm{g/mol}}{5.95\ \mathrm{g/cm^3}} = 12.08\ \mathrm{cm^3/mol}$$

Therefore, $\Delta V = (12.08 - 7.1)/7.1 = \underline{0.7}$ (70% volume increase by oxidation). This increase in volume by oxidation causes the oxides formed to be very porous. Continued oxidation can then take place via further diffusion of oxygen to the underlying metal.

5.2. Co-BASED ALLOYS

These materials are usually referred to as cobalt–chromium alloys. There are basically two types; one is the CoCrMo alloy, which is usually used to *cast* a product, and the other is the CoNiCrMo alloy, which is usually *wrought* by (hot) *forging*. The castable CoCrMo alloy has been in use for many decades in dentistry and recently in making artificial joints. The wrought CoNiCrMo alloy is a newcomer now used for making the stems of prostheses for heavily loaded joints such as the knee and hip.

5.2.1. Types and Compositions of Co-Based Alloys

ASTM lists four types of Co-based alloys that are recommended for surgical implant applications: (1) cast CoCrMo alloy (F76), (2) wrought CoCrWNi alloy (F90), (3) wrought CoNiCrMo alloy (F562), and (4) wrought CoNiCrMoWFe alloy (F563). The chemical compositions of the first three types are summarized in Table 5-3. At the present time, only two of the four alloys are used extensively in implant fabrications, the castable CoCrMo and the wrought CoNiCrMo alloy. As can be seen in Table 5-3, the compositions of the alloys are quite different.

5.2.2. Properties of Co-Based Alloys

The two basic elements of the Co-based alloys form a solid solution of up to 65 wt% Co and the remainder is Cr as shown in Figure 5-3. The molybdenum is added to produce finer grains, which results in higher strengths after casting or forging.

Table 5-3. Chemical Compositions of Co-Based Alloys[a]

	CoCrMo (F75)		CoCrWNi (F90)		CoNiCrMo (F562)	
Element	Min.	Max.	Min.	Max.	Min.	Max.
Cr	27.0	30.0	19.0	21.0	19.0	21.0
Mo	5.0	7.0	—	—	9.0	10.5
Ni	—	2.5	9.0	11.0	33.0	37.0
Fe	—	0.75	—	3.0	—	1.0
C	—	0.35	0.05	0.15	—	0.025
Si	—	1.00	—	1.00	—	0.15
Mn	—	1.00	—	2.00	—	0.15
W	—	—	14.0	16.0	—	—
P	—	—	—	—	—	0.015
S	—	—	—	—	—	0.010
Ti	—	—	—	—	—	1.0
Co			Balance			

[a] From *Annual Book of ASTM Standards*, Part 46, American Society for Testing and Materials, Philadelphia, 1980.

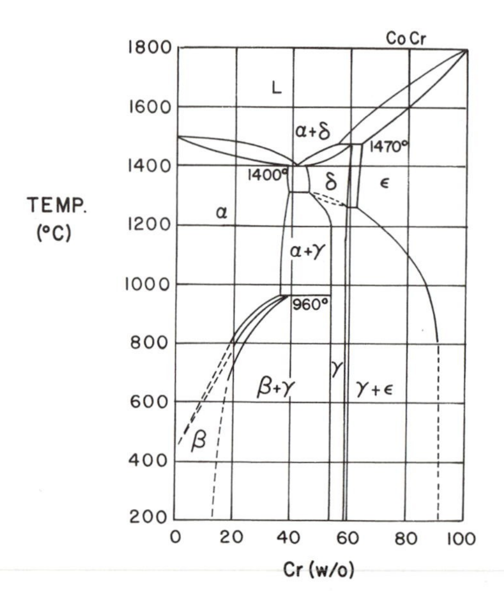

Figure 5-3. Phase diagram of Co–Cr. From C. J. Smithells (ed.), *Metals Reference Book*, Butterworths, London, 1976, p. 549.

One of the most promising wrought Co-based alloys is the CoNiCrMo alloy originally called MP35N (Standard Pressed Steel Co.), which contains approximately 35 wt% Co and Ni each. The alloy has a high degree of corrosion resistance to seawater (containing chloride ions) under stress. The cold-working can increase the strength of the alloy considerably as shown in Figure 5-4. However, there is

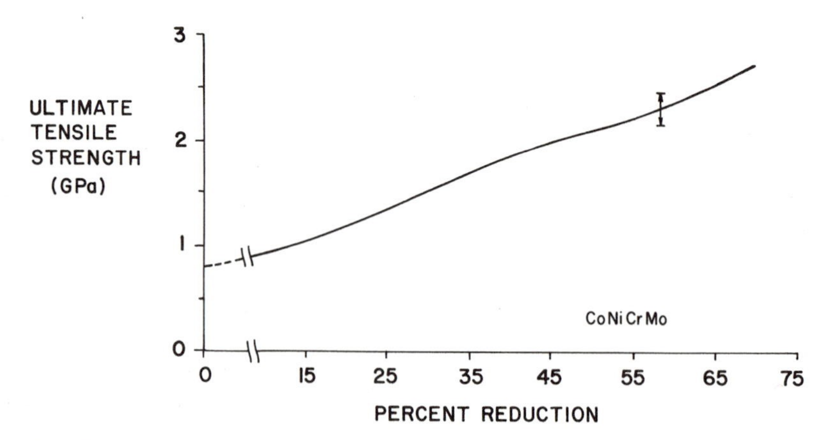

Figure 5-4. Relationship between ultimate tensile strength and the amount of cold-work for CoNiCrMo alloy. From T. M. Devine and J. Wulff, "Cast vs. Wrought Cobalt–Chromium Surgical Implant Alloys," *J. Biomed. Mater. Res.*, *9*, 151–167, 1975.

a considerable difficulty of cold-working especially when making large devices such as hip joint stems. Only hot-forging can be used to fabricate an implant with the alloy.

The abrasive wear properties of the wrought CoNiCrMo alloy are similar to those of the cast CoCrMo alloy (about 0.14 mm/year in joint simulation test); however, the former is not recommended for the bearing surfaces of a joint prosthesis because of its poor frictional properties with itself or other materials. The superior fatigue and ultimate tensile strength of the wrought CoNiCrMo alloy make it very suitable for the applications that require long service life without fracture or stress fatigue. Such is the case for the stems of the hip joint prostheses. This advantage is more appreciated when the implant has to be replaced with another one since it is quite difficult to remove the failed piece of implant embedded deep in the femoral medullary canal. Furthermore, the revision arthroplasty is usually inferior to the original in terms of its function owing to poorer fixation of the implant.

Table 5-4 shows the mechanical properties required of Co-based alloys. As is the case with other alloys, the increased strength is accompanied by decreased ductility. Both the cast and wrought alloys have excellent corrosion resistance.

Experimental determination of the rate of nickel release from the CoNiCrMo alloy and 316L stainless steel in 37°C Ringer's solution produced an interesting result. Although the cobalt alloy has more initial release of nickel ions into the solution, the rate of release was about the same (3×10^{-10} g/cm^2/day) for both alloys as shown in Figure 5-5. This is rather surprising since the nickel content of the CoNiCrMo alloy is about three times that of 316L stainless steel.

The modulus of elasticity for the cobalt-based alloys does not change with the changes in their ultimate tensile strength. The moduli range from 220 to 234 GPa, which are higher than other materials such as stainless steels. This may

Table 5-4. Mechanical Property Requirements of Co-Based Alloys[a]

Property	Cast CoCrMo (F76)	Wrought CoCrWNi (F90)	Wrought CoNiCrMo (F562)		
			Solution annealed	Cold worked and aged	Fully annealed
Tensile strength (MPa)	655	860	795–1000	1790	600
Yield strength (0.2% offset) (MPa)	450	310	240–655	1585	276
Elongation (%)	8	10	50.0	8.0	50
Reduction of area (%)	8	—	65.0	35.0	65
Fatigue strength (MPa)[b]	310	—	—	—	340

[a] From *Annual Book of ASTM Standards*, Part 46, American Society for Testing and Materials, Philadelphia, 1980.
[b] Data from M. Semlitsch, "Properties of Wrought CoNiCrMo Alloy Protasul-10, a Highly Corrosion and Fatigue Resistant Implant Material for Joint Endoprostheses," *Eng. Med.*, *9*, 201–207, 1980.

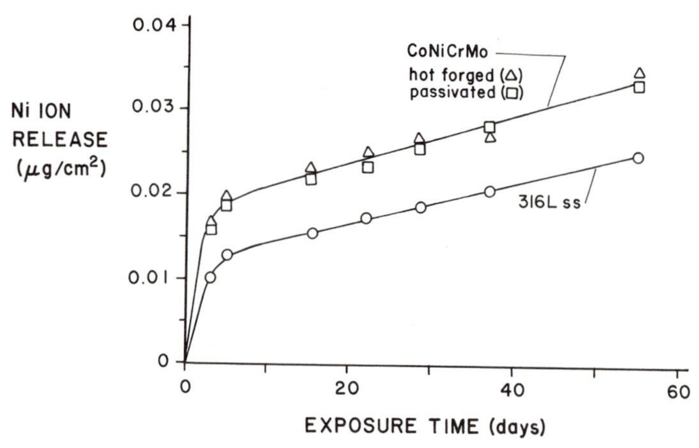

Figure 5-5. Nickel ion release versus time for hot-forged and passivated CoNiCrMo and 316L stainless steel in 37°C Ringer's solution. From Biophase Implant Material, Technical Information Publ. No. 3846, Richards Mfg. Co., Memphis, Tenn., 1980, p. 7.

have some implications of different load transfer modes to the bone although it is not established clearly what the effect of the increased modulus is.

5.2.3. Manufacturing of Implants Using Co-Based Alloys

The CoCrMo alloy is particularly susceptible to the work-hardening so that the normal fabrication procedure used with other metals cannot be employed. Instead the alloy is cast by a lost wax (or investment casting) method that involves the following steps (Figure 5-6):

1. A wax pattern of the desired component is made.
2. The pattern is coated with a refractory material, first by a thin coating with a slurry (suspension of silica in ethyl silicate solution) followed by complete investing after drying.
3. The wax is melted out in a furnace (100–150°C).
4. The mold is heated to a high temperature burning out any traces of wax or gas-forming materials.
5. Molten alloy is poured with gravitational or centrifugal force. The mold temperature is about 800–1000°C and the alloy is at 1350–1400°C.

Controlling the mold temperature will have an effect on the grain size of the final cast; coarse ones are formed at higher temperatures, which will decrease the strength. However, high processing temperature will result in larger carbide precipitates with greater distances between them resulting in a less brittle material. Again there is a complementary (trade-off) relationship between strength and toughness.

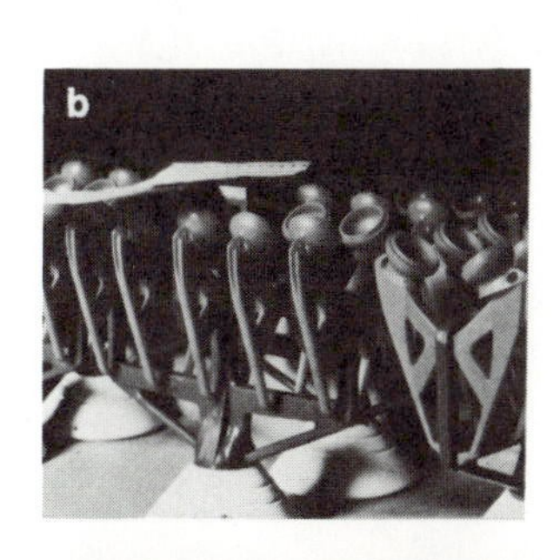

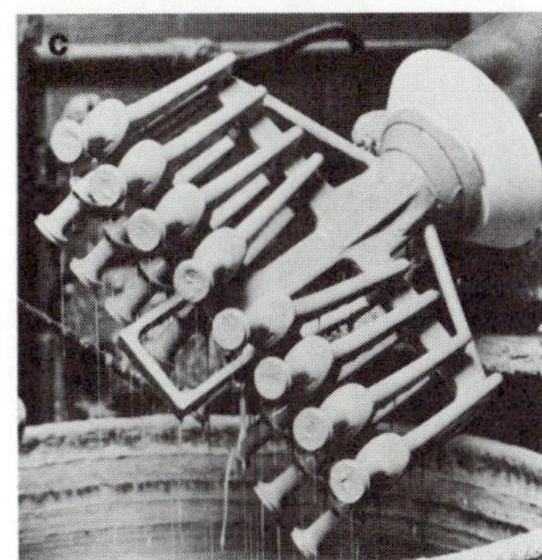

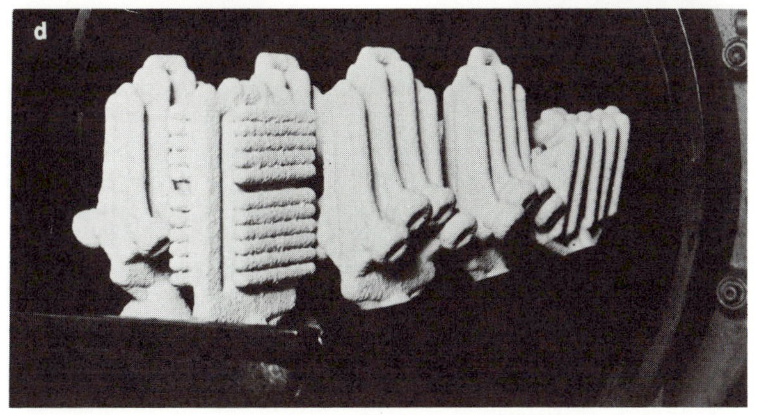

Figure 5-6. Lost wax casting of femoral joint prosthesis. (Courtesy of Howmedica, Inc., Rutherford, N.J.) (a) Injection of wax into a brass mold. (b) Wax patterns assembled for a ceramic coating (note the hollow part of the femoral head). (c) Application of the ceramic coating. (d) A hot pressure chamber retrieves the wax, leaving behind a ceramic coating. (e) Pouring molten metals into the preheated ceramic mold.

Figure 5-6. (*Continued*).

Example 5-2

Calculate the number of Co atoms released during a year from the femoral head of a hip joint prosthesis made of CoCrMo alloy. Assume that the wear rate is 0.14 mm/yr and that all of the atoms become ionized.

Answer

Assume a nominal diameter of the prosthetic femoral head of 28 mm. The surface area is $A = (4/3)\pi\ (1.4\ \text{cm})^2 = 8.21\ \text{cm}^2$. Half of this area is in contact with the socket portion of the joint. Therefore, the volume of wear material is $(1/2) \times (1.4\ \text{cm})^2 \times 0.014\ \text{cm/yr} = 0.055\ \text{cm}^3/\text{yr}$. Since the density of Co is $8.83\ \text{g/cm}^3$, and the atomic weight is 58.93, and the alloy is about 65% cobalt,

$$\frac{\text{atoms}}{\text{yr}} = \frac{0.65 \times 0.055\ \text{cm}^3/\text{yr} \times 8.83\ \text{g/cm}^3 \times 6.02 \times 10^{23}\ \text{atoms/mol}}{58.93\ \text{atoms/mol}}$$

$$= \underline{3.2 \times 10^{21}\ \text{atoms/yr}}, \text{ or } 10^{14}\ \text{atoms/sec.}$$

5.3. Ti AND Ti-BASED ALLOYS

Attempts to use titanium for implant fabrication date to the late 1930s. It was found that titanium was tolerated in cat femurs as was stainless steel and Vitallium® (CoCrMo alloy). Its lightness (4.5 g/cm^3 compared to 7.9 g/cm^3 for 316 stainless steel, 8.3 g/cm^3 for cast CoCrMo, and 9.2 g/cm^3 for wrought CoNiCrMo alloys) and good mechanochemical properties are salient features for implant application.

5.3.1. Compositions of Ti and Ti-Based Alloys

There are four grades of unalloyed titanium for surgical implant applications as given in Table 5-5. The impurity contents separate them; oxygen, iron, and nitrogen should be controlled carefully. Oxygen in particular has a great influence on the ductility and strength.

One titanium alloy (Ti6Al4V) is widely used to manufacture implants and its chemical requirements are given in Table 5-5. The main alloying elements of the alloy are aluminum (5.5–6.5 wt%) and vanadium (3.5-4.5 wt%).

5.3.2. Structure and Properties of Ti and Ti Alloys

Titanium is an allotropic material that exists as a hexagonal close-packed structure (α-Ti) up to 882°C and body-centered cubic structure (β-Ti) above that temperature. The addition of alloying elements to titanium enables it to have a wide range of properties:

1. Aluminum tends to stabilize the α phase, that is, increase the transformation temperature from α to β phase (Figure 5-7).

Table 5-5. Chemical Compositions of Titanium and Its Alloy (ASTM F67, F136)

Element	Grade 1	Grade 2	Grade 3	Grade 4	Ti6Al4V[a]
Nitrogen	0.03[b]	0.03	0.05	0.05	0.05
Carbon	0.10	0.10	0.10	0.10	0.08
Hydrogen	0.015	0.015	0.015	0.015	0.0125
Iron	0.20	0.30	0.30	0.50	0.25
Oxygen	0.18	0.25	0.35	0.40	0.13
Titanium			Balance		

[a] Aluminum 6.00 wt% (5.50–6.50), vanadium 4.00 wt% (3.50–4.50), and other elements 0.1 wt% maximum or 0.4 wt% total.

[b] All are maximum allowable weight percent.

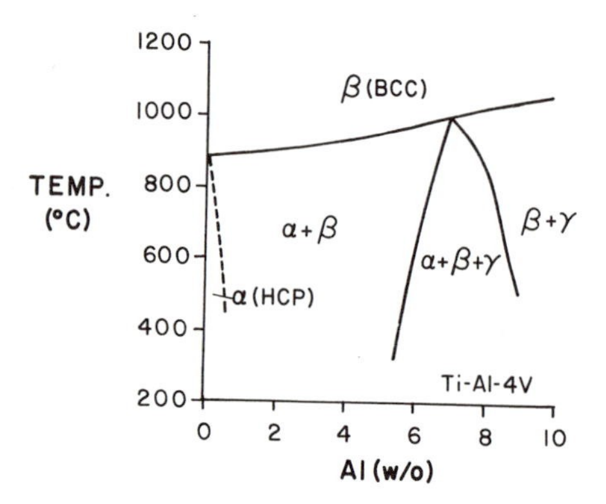

Figure 5-7. Part of phase diagram of Ti–Al–V at 4 wt% V. From C. J. E. Smith and A. N. Hughes, "The Corrosion Fatigue Behavior of a Titanium–6 w/o Aluminum–4 w/o Vanadium Alloy," *Eng. Med.*, *7*, 158–171, 1966.

2. Vanadium stabilizes the β phase by lowering the temperature of the transformation from α to β.

The α alloys have single-phase microstructure (Figure 5-8a), which promotes good weldability. The stabilizing effect of the high aluminum content of these groups of alloys results in excellent strength characteristics and oxidation resistance at high temperature (300–600°C). These alloys cannot be heat-treated for strengthening since they are single-phased.

The addition of a controlled amount of β-stabilizers causes the higher strength β phase to persist below the transformation temperature, which results in the two-phase system. As discussed in Section 3-4, the precipitates of β phase will appear by heat treatment in the solid solution temperature and subsequent quenching, followed by aging at a somewhat lower temperature. The aging cycle causes the precipitation of some fine α particles from the metastable β imparting α structure that is stronger than the annealed α–β structure (Figure 5-8b).

The higher percentage of β-stabilizing elements (13 wt% V in Ti13V11Cr3Al alloy) results in a microstructure that is substantially β which can be strengthened by the heat treatment (Figure 5-8c).

The mechanical properties of the commercially pure titanium and 6Al4V alloy are given in Table 5-6. The modulus of elasticity of these materials is about 110 GPa, which is half the value of Co-based alloys. From Table 5-6 one can see that the higher impurity content leads to higher strength and reduced ductility. The strength of the material varies from a value much lower than that of 316 stainless steel or the Co-based alloys to a value about equal to that of annealed 316 stainless steel of the cast CoCrMo alloy. However, when compared by the specific strength (strength per density) the titanium alloy excels over any other implant material as shown in Figure 5-9. Titanium, nevertheless, has poor shear strength, making it less desirable for bone screws, plates, and similar applications. It also tends to gall or seize when in sliding contact with itself or another metal.

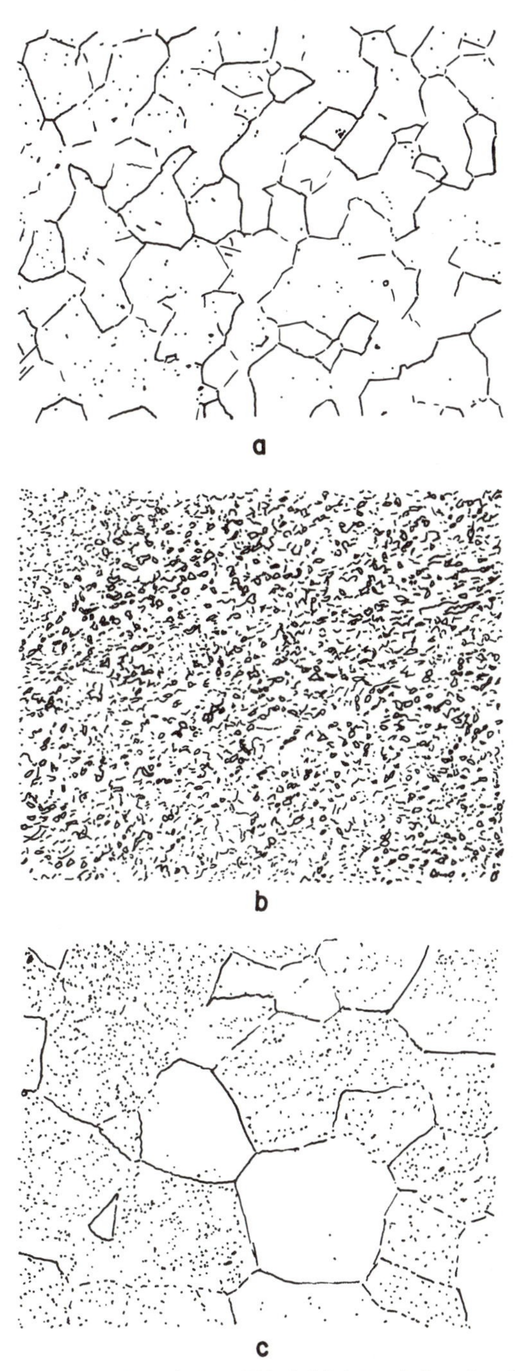

Figure 5-8. Microstructure of Ti alloys (all are 500×). (a) Annealed α alloy. (b) Ti6Al4V, α–β alloy, annealed. (c) β alloy, annealed. From G. H. Hille, "Titanium for Surgical Implants," *J. Mater.*, *1*, 373–383, 1966.

Table 5-6. Mechanical Properties of Ti and Ti Alloy (ASTM F136)

Properties	Grade 1	Grade 2	Grade 3	Grade 4	Ti6Al4V
Tensile strength (MPa)	240	345	450	550	860
Yield strength (0.2% offset) (MPa)	170	275	380	485	795
Elongation (%)	24	20	18	15	10
Reduction of area (%)	30	30	30	25	25

Titanium derives its resistance to corrosion by the formation of a solid oxide layer. Under *in vivo* conditions the oxide (TiO_2) is the only stable reaction product. The oxide layer forms a thin adherent film and passivates the material. Corrosion resistance mechanisms are discussed further in Section 5.6.

5.3.3. Manufacturing of Implants

Titanium is very reactive at high temperature and burns readily in the presence of oxygen. It therefore requires an inert atmosphere for high-temperature processing or is processed by vacuum melting. Oxygen diffuses readily in titanium and the dissolved oxygen embrittles the metal. As a result, any hot working or forging operation should be carried out below 925°C. Machining at room temperature is not the solution to all of the problems since the material also tends to gall or

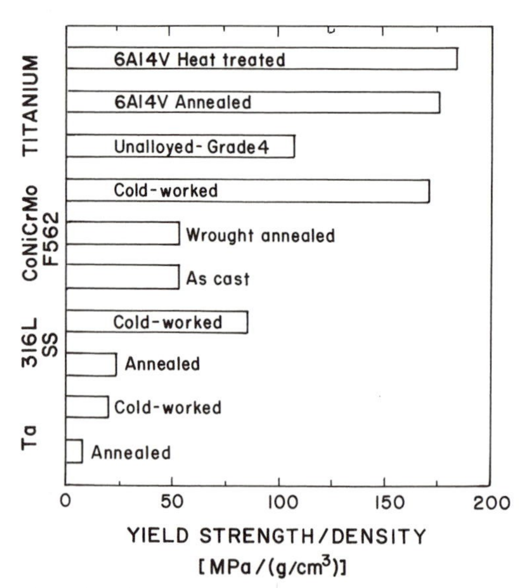

Figure 5-9. Yield strength-to-density ratio of some implant materials. From G. H. Hille, "Titanium for Surgical Implants," *J. Mater.*, *1*, 373–383, 1966.

seize the cutting tools. Very sharp tools with slow speeds and large feeds are used to minimize this effect. Electrochemical machining is an attractive means.

Example 5-3

The allotropic phase change of titanium at 882°C is from hcp structure ($a = 2.95$ Å, $C = 4.683$ Å) to bcc structure ($a = 3.32$ Å). Calculate the volume change in cm^3/g by heating above the transformation temperature. The density is $4.54\ g/cm^3$ and the atomic weight is 47.9 g.

Answer

$$\text{Area of hexagon} = a \times a\sqrt{\frac{3}{2}} \times \frac{1}{2} \times 6 = \frac{3\sqrt{3}}{2}a^2\ \text{Å}^2$$

$$V_{\text{hcp}} = \frac{3\sqrt{3}}{2}a^2c = 106.3\ \text{Å}^3 \text{ which contains 6 atoms}$$

$$V_{\text{bcc}} = a^3 = 36.59\ \text{Å}^3 \text{ which contains 2 atoms}$$

$$\text{hcp: } \frac{106.3\ \text{Å}^3}{6\ \text{atoms}} = \frac{106.3\ \text{Å}^3 \times 6.02 \times 10^{23}}{6 \times 47.9\ \text{g/mol atoms}} = 0.2227\ \text{cm}^3/\text{g}$$

$$\text{bcc: } \frac{36.59\ \text{Å}^3}{2\ \text{atoms}} = \frac{36.59\ \text{Å}^3 \times 6.02 \times 10^{23}}{2 \times 47.9\ \text{g/mol atoms}} = 0.23\ \text{cm}^3/\text{g}$$

The difference is $\underline{0.0073\ \text{cm}^3/\text{g}}$.

5.4. DENTAL METALS

5.4.1. Dental Amalgam

An *amalgam* is an alloy in which one of the component metals is mercury. The rationale for using amalgam as a tooth filling material is that since mercury is a liquid at room temperature, it can react with other metals such as silver and tin and form a plastic mass that can be packed into the cavity, and that hardens (sets) with time. To fill a cavity, the dentist mixes solid alloy, supplied in particulate form, with mercury in a mechanical triturator. The resulting material is readily deformable and is then packed into the prepared cavity. The solid alloy is composed of at least 65% silver, and not more than 29% tin, 6% copper, 2% zinc, and 3% mercury. The reaction during setting is thought to be

$$\gamma + \text{Hg} \rightarrow \gamma + \gamma_1 + \gamma_2 \qquad (5\text{-}1)$$

in which the γ phase is Ag_3Sn, the γ_1 phase is Ag_2Hg_3, and the γ_2 phase is Sn_7Hg, as shown in Figure 5-10.

The phase diagram for the Ag–Sn–Hg system shows that over a wide compositional range, all three phases are present. Dental amalgams typically contain 45%

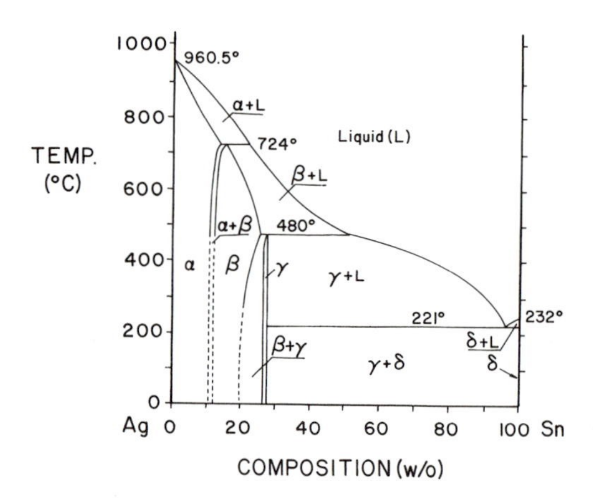

Figure 5-10. Phase diagram of Ag and Sn.

to 55% mercury, 35% to 45% silver, and about 15% tin, when fully set. Strength of the restoration increases during the setting process, so that the amalgam has attained one quarter of its final strength after 1 hour, and almost all of its final strength after 1 day.

5.4.2. Gold

Gold and gold alloys are useful metals in dentistry as a result of their durability, stability, and corrosion resistance. Gold fillings are introduced by two methods: casting and malleting. *Cast* restorations are made by taking a wax impression of the prepared cavity, making a mold from this impression in a material such as gypsum silica, which tolerates high temperature, and casting molten gold in the mold. The patient is given a temporary filling for the intervening time. Gold *alloys* are used for cast restorations, since they have mechanical properties that are superior to those of pure gold. Corrosion resistance is retained in these alloys provided they contain 75 wt% or more of gold and other noble metals. Copper, alloyed with gold, significantly increases its strength. Platinum also improves the strength, but no more than about 4% can be added, or the melting point of the alloy is elevated excessively. Silver compensates for the color of copper. A small amount of zinc may be added to lower the melting point and to scavenge oxides formed during melting. Gold alloys of different composition are available. Softer alloys containing more than 83% gold are used for inlays, which are not subjected to much stress. Harder alloys containing less gold are chosen for crowns and cusps, which are more heavily stressed.

Malleted restorations are built up in the cavity from layers of *pure* gold foil. The foils are degassed before use, and the layers are welded together by pressure at room temperature. In this type of welding, the metal layers are joined by thermal diffusion of atoms from one layer to the other. Since intimate contact is required in this procedure, it is particularly important to avoid contamination.

The pure gold is relatively soft, so this type of restoration is limited to areas not subjected to much stress.

5.4.3. Nickel–Titanium Alloys

The nickel-titanium alloys show an unusual property, i.e., after the material is deformed it can snap back to its previous shape following heating the material. This phenomenon is called *shape memory effect* (SME). This effect of Ni–Ti alloy was first observed by Buehler and Wiley at the U.S. Naval Ordnance Laboratory. The equiatomic Ni-Ti alloy (Nitinol®) exhibits an exceptional SME near room temperature: if it is plastically deformed below the transformation temperature, it reverts back to its original shape as the temperature is raised. The SME can be generally related to a diffusionless martensitic phase transformation that is also thermoelastic in nature, the thermoelasticity being attributed to the ordering in the parent and martensitic phases. The thermoelastic martensitic transformation exhibits the following general characteristics:

1. Martensite formation can be initiated by cooling the material below M_s, defined as the temperature at which the martensitic transformation begins. Martensite formation can also be initiated by applying a mechanical stress at a temperature above M_s.
2. M_s and A_s (the temperature at which the reverse austenitic transformation begins upon heating) temperatures can be increased by applying stresses below the yield point; the increase is proportional to the applied stress.
3. The material is more resilient than most metals.
4. The transformation is reversible.

Some possible applications of shape memory alloys are orthodontic dental archwires, intracranial aneurysm clips, a vena cava filter, contractile artificial muscles for an artificial heart, orthopedic implants and other medical devices.

In order to develop such devices, it is necessary to understand fully the mechanical and thermal behavior associated with the martensitic phase transformation. A widely known Ni–Ti alloy is 55-Nitinol (55 wt% or 50 at% Ni), which has a single phase and the "mechanical memory" plus other properties, e.g., high acoustic damping, direct conversion of heat energy into mechanical energy, good fatigue properties, and low-temperature ductility. Deviation from the 55-Nitinol (near stoichiometric NiTi) in the Ni-rich direction yields a second group of alloys that are also completely nonmagnetic but differ from 55-Nitinol in their capability of being thermally hardened to higher hardness levels. Shape recovery capability decreases and heat treatability increases rapidly as the Ni content approaches 60 wt%. Both 55- and 60-Nitinols have relatively low moduli of elasticity and can be tougher and more resilient than stainless steel, Ni–Cr, or Co–Cr based alloys.

Efficiency of 55-Nitinol shape recovery can be controlled by changing the final annealing temperatures during preparation of the alloy device. For the most efficient recovery, the shape is fixed by constraining the specimen in a desired

Table 5-7. Chemical Composition of Ni–Ti Alloy Wire

Element	Composition (wt%)
Ni	54.01
Co	0.64
Cr	0.76
Mn	0.64
Fe	0.66
Ti	Balance

configuration and heating to between 482 and 510°C. If the annealed wire is deformed at a temperature below the shape recovery temperature, shape recovery will occur upon heating, provided the deformation has not exceeded crystallographic strain limits (~8% strain in tension). The Ni–Ti alloys also exhibit good biocompatibility and corrosion resistance *in vivo.*

The mechanical properties of Ni–Ti alloys are especially sensitive to the stoichiometry of composition (typical composition is given in Table 5-7) and the individual thermal and mechanical history. Although much is known about the processing, mechanical behavior, and properties relating to the SME, considerably less is known about the thermomechanical and physical metallurgy of the alloy.

The differential scanning calorimeter (DSC) is a device that is capable of measuring the heat capacity of materials as a function of temperature. Figure 5-11 shows a typical DSC plot for Ni–Ti alloy and identifies several

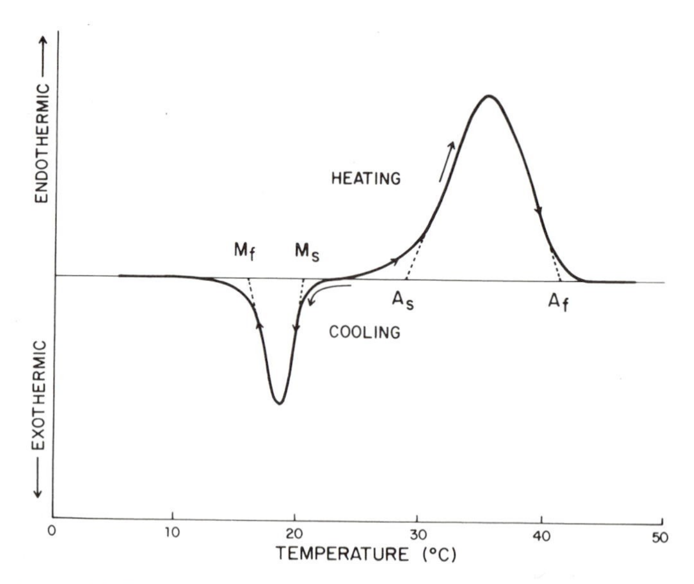

Figure 5-11. A typical DSC curve of the specific heat (thermal energy) versus temperature. From J. H. Lee, J. B. Park, S. F. Andreasen, and R. S. Lakes, "Thermomechanical Studies of NiTi Alloys," *J. Biomed. Mater. Res. 22*, 573–588, 1989.

relevant parameters. The Ni–Ti alloys generally display two peaks that are associated with the martensitic phase transformation temperature in cooling (M_s and M_f) and heating (austenitic phase transformation temperature; A_s and A_f) where subscripts s and f indicate the starting and finishing of transformations, respectively. The area under the specific heat versus temperature curve can be used to calculate the amount of thermal energy used for the phase transformation.

The typical curves for bending moment versus bend angle are shown in Figure 5-12. Ni–Ti alloy wire specimens were tested at 0°C and room temperature. As can be seen from the curves, the samples deformed at room temperature recovered almost completely to their original states, indicating that the transformation temperature is close to room temperature. From these curves the elastic stiffnesses of the alloy were calculated using the straight portions of the curves and are given in Table 5-8. The results also indicate that the elastic modulus is higher at higher temperature.

Figure 5-13 shows the typical microstructure of a Ni–Ti specimen at room temperature. The optical microscopic picture of the cross section of the wire shown in Figure 5-13a illustrates the evenly dispersed nonmetallic inclusions in the Ni–Ti matrix. The inclusions are presumed to be primarily titanium carbonitrides, with a few nickel-titanium oxides. Panels b and c show SEM photomicrographs of the unbent and bent portions of specimens, respectively, at room temperature. From these photomicrographs one can see that long pores are aligned

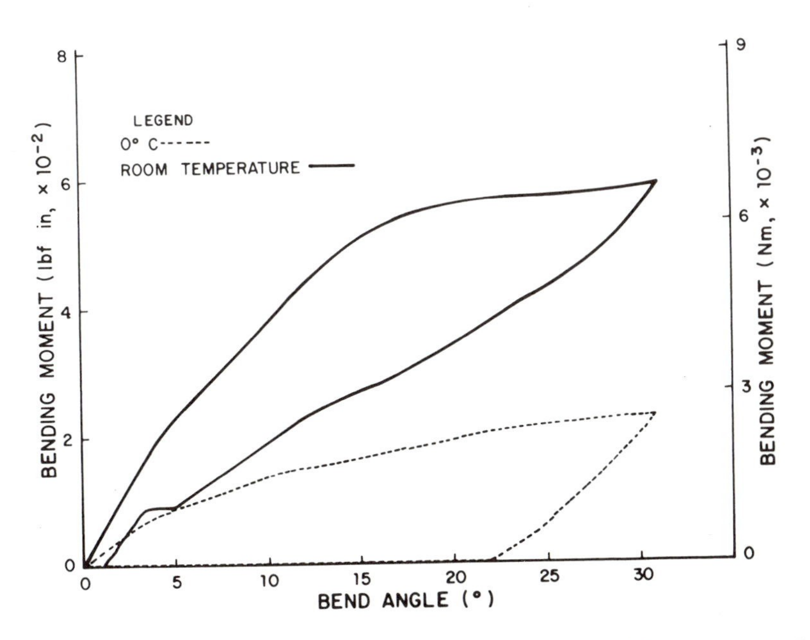

Figure 5-12. The bending moment versus bending angle curves for a NiTi specimen at 0°C and room temperature. From J. H. Lee, J. B. Park, S. F. Andreasen, and R. S. Lakes, "Thermomechanical Studies of NiTi Alloys," *J. Biomed. Mater. Res. 22*, 573–588, 1989.

Table 5-8. Elastic Properties of Ni–Ti Alloy Wire

Test temp. (°C)	Elastic stiffness (E_b)[a] lbf in/degree, $\times 10^{-3}$	Elastic stiffness (E_b)[a] N m/degree, $\times 10^{-4}$	Young's modulus (E) (GPa)
0	1.4	1.58	31
Room temp.	4.3	4.86	30

[a] $E_b = \Delta I_b/\Delta\omega$ (bending moment/angle of bending).

with the longitudinal direction of the specimen. The long pores appear to be created by the polishing and chemical etching steps used to prepare the specimen. However, Figure 5-13c shows transformed martensitic structure near the surface in contrast to the undeformed structure shown in Figure 5-13b, which does not show martensitic structure at room temperature.

5.5. OTHER METALS

Several other metals have been used for a variety of specialized implant applications.

Tantalum has been subjected to animal implant studies and has been shown very biocompatible. Due to its poor mechanical properties (Table 5-9) and its high density (16.6 g/cm^3), it is restricted to few applications such as wire sutures for plastic surgeons and neurosurgeons and a radioisotope for bladder tumors.

Platinum and other noble metals in the platinum group are extremely corrosion resistant but have poor mechanical properties. They are mainly used as alloys for electrodes such as pacemaker tips because of their high resistance to corrosion and low threshold potentials.

5.6. CORROSION OF METALLIC IMPLANTS

Corrosion is the unwanted chemical reaction of a metal with its environment, resulting in its continued degradation to oxides, hydroxides, or other compounds.

Table 5-9. Mechanical Properties of Tantalum (ASTM F560)

Properties	Fully annealed	Cold-worked
Tensile strength (MPa)	205	515
Yield strength (0.2%) offset) (MPa)	140	345
Elongation (%)	20–30	2
Young's modulus (GPa)	—	190

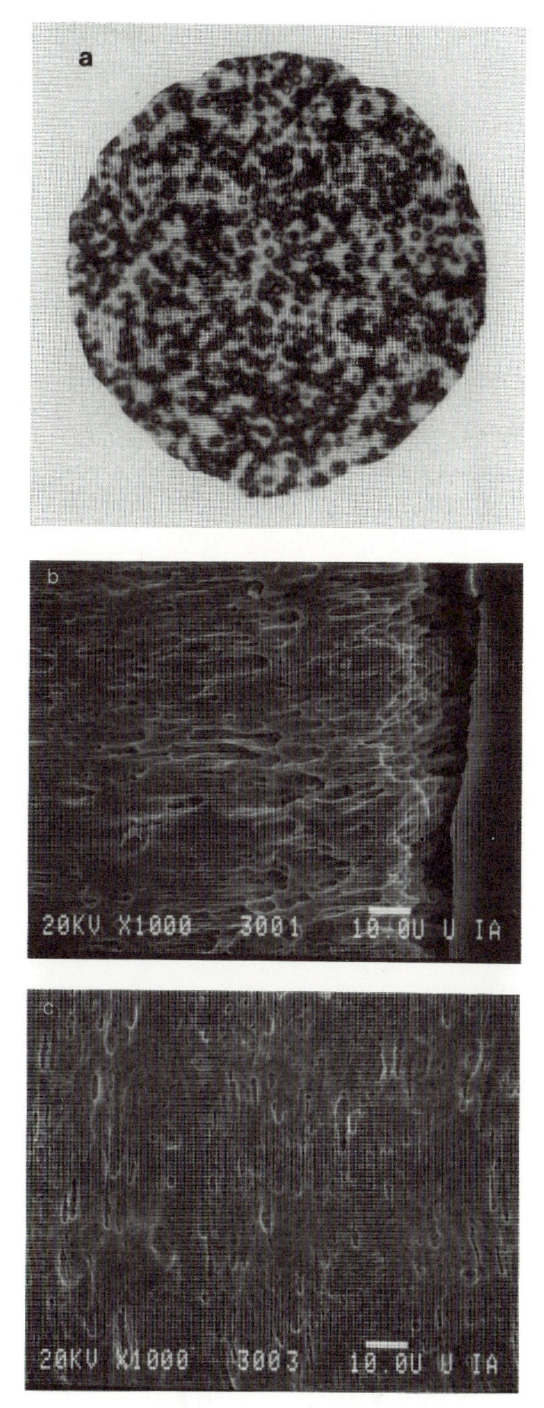

Figure 5-13. Microstructure of a NiTi specimen at room temperature. (a) Optical micrograph of the cross section (100×) showing nonmetallic inclusions (black particulates) and NiTi matrix (white background). (b) Scanning electron micrograph of unbent specimen at a longitudinal section (1000×) showing elongated pores in the wire-drawing direction. (c) Scanning electron micrograph of bent specimen at a longitudinal section (1000×). Note the martensites formed at about 45° to the wire-drawing direction (top to bottom in this photo).

Tissue fluid in the human body contains water, dissolved oxygen, proteins, and various ions such as chloride and hydroxide. As a result, the human body presents a very aggressive environment to metals used for implantation. Corrosion resistance of a metallic implant material is consequently an important aspect of its biocompatibility.

5.6.1. Electrochemical Aspects

The lowest free energy state of many metals in an oxygenated and hydrated environment is that of the *oxide.* Corrosion occurs when metal atoms become ionized and go into solution, or combine with oxygen or other species in solution to form a compound that flakes off or dissolves. The body environment is very aggressive in terms of corrosion since it is not only aqueous but also contains chloride ions and proteins. A variety of chemical reactions occur when a metal is exposed to an aqueous environment, as shown in Figure 5-14. The electrolyte, which contains ions in solution, serves to complete the electric circuit. In the human body, the required ions are plentiful in the body fluids. Anions are negative ions, which migrate toward the anode, and cations are positive ions, which migrate toward the cathode. The electrical component V in Figure 5-14 can be a voltmeter with which to measure the potential produced; or it can be a battery in which case the cell is an electroplating cell; or it can be tissue resistance, in which case the electrochemical cell is an unwanted corrosion cell for a biomaterial in the body. In the body, an external electrical driving source may be present in the form of a cardiac pacemaker, or an electrode used to stimulate bone growth. At the anode, or positive electrode, the metal oxidizes. The following reactions involving the metal M may occur:

$$M \rightarrow M^{+n} + ne^- \tag{5-2}$$

At the cathode, or negative electrode, the following reduction reactions are important:

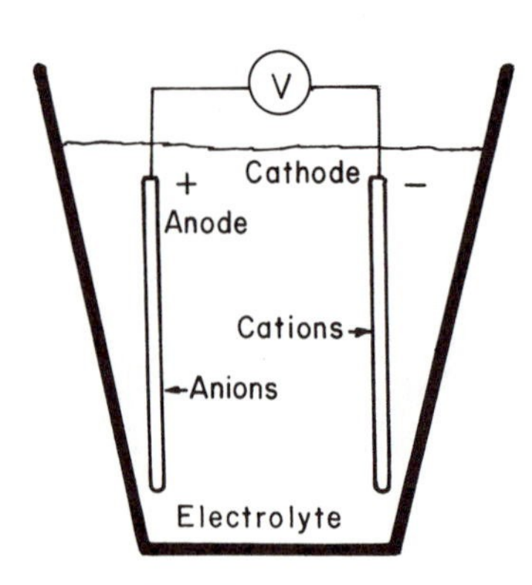

Figure 5-14. Electrochemical cell.

$$M^+ + e^- \rightarrow M \tag{5-3}$$

$$M^{2+} + OH^- + 2e^- \rightarrow MOH \tag{5-4}$$

$$2H_3O^+ + 2e^- \rightarrow H_2\uparrow + 2H_2O \tag{5-5}$$

$$\tfrac{1}{2}O_2 + H_2O + 2e^- \rightarrow 2OH^- \tag{5-6}$$

Consider as an example the corrosion of a metal such as iron. Metallic iron goes into solution in ionized form as follows:

$$Fe + 2H_2O \rightarrow Fe^{2+} + H_2\uparrow + 2OH^- \tag{5-7}$$

In the presence of oxygen, rust may be formed in the following reactions:

$$4Fe^{2+} + O_2 + 2H_2O \rightarrow Fe^{3+} + 4OH^- \tag{5-8}$$

$$4Fe^{3+} + 12OH^- \rightarrow 4Fe(OH)_3\downarrow \tag{5-9}$$

If less oxygen is available, Fe_3O_4, magnetite, may precipitate rather than ferric hydroxide.

The tendency of metals to corrode is expressed most simply in the standard electrochemical series of Nernst potentials, shown in Table 5-10. These potentials are obtained in electrochemical measurements in which one electrode is a standard hydrogen electrode formed by bubbling hydrogen through a layer of finely divided platinum black. The potential of this reference electrode is defined to be zero. Noble metals are those that have a potential higher than that of a standard hydrogen electrode; base metals have lower potentials.

If two dissimilar metals are present in the same environment, the one that is most negative in the galvanic series will become the anode, and bimetallic (or galvanic) corrosion will occur. Galvanic corrosion can be much more rapid than the corrosion of a single metal. Consequently, implantation of dissimilar metals (mixed metals) is to be avoided. Galvanic action can also result in corrosion within a single metal, if there is inhomogeneity in the metal or in its environment, as shown in Figure 5-15.

Table 5-10. Standard Electrochemical Series

Reaction	ΔE^0 (volts)
$Li \leftrightarrow Li^+$	−3.045
$Na \leftrightarrow Na^+$	−2.714
$Al \leftrightarrow Al^+$	−1.66
$Ti \leftrightarrow Ti^{3+}$	−1.63
$Fe \leftrightarrow Fe^{2+}$	−0.44
$H_2 \leftrightarrow 2H^+$	0.000
$Ag \leftrightarrow Ag^+$	+0.799
$Au \leftrightarrow Au^+$	+1.68

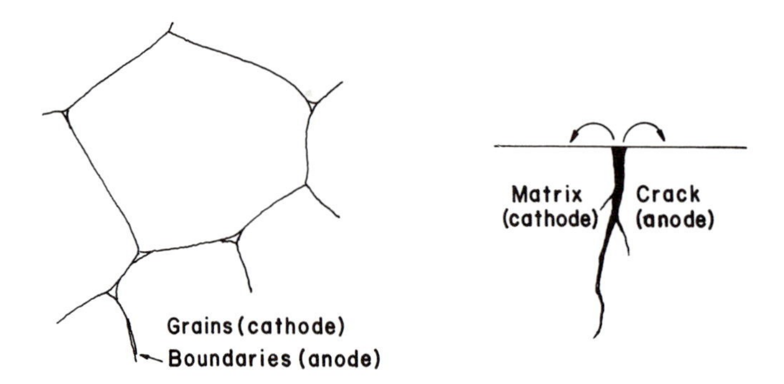

Figure 5-15. Micro-corrosion cells. (a) Grain boundaries are anodic with respect to the grain interior. (b) Crevice corrosion due to oxygen-deficient zone in metal's environment.

The potential difference E actually observed depends on the concentration of the metal ions in solution according to the Nernst equation,

$$E = E_0 + (RT/nF)\ln[M^{n+}] \tag{5-10}$$

in which R is the gas constant, E_0 is the standard electrochemical potential, T is the absolute temperature, F is Faraday's constant, 96,485 coulombs/mole, and n is the number of moles of ions.

The order of nobility observed in actual practice may differ from that predicted thermodynamically. The reasons are that some metals become covered with a *passivating* film of reaction products, which protects the metal from further attack. The dissolution reaction may be strongly irreversible so that a potential barrier must be overcome. In this case, corrosion may be inhibited even though it remains energetically favorable. Finally, the corrosion reactions may proceed slowly: the kinetics are not determined by the thermodynamics.

5.6.2. Pourbaix Diagrams in Corrosion

The Pourbaix diagram is a plot of regions of corrosion, passivity, and immunity as they depend on electrode potential and pH. The Pourbaix diagrams are derived from the Nernst equation and from the solubility of the degradation products and the equilibrium constants of the reaction. For the sake of definition, the *corrosion region* is set arbitrarily at a concentration of greater than 10^{-6} g atom per liter (molar) or more of metal in the solution at equilibrium. This corresponds to about 0.06 mg/liter for metals such as iron and copper, and 0.03 mg/liter for aluminum. *Immunity* is defined as equilibrium between metal and its ions at less than 10^{-6} molar. In the region of immunity, the corrosion is energetically impossible. Immunity is also referred to as cathodic protection. In the passivation domain, the stable solid constituent is an oxide, hydroxide, a hydride, or a salt of the metal. *Passivity* is defined as equilibrium between a metal

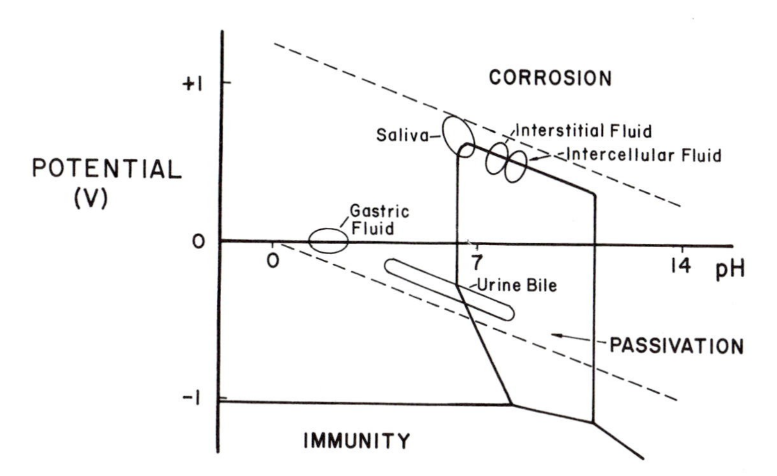

Figure 5-16. Pourbaix diagram for chromium, showing regions associated with various body fluids. From J. H. Dumbleton and J. Black, *An Introduction to Orthopaedic Materials*, Charles C. Thomas, Springfield, Ill., 1975.

and its reaction products (oxides, hydroxides, etc.) at a concentration of 10^{-6} molar or less. This situation is useful if reaction products are adherent. In the biomaterials setting, passivity may or may not be adequate: disruption of a passive layer may cause an increase in corrosion. The equilibrium state may not occur if reaction products are removed by the tissue fluid. Materials differ in their propensity to reestablish a passive layer that has been damaged. This layer of material may protect the underlying metal if it is firmly adherent and nonporous; in that case further corrosion is prevented. Passivation can also result from a concentration polarization due to a buildup of ions near the electrodes. This is not likely to occur in the body since the ions are continually replenished. Cathodic depolarization reactions can aid in the passivation of a metal by virtue of an energy barrier that hinders the kinetics. Equations (5-5) and (5-6) are examples.

In the diagrams shown in Figures 5-16 to 5-19, there are two diagonal lines. The top ("oxygen") line represents the upper limit of the stability of water and is associated with oxygen-rich solutions or electrolytes near oxidizing materials.

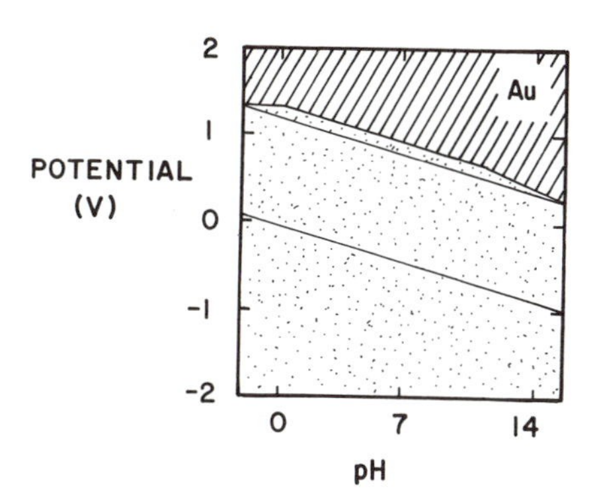

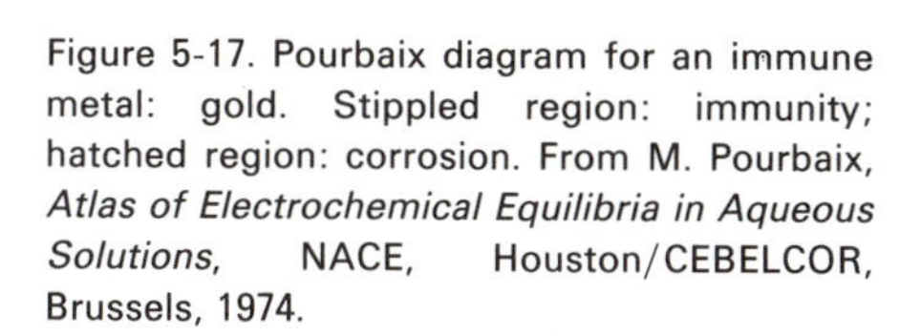
Figure 5-17. Pourbaix diagram for an immune metal: gold. Stippled region: immunity; hatched region: corrosion. From M. Pourbaix, *Atlas of Electrochemical Equilibria in Aqueous Solutions*, NACE, Houston/CEBELCOR, Brussels, 1974.

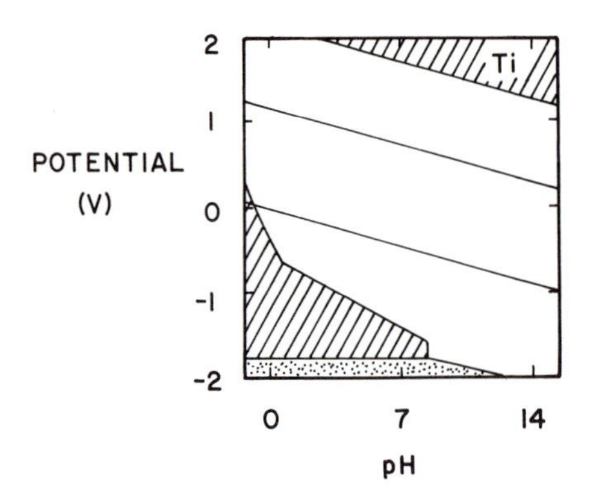

Figure 5-18. Pourbaix diagram for a passive metal: titanium. Stippled region: immunity; hatched region: corrosion; unmarked region: passivity. From M. Pourbaix, *Atlas of Electrochemical Equilibria in Aqueous Solutions*, NACE, Houston/CEBELCOR, Brussels, 1974.

In the region above this line, oxygen is evolved according to $2H_2O \rightarrow O_2 + 4H^+ + 4e^-$. In the human body, saliva, intracellular fluid, and interstitial fluid occupy regions near the oxygen line, since they are saturated with oxygen. The lower ("hydrogen") diagonal line represents the lower limit of the stability of water. Hydrogen gas is evolved according to Eq. (5-5). Aqueous corrosion occurs in the region between these diagonal lines on the Pourbaix diagram. In the human body, urine, bile, the lower gastrointestinal tract, and secretions of ductless glands occupy a region somewhat above the hydrogen line.

The significance of Pourbaix diagrams is as follows. Different parts of the body have different pH values and oxygen concentrations. Consequently, a metal that performs well (is immune or passive) in one part of the body may suffer an unacceptable amount of corrosion in another part. Moreover, pH can change dramatically in tissue that has been injured or infected. In particular, normal tissue fluid has a pH of about 7.4, but in a wound it can be as low as 3.5, and in an infected wound the pH can increase to 9.0.

Pourbaix diagrams are useful, but do not tell the whole story; there are some limitations. Diagrams are made considering equilibrium among metal, water, and reaction products. The presence of other ions, e.g., chloride, may result in very much different behavior and large molecules in the body may also change the

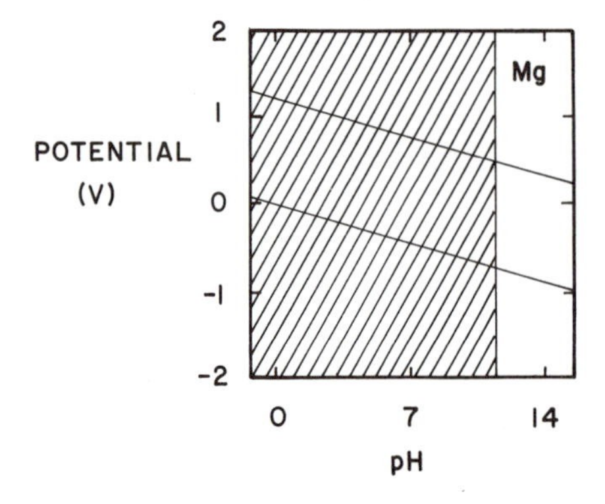

Figure 5-19. Pourbaix diagram for magnesium. Hatched region: corrosion; unmarked region: passivity. From M. Pourbaix, *Atlas of Electrochemical Equilibria in Aqueous Solutions*, NACE, Houston/CEBELCOR, Brussels, 1974.

situation. Prediction of "passivity" may in some cases be optimistic, since reaction rates are not considered.

5.6.3. Rates of Corrosion and Polarization Curves

The regions in the Pourbaix diagram specify whether corrosion will take place, but they do not determine the rate. The rate, expressed as an electric current density (current per unit area), depends on electrode potential as shown in the polarization curves in Figure 5-20. From such curves, it is possible to calculate the number of ions per unit time liberated into the tissue, as well as the depth of metal removed by corrosion in a given time. An alternative experiment is one in which the weight loss of a specimen of metal due to corrosion is measured as a function of time.

The rate of corrosion also depends on the presence of synergistic factors, such as those of mechanical origin. For example, in corrosion fatigue, repetitive deformation of a metal in a corrosive environment results in acceleration of both the corrosion and the fatigue microdamage. Since the body environment involves both repeated mechanical loading and a chemically aggressive environment, fatigue testing of implant materials should always be performed under physiological environmental conditions: under Ringer's solution at body temperature. In *fretting corrosion*, rubbing of one part on another disrupts the passivation layer, resulting in accelerated corrosion. In *pitting*, the corrosion rate is accelerated in a local region. Stainless steel is vulnerable to pitting. Localized corrosion can

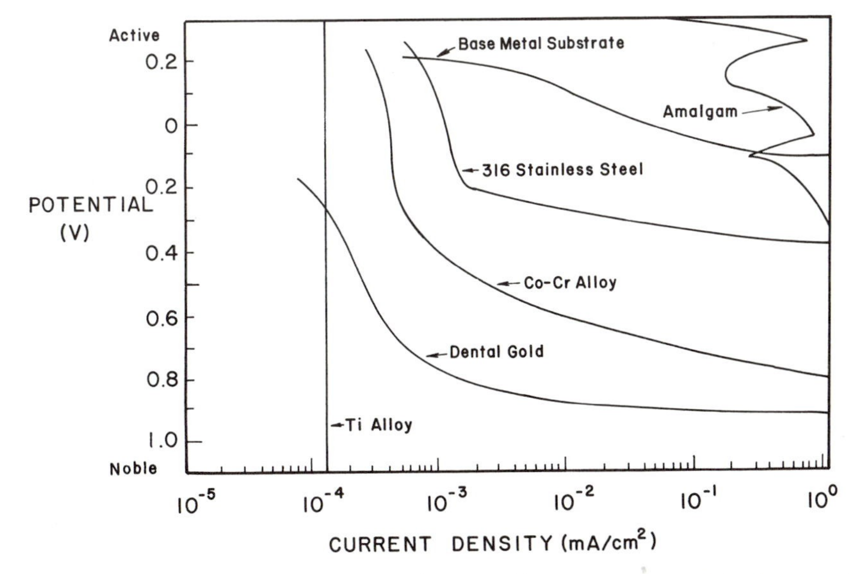

Figure 5-20. Potential–current density curves for some biomaterials. From E. H. Greener, J. K. Harcourt, and E. P. Lautenschlager, *Materials Science in Dentistry*, Williams and Wilkins, Baltimore, 1972.

occur if there is inhomogeneity in the metal or in the environment. *Grain boundaries* in the metal may be susceptible to the initiation of corrosion, as a result of their higher energy level. *Crevices* are also vulnerable to corrosion, since the chemical environment in the crevice may differ from that in the surrounding medium. The area of contact between a screw and a bone plate, for example, can suffer crevice corrosion.

Example 5-4

Consider a restoration made of dental amalgam. Suppose that the corrosion current density is 100 μA/cm^2. Is this a reasonable value? This material is an alloy; assume the density is 11 g/cm^3 and that the mean atomic weight is 150 amu.

(a) How many univalent ions will be released per year from a restoration surface 3.16 mm by 3.16 mm?
(b) If the corrosion is uniform, how many millimeters of depth will be lost per year? Discuss your answer.

Answer

(a) Ion flow,

$$J = \left[10^{-4}\frac{\text{A}}{\text{cm}^2}\right]\left[\frac{\text{coul} \times e^-}{\text{sec} \times 1.6 \times 10^{-19}\ \text{coul}}\right]$$

$[0.1\ \text{cm}^2][3.15 \times 10^7\ \text{sec/yr}] = \underline{1.97 \times 10^{21}\ e^-/\text{yr}}$

(b) Depth loss,

$$L = J[1\ \text{ion}/e^-\ 150\ \text{g/mol}]\left[\frac{1}{6.02 \times 10^{23}\ \text{ions/mol}}\right]$$

$[1/(11\ \text{g/cm}^3)][1/0.1\ \text{cm}^2] = \underline{0.45\ \text{cm/yr}}$

(Depth loss is far in excess of what is known to occur in dental amalgams. However, the value of 100 μA/cm^2 is a conservative estimate of the corrosion current density from the polarization curve in Figure 5-19. Surface changes that depend upon time may occur *in vivo*.)

5.6.4. Corrosion of Available Metals

Choice of a metal for implantation should take into account the corrosion properties discussed above. Metals that are in current use as biomaterials include gold, cobalt–chromium alloys, type 316 stainless steel, titanium, nickel–titanium alloys, and silver–mercury amalgam. This section deals with the corrosion aspects of these metals. Their composition and physical properties are discussed in Section 5.4.

The noble metals are immune to corrosion and would be ideal materials if corrosion resistance were the only concern. Gold is widely used in dental restorations and in that setting it offers superior performance and longevity. Gold is not, however, used in orthopedic applications as a result of its high density, insufficient strength, and high cost.

Titanium is a base metal in the context of the electrochemical series; however, it forms a robust passivating layer and, as shown in Figure 5-18, it remains passive under physiological conditions. Corrosion currents in normal saline are very low: 10^{-8} A/cm^2. Titanium implants remain virtually unchanged in appearance. Ti offers superior corrosion resistance but is not as stiff or strong as steel.

Cobalt–chromium alloys, like titanium, are passive in the human body. They are widely in use in orthopedic applications. They do not exhibit pitting.

Stainless steels contain enough chromium to confer corrosion resistance by passivity. The passive layer is not as robust as in the case of titanium or the cobalt–chromium alloys. Only the most corrosion resistant of the stainless steels are suitable for implants. These are the austenitic types 316, 316L, and 317, which contain molybdenum. Even these types of stainless steel are vulnerable to pitting and to crevice corrosion around screws.

Dental amalgam is an alloy of mercury, silver, and tin. Although the phases are passive at neutral pH, the transpassive potential for the γ_2 phase is easily exceeded, due to interphase galvanic couples or potentials due to differential aeration under dental plaque. Amalgam, therefore, often corrodes and is the most active (corrosion prone) material used in dentistry.

5.6.5. Minimization of Corrosion: Case Studies

Although laboratory investigations are essential in the choice of a metal, clinical evaluation in follow-up is also essential. Corrosion of an implant in the clinical setting can result in symptoms such as local pain and swelling in the region of the implant, with no evidence of infection; cracking or flaking of the implant as seen on X-ray films, and excretion of excess metal ions. At surgery, gray or black discoloration of the surrounding tissue may be seen, and flakes of metal may be found in the tissue. Corrosion also plays a role in the mechanical failures of orthopedic implants. Most of these failures are due to fatigue, and the presence of a saline environment certainly exacerbates fatigue. The extent to which corrosion influences fatigue in the body is not precisely known. Some specific case histories follow.

Case 1. A total hip replacement broke after 1.5 years of service.

The prosthesis was X-rayed and found to be broken high in the femoral stem area. The femoral implant was retrieved at surgery and analyzed. It was found to be made of cast stainless steel. Cementation of the implant was not optimal but was adequate. The choice of material was considered poor for this demanding application. Cast stainless steel is mechanically inferior to the wrought steels commonly used in hip nails. Consequently, scratches on the stem during implantation could initiate corrosion fatigue cracks. The stem was found to have failed by fatigue followed by overload fracture. Moreover, in this implant, excessive inclusion content and porosity were found in the area of the failure, indicating poor manufacturing processes.

Case 2. A patient's arm was X-rayed. A bone plate, shown in Figure 5-21, had been left in place for 30 years.

The screws had lost their clear outline due to corrosion and the irritating effect of the corrosion products resulted in osseous proliferation. The plate was found to be vanadium steel, a metal that was considered suitable in the 1920s but that has since been abandoned for implants.

Case 3. A radiograph of a mold arthroplasty revealed that the surgeon had used an ordinary iron nail (Figure 5-22) to reattach the osteotomized greater trochanter.

The cup was found to be made of CoCrMo. The nail was removed and found to be grossly corroded. The mold technique is rarely used today, and surgeons would now not even think of using a carpenter's nail.

Case 4. A patient experienced pain and disability in a repaired shoulder fracture (Figure 5-23).

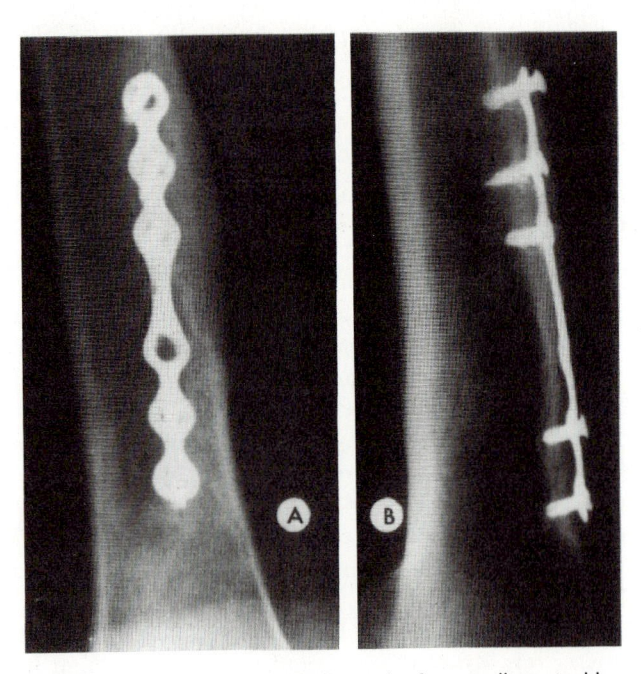

Figure 5-21. Corrosion of an obsolete metal. Radiograph of a vanadium steel bone plate in place 30 years. The screws have lost their clear outline as a result of corrosion. From C. O. Bechtol, A. B. Ferguson, Jr., and P. G. Laing, *Metals and Engineering in Bone and Joint Surgery*, Williams and Wilkins, Baltimore, 1959.

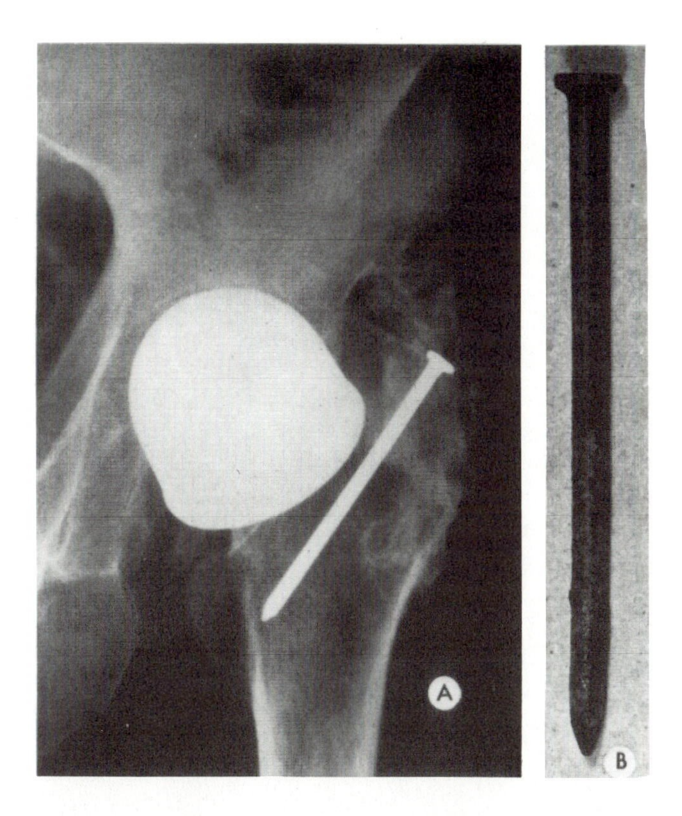

Figure 5-22. Corrosion due to use of improper metals. (A) Radiograph of a CoCrMo mold hip arthroplasty. An ordinary iron nail was used to reattach the osteotomized greater trochanter. (B) The retrieved nail is grossly rusty. From C. O. Bechtol, A. B. Ferguson, Jr., and P. G. Laing, *Metals and Engineering in Bone and Joint Surgery*, Williams and Wilkins, Baltimore, 1959.

The screws were removed and examined. One was found to be CoCrMo and the others of stainless steel. Bimetallic corrosion had resulted. Such cases can be avoided by better efforts by both manufacturers and surgeons to avoid mixed metals.

Case 5. A broken drill bit was found in a radiograph of a repaired broken hip (Figure 5-24).

Removal of the metal revealed severe corrosion. Mixed metals had been implanted unintentionally.

Experience in the orthopedic setting suggests that corrosion is minimized by the following:

1. Use appropriate metals.
2. Avoid implantation of different types of metal in the same region. In the manufacturing process, provide matched parts from the same batch of the same variant of a given alloy.

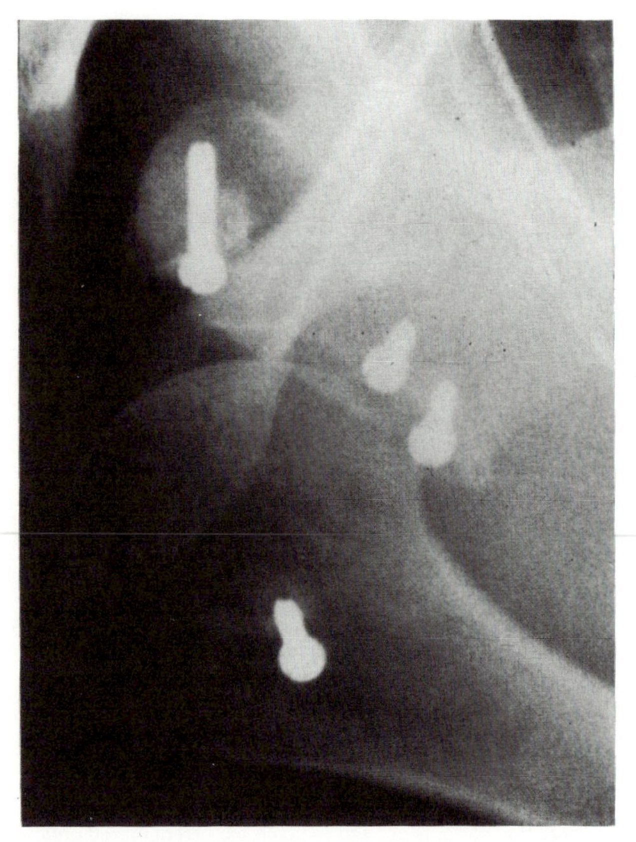

Figure 5-23. Mixed metals. Radiograph of a repaired shoulder. This patient experienced pain and disability. Examination of the screws following removal revealed one to be made of CoCrMo and the others of stainless steel.

3. Design the implant to minimize pits and crevices.
4. In surgery, avoid transfer of metal from tools to the implant or tissue. Avoid contact between metal tools and the implant, unless special care is taken.
5. Recognize that a metal that resists corrosion in one body environment may corrode in another part of the body.

Experience in the dental setting has led to the following suggestions for the minimization of corrosion.

1. Avoid using mixed metals to restore teeth in apposition, and if possible, in the same mouth.
2. Use an insulating base when seating a metallic restoration to minimize electrical conduction below the restoration.

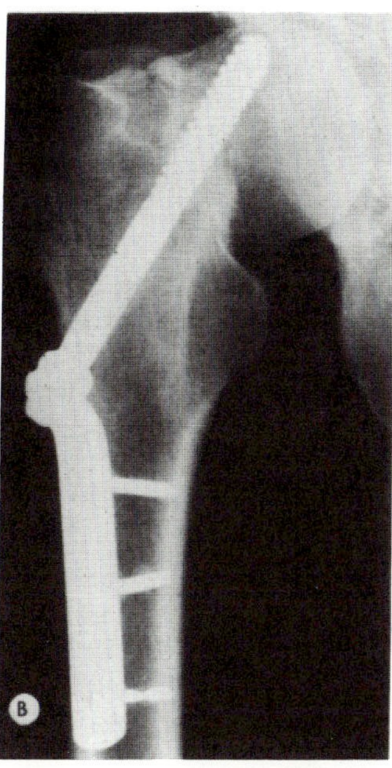

Figure 5-24. Unintentional implantation of mixed metals. (A) Broken and corroded screws and a broken drill bit removed from a patient. (B) Radiograph of fracture fixation device showing screws and drill. From C. O. Bechtol, A. B. Ferguson, Jr., and P. G. Laing, *Metals and Engineering in Bone and Joint Surgery*, Williams and Wilkins, Baltimore, 1959.

3. Avoid conditions that lead to plaque buildup, since regions covered by plaque will experience a reduced pH. This could result in corrosion.

PROBLEMS

5-1. A 1-mm-diameter eutectoid steel is coated with 1-mm-thick aluminum. Using the following data, answer (a)–(d) below.

Material	Young's modulus (GPa)	Yield strength (MPa)	Density (g/cm^3)	Expansion coefficient (/°C)
Eutectoid steel	205	300	7.84	10.8×10^{-6}
Aluminum	70	100	2.7	22.5×10^{-6}

(a) If the composite bar is loaded in tension, which metal will yield first?
(b) How much load can the composite carry in tension without plastic deformation?
(c) What is the Young's modulus of the composite?
(d) What is the density of the composite?

5-2. Cu has the following characteristics:

Atomic no.	29
Atomic mass (amu)	63.54
Melting point (°C)	1084.5
Density (g/cm^3)	8.96
Crystal structure	fcc
Atomic radius (nm)	0.1278
Ionic radius (nm)	0.096

(a) Calculate the density of Cu.
(b) Calculate the packing efficiency (or factor) of Cu.
(c) Would the grain size increase, decrease, or remain the same if the Cu was cooled more slowly than the original Cu after remelting?
(d) Would the grain size increase, decrease, or remain the same if the metal of (c) was annealed?

5-3. Silver has the following characteristics:

Atomic no.	47	Crystal structure	fcc
Atomic mass (amu)	107.87	Atomic radius (nm)	0.1444
Melting point (°C)	961.9	Ionic radius (nm)	0.126
Density (g/cm^3)	10.5		

A bioengineer is trying to construct a binary phase diagram of Cu–Ag and the following additional information was given:

Eutectic temperature	779.4°C
Eutectic composition	28.1 Cu–71.9 Ag
Maximum solubility of Cu in Ag	8.8%
Maximum solubility of Ag in Cu	8.0%

(a) Construct the binary phase diagram by using the information on Cu in Problem 5-2.
(b) Label all the phases in the phase diagram.

5-4. Sterling silver is made of 92.5% Ag and 7.5% Cu.
(a) Construct a temperature versus % fraction chart for liquid phases. Indicate specific temperatures where the phases start to change.
(b) Calculate the density of the sterling silver by using the mixture rule.
(c) When the sterling silver was equilibrated at 780°C, what phase(s) exist? What are their compositions?
(d) Can the sterling silver be age-hardened? Give a reason.

5-5. In general, pearlite steels are susceptible to corrosion. Explain why this is to be expected and how it may be prevented.

5-6. Type 316L stainless steel has a maximum carbon content of 0.03%. Welding of finished components is acceptable for this steel but not for type 316. Why? Explain how you would expect the mechanical properties of each metal to differ.

5-7. Calculate the densities of cast and wrought cobalt alloys (CoNiCrMo). Densities for Co, Cr, Mo, and Ni are 8.8, 7.9, 10.22, and 8.91 g/cm^3, respectively. Use the mixture rule.

5-8. The phase diagram of dental amalgam is shown in Figure 5-10.

(a) Calculate the theoretical weight percentage of the silver and tin of the γ phase (Ag_3Sn).
(b) What is the eutectic temperature?
(c) Can you age-harden the amalgam?

5-9. Suppose a titanium alloy (Ti6Al4V) implant experiences a corrosion current density of 0.1 $\mu A/cm^2$.

(a) Determine the depth of metal removed each year.
(b) Suppose no titanium were excreted and that all of the metal became lodged in a 1-cm thickness of tissue near the implant. Determine the average concentration of metal in the tissue, in parts per million by weight, in one year.

5-10. Magnesium is used in airplanes. Is it sensible to use it in an implant? Explain.

5-11. A patient is to be treated for a fracture from a hunting accident. Lead shotgun pellets remain in the wound. The surgeon decides a bone plate should be used. What is your recommendation?

5-12. Chromium is considered as a candidate material for a staple to be exposed to stomach contents. Is this suitable? Why?

5-13. An amalgam filling has been in service for 30 years and has corroded. The filling surface has become black and the patient complains of tooth sensitivity. A gap has formed at the margin of the restoration, allowing leakage. Should the filling be drilled out and replaced by gold?

5-14. A patient with a stainless steel bone plate complains of pain. The plate has been in service for 30 years and the original fracture has long since healed. Should the plate be removed and replaced by gold?

5-15. Aluminum and iron are used as electrodes in an electrochemical cell. A voltmeter is connected across the electrodes. What potential difference is observed?

5-16. The ASTM (American Society for Testing and Materials) issued standard specifications for thermomechanically processed CoCrMo alloy for surgical implants (F799-82). What is the rationale of the new standard and what are the differences from the earlier standard (F76)?

SYMBOLS/DEFINITIONS

Greek Letters

γ	Silver tin intermetallic compound (Ag_3Sn).
γ_1	Silver mercury intermetallic compound (Ag_2Hg_3).
γ_2	Tin mercury intermetallic compound (Sn_7Hg).
ω	Bending angle of the three-point mechanical bend test.

Latin Letters

A_s	Temperature for the start of reverse transformation, NiTi (III)–NiTi (II). NiTi (II) and NiTi (III) designate structural variations (phases) produced by diffusionless (martensitic) transformation.
E	Observed electrochemical potential difference.
E_0	Standard electrochemical potential difference, for negligible concentration.
F	Faraday's constant, 96,485 coulombs per mole.
M_f	Temperature where martensitic transformation ceases.
M_s	Martensitic transformation starting point in cooling.
n	Number of electron charges removed from ion.
n	Number of moles.
R	Gas constant.
T	Absolute temperature in K.

Terms

Age harden: Hardening of alloys by precipitating second phase during aging process.

Amalgam: An alloy obtained by mixing silver tin alloy with mercury.

Anode: Positive electrode in an electrochemical cell.

Cathode: Negative electrode in an electrochemical cell.

Corrosion: Unwanted reaction of metal with environment. In a Pourbaix diagram, it is the region in which the metal ions are present at a concentration of more than 10^{-6} molar.

Crevice corrosion: A form of localized corrosion in which concentration gradients around preexisting crevices in the material drive corrosion processes.

Galvanic corrosion: Dissolution of metal driven by macroscopic differences in electrochemical potential, usually as a result of dissimilar metals in proximity.

Galvanic series: Table of electrochemical potentials associated with the ionization of metal atoms. These are called Nernst potentials.

Immunity: Resistance to corrosion by an energetic barrier. In a Pourbaix diagram, it is the region in which the metal is in equilibrium with its ions at a concentration of less than 10^{-6} molar.

Nernst potential: Standard electrochemical potential measured with respect to a standard hydrogen electrode.

Noble: Type of metal with a positive standard electrochemical potential. Noble metals resist corrosion by immunity.

Passivation: Production of corrosion resistance by a surface layer of reaction products.

Passivity: Resistance to corrosion by a surface layer of reaction products. In a Pourbaix diagram, it is the region in which the metal is in equilibrium with its reaction products at a concentration of less than 10^{-6} molar.

Pitting: A form of localized corrosion in which pits form on the metal surface.

Pourbaix diagram: Plot of electrical potential versus pH for a material in which the regions of corrosion, passivity, and immunity are identified.

Shape memory effect (SME): Thermoelastic behavior of some alloys that can revert back to their original shape when the temperature is greater than the phase transformation temperature of the alloy.

BIBLIOGRAPHY

C. O. Bechtol, A. B. Ferguson, and P. G. Laing, *Metals and Engineering in Bone and Joint Surgery*, Balliere, Tindall and Cox, London, 1959.

T. W. Duerig, K. N. Melton, D. Stockel, and C. M. Wayman (eds.), *Engineering Aspects of Shape Memory Alloys*, Butterworths-Heinemann, London, 1990.

J. H. Dumbleton and J. Black, *An Introduction to Orthopedic Materials*, Chapter 9, Charles C. Thomas, Springfield, IL. 1975.

E. H. Greener, J. K. Harcourt, and E. P. Lautenschlager, *Materials Science in Dentistry*, Williams and Wilkins, Baltimore, 1972.

B. Harris and A. R. Bunsell, *Structure and Properties of Engineering Materials*, Chapters 7–9, Longmans, London, 1977.

S. N. Levine (ed.), *Materials in Biomedical Engineering*, Annals of the New York Academy of Sciences, Vol. 146, 1968.

D. C. Mears, *Materials and Orthopedic Surgery*, Chapter 5, Williams and Wilkins, Baltimore, 1979.

J. Perkins (ed.), *Shape Memory Effects in Alloys*, Plenum Press, New York, 1975.

F. B. Puckering (ed.), *The Metallurgical Evolution of Stainless Steels*, Chapter 1, p. 42, American Society for Metals and the Metals Society, Metals Park, Ohio, 1979.

L. L. Shreir (ed.), *Corrosion*, 2nd ed., Butterworths, London, 1976.

L. H. Van Vlack, *Materials Science for Engineers*, Chapters 6 and 22, Addison-Wesley Publ. Co., Reading, Mass., 1970.

A. Weinstein, E. Horowitz, and A. W. Ruff (eds.), *Retrieval and Analysis of Orthopedic Implants*, NBS, U.S. Department of Commerce, 1977.

D. F. Williams and R. Roaf, *Implants in Surgery*, Chapters 6 and 8, W. B. Saunders Co., Philadelphia, 1973.

CHAPTER 6

CERAMIC IMPLANT MATERIALS

Ceramics are refractory, polycrystalline compounds, usually inorganic, including silicates, metallic oxides, carbides, and various refractory hydrides, sulfides, and selenides. Oxides such as Al_2O_3, MgO, and SiO_2 contain metallic and nonmetallic elements, whereas others such as $NaCl$, $CsCl$, and ZnS are ionic salts. Exceptions are diamond and carbonaceous structures like graphite and pyrolized carbons, which are covalently bonded. Important factors influencing the structure-property relationship of the ceramic materials are radius ratios (Section 2.2.2) and the relative *electronegativity* between the positive and negative ions.

Recently, ceramic materials have been given a lot of attention as candidates for implant materials since they possess certain highly desirable characteristics for some applications. Ceramics have been used for some time in dentistry for dental crowns owing to their inertness to the body fluids, high compressive strength, and good esthetic appearance.

Some carbons also found use as implants especially for blood interfacing applications such as heart valves. Due to their high specific strength as fibers and their biocompatibility, they are also being used as a reinforcing component for composite implant materials and tensile loading applications such as artificial tendon and ligament replacements. Although the black color can be a drawback in some dental applications, if used as implants, they have desirable qualities such as good biocompatibility and ease of fabrication.

6.1. STRUCTURE–PROPERTY RELATIONSHIP OF CERAMICS

6.1.1. Atomic Bonding and Arrangement

As discussed in Chapter 2, when (neutral) atoms such as sodium (metal) and chlorine (nonmetal) are ionized the sodium will lose an electron and chlorine will gain an electron:

$$\begin{aligned} \mathrm{Na} &\Leftrightarrow \mathrm{Na}^+ + e^- \\ e^- + \mathrm{Cl} &\Leftrightarrow \mathrm{Cl}^- \end{aligned} \tag{6-1}$$

Thus, the sodium and chlorine can make an ionic compound by the strong affinity of the positive and negative ions. The negatively charged ions are much larger than the positively charged ions due to the gain and loss of electrons as given in Table 6-1. The radius of an ion varies according to the coordination numbers, the higher the coordination number the larger the radius. For example, oxygen ion (O^{2-}) has a radius of 1.28, 1.40, and 1.44 Å for coordination numbers 4, 6, and 8, respectively.

Ceramics can be classified according to their structural compounds of which A_mX_n is an example. The A represents a metal and X a nonmetal element, and m and n are integers. The simplest case of this system is AX, of which there are three types as shown in Figure 6-1. The difference between these structures is due to the relative size of the ions (*minimum radius ratio*). If the positive and negative ions are about the same size ($r_A/R_X > 0.732$) the structure becomes a simple cubic (CsCl structure). The face-centered cubic structure arises when the relative sizes of the ions are quite different since the positive ions can be fitted in the tetragonal or octagonal spaces created among larger negative ions. These are summarized in Table 6-2. The aluminum and chromium oxide belong to A_2X_3 type structure. The O^{2-} ions form hexagonal close packing while the positive ions (Al^{3+}, Cr^{3+}) fill in two-thirds of the octahedral sites leaving one-third vacant.

6.1.2. Physical Properties

Ceramics are generally hard; in fact, the measurement of hardness is calibrated against ceramic materials. Diamond is the hardest with a hardness index on Moh's scale of 10 and talc ($Mg_3Si_4O_{10}COH$) is the softest (Moh's hardness 1), and others such as alumina (Al_2O_3; hardness 9), quartz (SiO_2; hardness 8) and apatite ($Ca_5P_3O_{12}F$; hardness 5) are intermediate. Other characteristics of ceramic materials are their high melting temperatures, and low conductivity of electricity and heat. These characteristics reflect the nature of the chemical bonding in ceramics.

Unlike metals and polymers, ceramics are difficult to shear plastically due to the ionic nature of bonding as shown in Figure 6-2. In order to shear, the planes of atoms should slip past each other. However, for ceramic materials the ions of like electric charge repel each other; moving the plane of atoms is thus very difficult. This makes the ceramics nonductile and creep at room temperature is almost zero. The ceramics are also very sensitive to notches or microcracks since instead of undergoing plastic deformation (or yield) they will fracture elastically once the crack propagates. This is also why the ceramics have low tensile strength compared to compressive strength as discussed in Section 3.1.2. If the ceramic is made flawfree, then it becomes very strong even in tension.

Table 6-1. Atomic and Ionic Radii (Å) of Some Elements[a]

Group I			Group II			Group IV			Group VI		
Element	Atomic radius[b]	Ionic radius	Element	Atomic radius[b]	Ionic radius	Element	Atomic radius[b]	Ionic radius	Element	Atomic radius[b]	Ionic radius
Li^+	1.52	0.68	Be^{2+}	1.11	0.31	O^{2-}	0.74	1.40	F^-	0.71	1.36
Na^+	1.86	0.95	Mg^{2+}	1.60	0.65	S^{2-}	1.02	1.84	Cl^-	0.99	1.81
K^+	2.27	1.33	Ca^{2+}	1.97	0.99	Se^{2-}	1.16	1.98	Br^-	1.14	1.95

[a] From M. J. Starfield and A. M. Shrager, *Introductory Materials Science*, McGraw-Hill Book Co., New York, p. 64.
[b] Covalent.

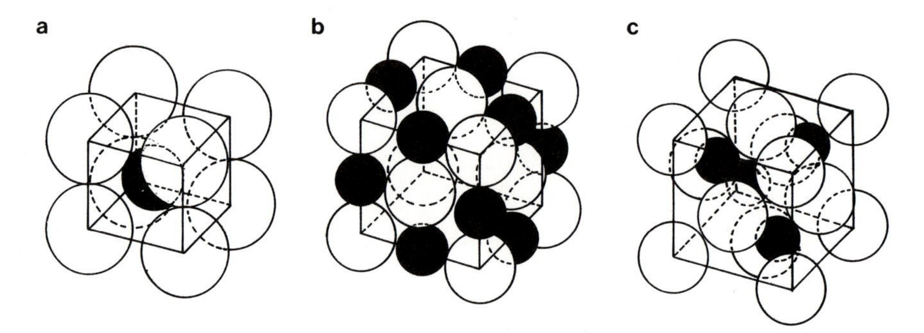

Figure 6-1. AX structures of ceramics. The black spheres represent positive ions (A^+) and the white ones represent negative ions (X^-). (a) AX structure (CsCl); (b) AX structure (NaCl); (c) AX structure (ZnS).

Table 6-2. Selected A_mX_n Structures

Prototype compound	Lattice of A (or X)	CN of A (or X) sites	Available sites filled	Minimum r_A/R_X	Other compounds
CsCl	Simple cubic	8	All	0.732	Csl
NaCl	fcc	6	All	0.414	MgP, MnS, LiF
ZnS	fcc	4	1/2	0.225	β-SiC, CdS, AlP
Al_2O_3	hcp	6	2/3	0.414	Cr_2O_3, Fe_2O_3

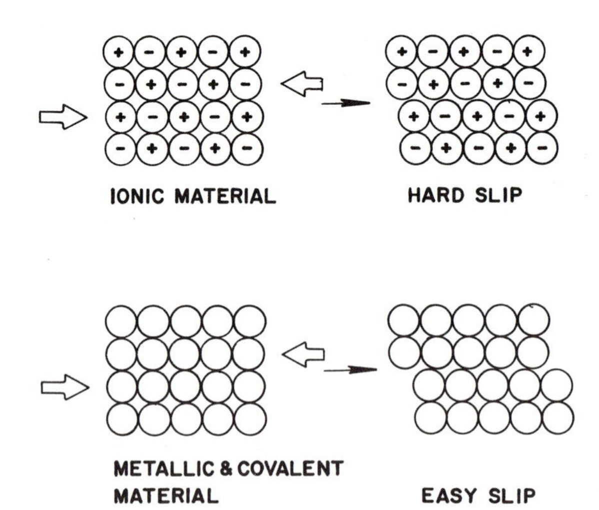

Figure 6-2. Schematic two-dimensional illustration of slips in ionic and nonionic bond materials.

Glass fibers made this way have tensile strengths twice that of a high-strength steel ($\approx$7 GPa).

Example 6-1

A piece of window glass fails at 70 MPa (10^4 psi). Calculate the largest size elliptic crack that is responsible for the low strength. The stress concentration factor (σ/σ_a) can be expressed as $2\sqrt{c/r}$, where c is the crack depth ($2c$ if away from surface) and r is the crack tip radius.

Answer

Assuming the crack tip radius has the dimension of an oxygen ion (1.4 Å) and the theoretical strength of glass is 7 GPa,

$$c = \frac{r(\sigma/\sigma_a)^2}{4}$$

$$= \frac{(1.4)(7000/70)^2}{4}$$

$$= 3.5 \times 10^3 \text{ Å}, \text{ or } 0.35\ \mu\text{m}$$

6.2. ALUMINUM OXIDES

The main source of high-purity alumina (aluminum oxide) is bauxite and native corundum. The commonly available (alpha, α) alumina can be prepared by calcining alumina trihydrate resulting in calcined alumina. Chemical composition and density of commercially available "pure" calcined aluminas are given in Table 6-3. The American Society for Testing and Materials (ASTM) specifies 99.5% pure alumina and less than 0.1% of combined SiO_2 and alkali oxides (mostly Na_2O) for implant use.

Alpha alumina has a rhombohedral crystal structure ($a = 4.758$ Å, and $c = 12.991$ Å). Natural alumina is known as sapphire or ruby (depending on the types of impurities, which give rise to color). The single-crystal form of alumina

Table 6-3. Chemical Composition of Calcined Aluminas[a]

Chemical	Composition (wt%)
Al_2O_3	99.6
SiO_2	0.12
Fe_2O_3	0.03
Na_2O	0.04

[a] From W. H. Gitzen (ed.), *Alumina as a Ceramic Material*, American Ceramic Society, Columbus, Ohio, 1970.

Table 6-4. Physical Property Requirements of Alumina Implants[a]

Property	Value
Flexural strength	>400 MPa (58,000 psi)
Elastic modulus	380 GPa (55.1×10^6 psi)
Density (g/cm^3)	3.8–3.9

[a] *Annual Book of ASTM Standards*, Part 46, American Society for Testing and Materials, Philadelphia, 1980.

has been used successfully to make implants. Single-crystal alumina can be made by feeding fine alumina powders onto the surface of a seed crystal that is slowly withdrawn from an electric arc or oxyhydrogen flame as the fused powder builds up. Alumina single crystals of up to 10-cm diameter have been grown by this method.

The strength of polycrystalline alumina depends on the porosity and grain sizes. Generally the smaller the grains and porosity, the higher the strength will be. The ASTM standard (F603-78) requires a flexural strength of greater than 400 MPa and elastic modulus of 380 GPa as given in Table 6-4.

Alumina in general is quite hard (Moh's hardness 9); the hardness varies from 2000 kg/mm^2 (19.6 GPa) to 3000 kg/mm^2 (29.4 GPa). This high hardness permits one to use alumina as an abrasive (emery) and as bearings for watch movements. The high hardness is accompanied by low friction and wear; these are major advantages of using alumina as a joint replacement material in spite of its brittleness.

6.3. CALCIUM PHOSPHATE

Calcium phosphate has been used in the form of artificial bone. Recently, this material has been synthesized and used for manufacturing various forms of implants as well as for solid or porous coatings on other implants.

6.3.1. Structure of Calcium Phosphate

Calcium phosphate can be crystallized into salts, hydroxyapatite and β-whitlockite depending on the Ca/P ratio, presence of water, impurities, and temperature. In a wet environment and at lower temperature (<900°C) it is more likely that the (hydroxyl or hydroxy) apatite will form, whereas in a dry atmosphere and at higher temperature the β-whitlockite will be formed. Both forms are very tissue compatible and are used for bone substitute in a granular form or a solid block. We will consider the apatite form of calcium phosphate since it is regarded to be more closely related to the mineral phase of bone and teeth.

The mineral part of bone and teeth is made of a crystalline form of calcium phosphate similar to hydroxyapatite [$Ca_{10}(PO_4)_6(OH)_2$]. The apatite family of minerals, $A_{10}(BO_4)_6X_2$, crystallizes into hexagonal rhombic prisms and has unit cell dimensions $a = 9.432$ Å and $c = 6.881$ Å. The atomic structure of hydroxyapatite projected down the c axis onto the basal plane is given in Figure 6-3. Note that the hydroxyl ions lie on the corners of the projected basal plane and they occur at equidistant intervals one-half of the cell (3.44 Å) along columns perpendicular to the basal plane and parallel to the c axis. Six of the ten calcium ions in the unit cell are associated with the hydroxyls in these columns, resulting in strong interactions among them.

The ideal Ca/P ratio of hydroxyapatite is 10:6 and the calculated density is 3.219 g/cm^3. It is interesting to note that the substitution of OH with F will give greater chemical stability due to the closer coordination of F (symmetric shape) as compared to the hydroxyl (nonsymmetric, two atoms) by the nearest calcium. This is one of the reasons for the better caries resistance of teeth following fluoridation.

6.3.2. Properties of Hydroxyapatite

There is a wide variation of mechanical properties of synthetic calcium phosphates as given in Table 6-5. This variation of properties is the result of the variation in the structure of polycrystalline calcium phosphates, in turn the result of differences in manufacturing processes. Depending on the final firing conditions, the calcium phosphate can be calcium hydroxyapatite or β-whitlockite. In many instances, however, both types of structure exist in the same final product.

Polycrystalline hydroxyapatite has a high elastic modulus (40–117 GPa). Hard tissues such as bone, dentin, and dental enamel are natural composites that contain hydroxyapatite (or a similar mineral) as well as protein, other organic materials, and water. Enamel is the stiffest hard tissue with an elastic modulus of 74 GPa and it contains the most mineral. Dentin (E = 21 GPa) and compact bone (E = 12–18 GPa) contain comparatively less mineral. The Poisson's ratio for the mineral or synthetic hydroxyapatite is about 0.27, which is close to that of bone ($\approx$0.3).

Among the most interesting properties of hydroxyapatite as a biomaterial is its excellent biocompatibility. Indeed, it appears to form a direct chemical bond with hard tissues. In an experimental trial, new lamellar cancellous bone was formed around implanted hydroxyapatite granules in the marrow cavity of rabbits after 4 weeks as shown in Figure 6-4.

6.3.3. Manufacturing of Hydroxyapatite

Many different methods have been developed to make precipitates of hydroxyapatite from an aqueous solution of $Ca(NO_3)_2$ and NaH_2PO_4. One method uses precipitates that are filtered and dried to form a fine particle powder. After calcination for about 3 hours at 900°C to promote crystallization, the powder

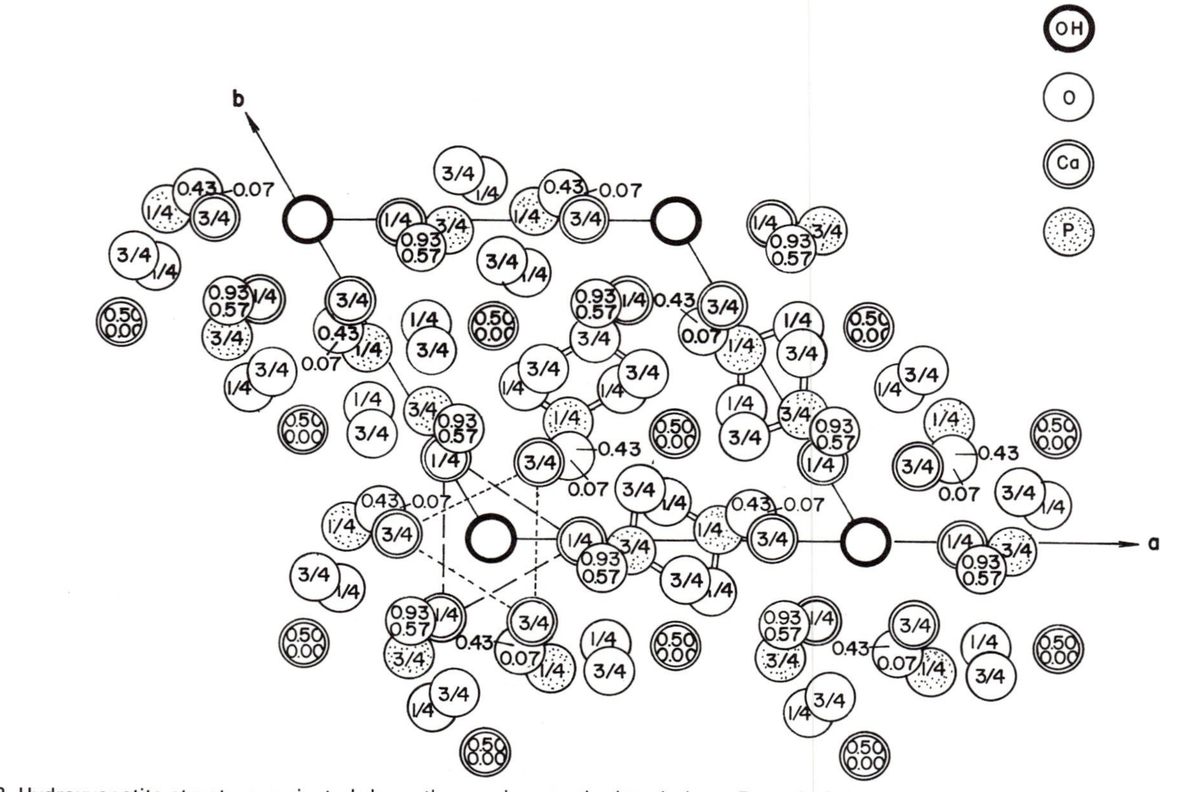

Figure 6-3. Hydroxyapatite structure projected down the *c* axis onto the basal plane. From A. S. Posner, A. Perloff, and A. D. Diorio, "Refinement of the Hydroxyapatite Structure," *Acta Crystallogr.*, *11*, 308–309, 1958.

Table 6-5. Physical Properties of Calcium Phosphate[a]

Property	Value
Elastic modulus (GPa)	40–117
Compressive strength (MPa)	294
Bending strength (MPa)	147
Hardness (Vickers, GPa)	3.43
Poisson's ratio	0.27
Density (theoretical, g/cm^3)	3.16

[a] Data from, among other sources, D. E. Grenoble, J. L. Katz, K. L. Dunn, R. S. Gilmore, and K. L. Murty, "The Elastic Properties of Hard Tissues and Apatites," *J. Biomed. Mater. Res.*, *6*, 221–233, 1972.

is pressed into a final form and sintered at about 1050 to 1200°C for 3 hours. Above 1250°C the hydroxyapatite shows a second phase precipitation along the grain boundaries.

Example 6-2

Calculate the theoretical density of hydroxyapatite crystal $[Ca_{10}(PO_4)_6(OH)_2]$.

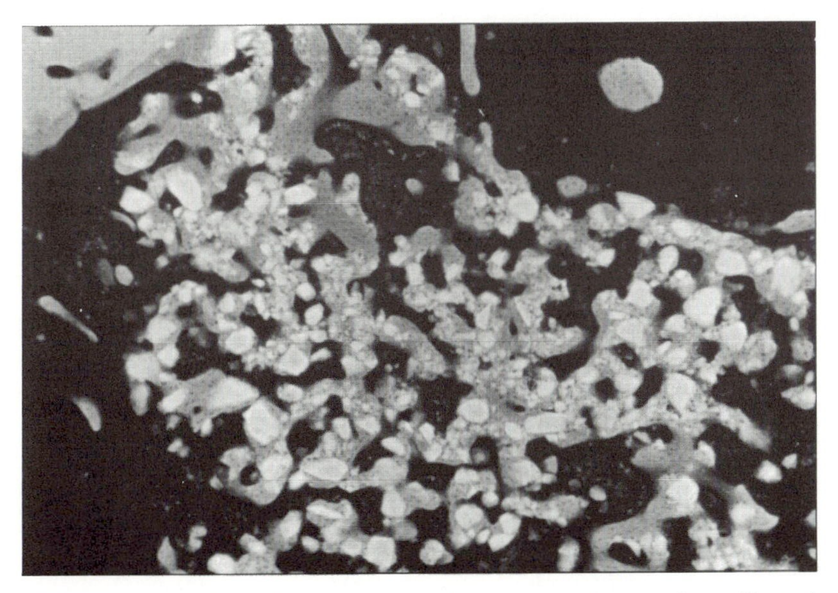

Figure 6-4. X-ray microradiographic picture shows the hydroxyapatite granules and bony tissues surrounding them after 4 weeks in rabbit marrow cavity (40×). The mottled areas are new bones and white areas are implants. From S. Niwa, K. Sawai, S. Takahashi, H. Tagai, M. Ono, and Y. Fukuda, Experimental Studies on the Implantation of Hydroxyapatite in the Medullary Canal of Rabbits, *Trans. First World Biomaterials Congress*, Baden, Austria, April 8–12, 1980.

Answer

From Figure 6-3 one can see that there are 10 Ca atoms in the hexagonal unit cell prism, 4 inside (2 for 1/2, 2 for 1/4, 3/4 position), 2 for top and bottom (0 and 1 position), and 4 for sides (1/4, 3/4 position). Therefore,

$$\rho = \frac{(10 \times 40 + 6 \times 31 + 26 \times 16 + 2 \times 1)}{9.432 \times (\sqrt{3}/2) \times 9.432 \times 6.881 \times 10^{-24} \times 6.02 \times 10^{23}}$$

$$= \underline{3.16\ \text{g/cm}^3}$$

[This is very close to the value given in the literature (McConell, 1973; see the Bibliography).]

6.4. GLASS-CERAMICS

Glass-ceramics are polycrystalline ceramics made by controlled crystallization of glasses developed by S. D. Stookey of Corning Glass Works in the early 1960s. They were first utilized in photosensitive glasses in which small amounts of copper, silver, and gold are precipitated by ultraviolet light irradiation. These metallic precipitates help to nucleate and crystallize the glass into a fine-grained ceramic that possesses excellent mechanical and thermal properties. Bioglass® and Ceravital® are two glass-ceramics developed for implants.

6.4.1. Formation of Glass-Ceramics

The formation of glass-ceramics is influenced by the nucleation and growth of small (<1 μm diameter) crystals as well as the size distribution of these crystals. It is estimated that about 10^{12} to 10^{15} nuclei per cm^3 are required to achieve such small crystals. In addition to the metallic agents mentioned (Cu, Ag, and Au), Pt groups, TiO_2, ZrO_2, and P_2O_5 are widely used for this purpose. The nucleation of glass is carried out at temperatures much lower than the melting temperature. During processing the melt viscosity is kept in the range of 10^{11} and 10^{12} poise for 1-2 hours. In order to obtain a larger fraction of the microcrystalline phase, the material is further heated to an appropriate temperature for maximum crystal growth. Deformation of the product, phase transformation within the crystalline phases, or redissolution of some of the phases are to be avoided. The crystallization is usually more than 90% complete with grain sizes of 0.1 to 1 μm. These grains are much smaller than those of the conventional ceramics. Figure 6-5 shows a schematic representation of the temperature–time cycle for a glass-ceramic.

The glass-ceramics developed for implantation are SiO_2–CaO–Na_2O–P_2O_5 and Li_2O–ZnO–SiO_2 systems. There are two different groups experimenting with the SiO_2–CaO–Na_2O–P_2O_5 glass-ceramic. One group varied the compositions (except for P_2O_5) as given in Table 6-6 in order to obtain the best composition to induce direct bonding with bone. The bonding is related to the simultaneous formation of a calcium phosphate and SiO_2-rich film layer on the surface as

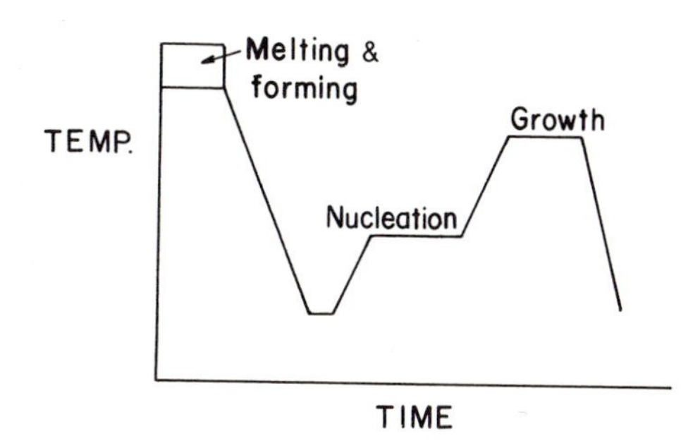

Figure 6-5. Temperature–time cycle for a glass-ceramic. From W. D. Kingery, H. K. Bowen, and D. R. Uhlmann, *Introduction to Ceramics*, 2nd ed., J. Wiley and Sons, New York, 1976, p. 368.

Table 6-6. Compositions of Bioglass® and Ceravital® Glass-Ceramics[a]

Type	Code	SiO_2	CaO	Na_2O	P_2O_5	MgO	K_2O
Bioglass	42S5.6	42.1	29.0	26.3	2.6	—	—
	(45S5) 46S5.2	46.1	26.9	24.4	2.6	—	—
	49S4.9	49.1	25.3	23.0	2.6	—	—
	52S4.6	52.1	23.8	21.5	2.6	—	—
	55S4.3	55.1	22.2	20.1	2.6	—	—
	60S3.8	60.1	19.6	17.7	2.6	—	—
Cervital[b]	Bioactive	40.0–50.0	30.0–35.0	5.0–10.0	10.0–15.0	2.5–5.0	0.5–3.0
	Nonbioactive[c]	30.0–35.0	25.0–30.0	3.5–7.5	7.5–12.0	1.0–2.5	0.5–2.0

[a] Data from (1) M. Ogino, F. Ohuchi, and L. L. Hench, "Compositional Dependence of the Formation of Calcium Phosphate Film on Bioglass," *J. Biomed. Mater. Res.*, *12*, 55–64, 1980; and (2) B. A. Blencke, H. Bromer, and K. K. Deutscher, "Compatibility and Long-term Stability of Glass-ceramic Implants," *J. Biomed. Mater. Res.*, *12*, 307–318, 1978.
[b] The Ceravital composition is in weight % while the Bioglass compositions are in mol %.
[c] In addition, Al_2O_3 (5.0–15.0), TiO_2 (1.0–5.0), and Ta_2O_5 (5.0–15.0) are added.

exhibited by the 46S5.2-type Bioglass®. If a SiO_2-rich layer forms first and a calcium phosphate film develops later (46–55 mol% SiO_2 samples) or no phosphate film is formed (60 mol% SiO_2), then no direct bonding with bone is observed. The approximate region of the SiO_2-CaO-Na_2O system for the tissue-glass-ceramic reaction is shown in Figure 6-6. As can be seen the best region (Region A) for good tissue bonding is the composition given for 46S5.2-type Bioglass® (Table 6-6).

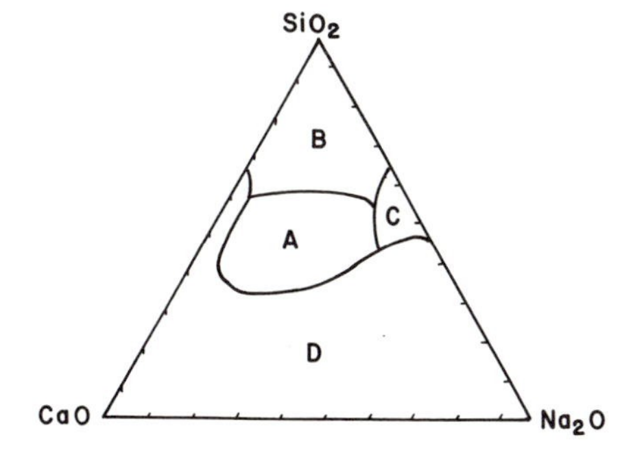

Figure 6-6. Approximate regions of the tissue–glass-ceramic bonding for the SiO_2–CaO–Na_2O system. A: Bonding within 30 days. B: Nonbonding; reactivity is too low. C: Nonbonding; reactivity is too high. D: Bonding; does not form glass. From L. L. Hench and E. C. Ethridge, *Biomaterials: An Interfacial Approach*, Academic Press, New York, 1982, p. 147.

The composition of Ceravital® is similar to the Bioglass® in SiO_2 content but differs somewhat in others as given in Table 6-6. Also, Al_2O_3, TiO_2, and Ta_2O_5 are used for the Ceravital glass-ceramic in order to control the dissolution rate. The mixtures were melted in a platinum crucible at 1500°C for 3 hours and annealed, then cooled. The nucleation and crystallization temperatures were 680 and 750°C, respectively, for 24 hours each. When the size of crystallites was about 4 Å and did not exhibit the characteristic needle structure, the process was stopped to obtain fine grain structure.

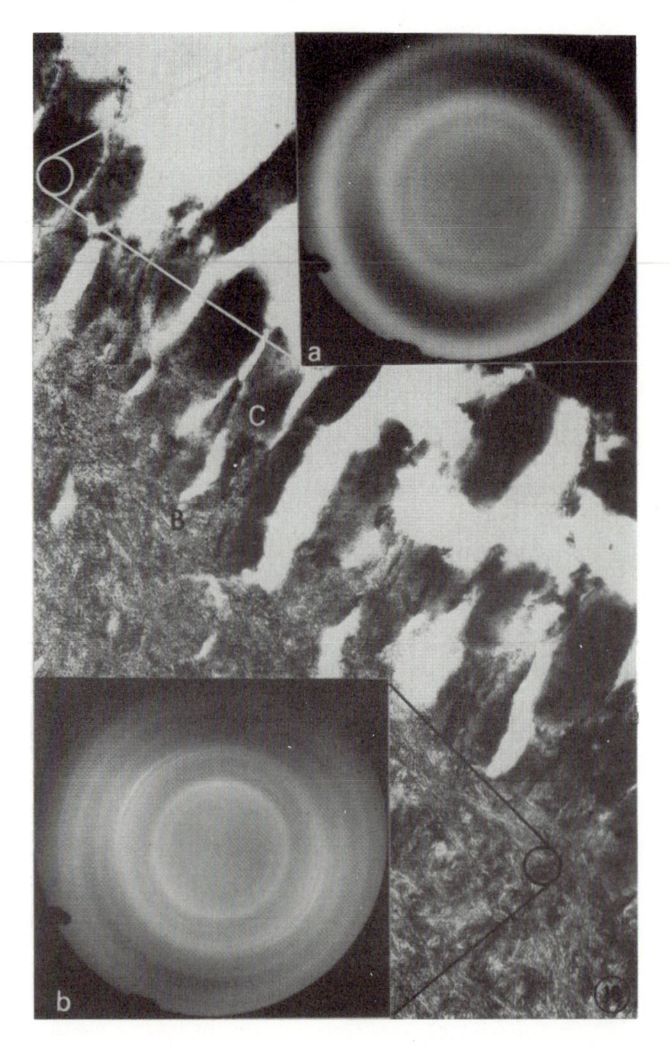

Figure 6-7. Transmission electron micrograph of well-mineralized bone (b) juxtaposed to the glass-ceramic (c), which was fractured during sectioning. ×51,500. Insert a is the diffraction pattern from ceramic area and b is from bone area. From C. A. Beckham, T. K. Greenlee, Jr., and A. R. Crebo, "Bone Formation at a Ceramic Implant Interface," *Calcif. Tissue Res.*, *8*, 165–171, 1971.

6.4.2. Properties of Glass-Ceramics

Glass-ceramics have several desirable properties compared to glasses and ceramics. The thermal coefficient of expansion is very low, typically 10^{-7} to 10^{-5} per degree C and in some cases it can be made even negative. Due to the controlled grain size and improved resistance to surface damage, the tensile strength of these materials can be increased at least a factor of two from about 100 MPa to 200 MPa. The resistance to scratching and abrasion is close to that of sapphire.

In an experimental trial, Bioglass® glass-ceramic was implanted in the femur of rats for 6 weeks; a transmission electron micrograph (Figure 6-7) showed intimate contacts between the mineralized bone and the Bioglass®. The mechanical strength of the interfacial bond between bone and Bioglass® ceramic is the same order of magnitude as the strength of the bulk glass-ceramic ($850\ kg/cm^2$ or 83.3 MPa), which is about three-fourths that of the host bone strength.

The main drawback of the glass-ceramic is its brittleness as is the case with other glasses and ceramics. Also, due to the restrictions on the composition for biocompatibility (or osteogenicity) the mechanical strength cannot be substantially improved as for other glass-ceramics. Therefore, they cannot be used for making major load-bearing implants such as joint implants. However, they can be used as fillers for bone cement, dental restorative composites, and coating material.

Example 6-3

From the phase diagram of Al_2O_3–SiO_2 shown below,

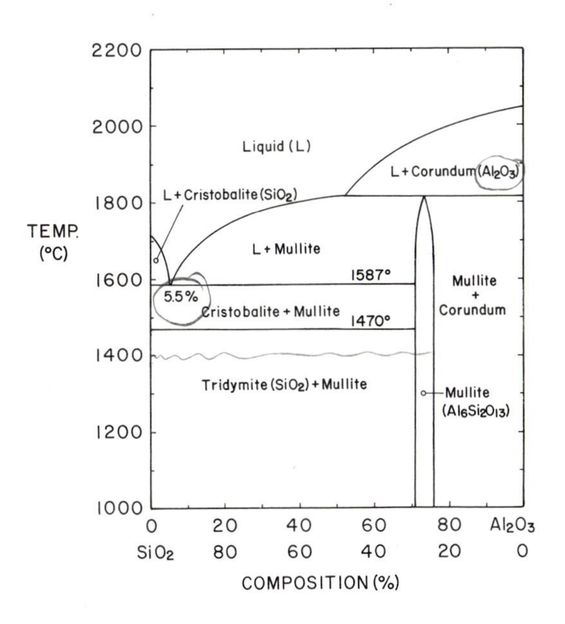

(a) Determine the exact wt% of Al_2O_3 for mullite, which has $3Al_2O_3 \cdot 2SiO_2$ composition.
(b) Determine the amount of liquid in 50 wt% Al_2O_3–50 wt% SiO_2 at 2815°F.

Answer

(a)
$$\frac{3Al_2O_3}{3Al_2O_3 + 2SiO_2} = \frac{6(27) + 9 \times 16}{6 \times 27 \times 9 \times 16 + 2 \times 28 + 4 \times 16}$$

$$= \frac{162 + 144}{306 + 56 + 64}$$

$$= \frac{306}{426}$$

$$= \underline{0.718\ (71.8\%)}$$

(b) Using the lever rule (see Section 5.1.1):

$$\%L = \frac{71.8 - 50}{71.8 - 5.5} = \frac{21.8}{66.3} = \underline{0.329\ (32.9\%)}$$

6.5. OTHER CERAMICS

In addition to the ceramic materials discussed so far, many others have been studied. These are titanium oxide (TiO_2), barium titanate ($BaTiO_2$), tricalcium phosphate [$Ca_3(PO_4)_2$], and calcium aluminate ($CaO \cdot Al_2O_3$). Titanium oxide was tested as a potential component of bone cement or as a blood interfacing material. Porous calcium aluminate was used to induce tissue ingrowth into the pores with the aim of achieving better implant fixation. However, this material loses its strength considerably after *in vivo* and *in vitro* aging as shown in Figure 6-8. The tricalcium phosphate together with calcium aluminate were tried as biodegradable implants in the hopes of regenerating new bone.

Barium titanate with a textured surface was used in experimental trials to achieve improved fixation of implants to bone. This material is piezoelectric (following a polarization procedure). Therefore, mechanical loads on the implant will generate electrical signals that are capable of stimulating bone healing and ingrowth. These loads on the implant arise during the use of the implanted limb. Alternatively, the implant can be exposed to ultrasound to generate the electrical signals.

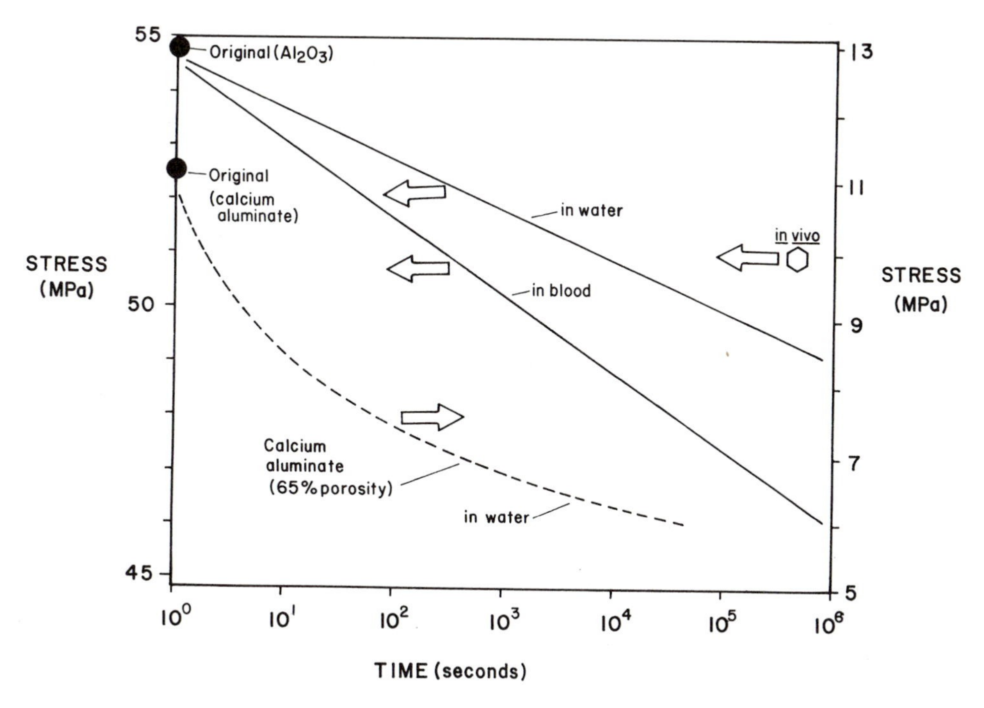

Figure 6-8. Aging effect on the strength of calcium aluminate *in vitro* and *in vivo*. From G. S. Schnittgrund, G. H. Kenner, and S. D. Brown, "In Vivo and in Vitro Changes in Strength of Orthopedic Calcium Aluminate," *J. Biomed. Mater. Res. Symp.*, No. 4, 435–452, 1973.

6.6. CARBONS

Carbons can be made in many allotropic forms: crystalline diamond, graphite, noncrystalline glassy carbon, and quasi-crystalline pyrolytic carbon. Among these, only pyrolytic carbon is widely utilized for implant fabrication; it is normally used as a surface coating. It is also possible to coat surfaces with diamond. This technique has the potential to revolutionize medical device manufacturing; however, it is not yet commercially available.

6.6.1. Structure of Carbons

The crystalline structure of carbon as used in implants is similar to the graphite structure as shown in Figure 6-9. The planar hexagonal arrays are formed by strong covalent bonds in which one of the valence electrons per atom is free to move, resulting in high but anisotropic electric conductivity. The bonding between layers is stronger than the van der Waals force; therefore, *cross-links* between them were suggested. Indeed, the remarkable lubricating property of graphite cannot be realized unless the cross-links are eliminated.

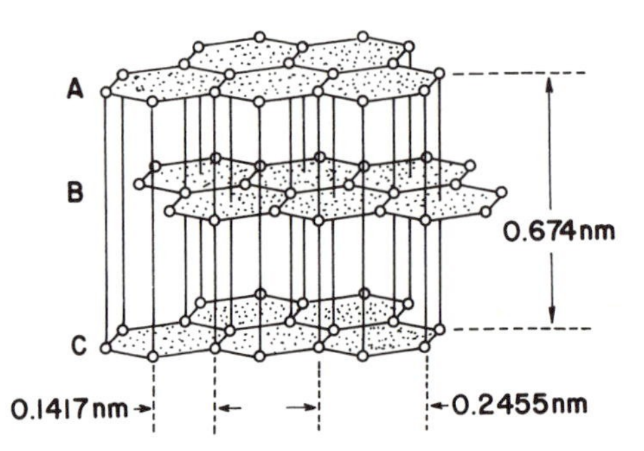

Figure 6-9. Crystal structure of graphite. From E. I. Shobert, II, *Carbon and Graphite*, Academic Press, New York, 1964.

The poorly crystalline carbons are thought to contain unassociated or unoriented carbon atoms. The hexagonal layers are not perfectly arranged as shown in Figure 6-10. The strong bonding within layers and the weaker bonding between layers cause the properties of individual crystallites to be highly anisotropic. However, if the crystallites are randomly dispersed, then the aggregate becomes isotropic.

6.6.2. Properties of Carbon

The mechanical properties of carbon, especially pyrolytic carbon, are largely dependent on the density as shown in Figures 6-11 and 6-12. The increased mechanical properties are directly related to the increased density, which indicates that the properties depend mainly on the aggregate structure of the material.

Graphite and glassy carbon have much lower mechanical strength than pyrolytic carbon (Table 6-7). However, the average modulus of elasticity is almost the same for all carbons. The strength of pyrolytic carbon is quite high compared

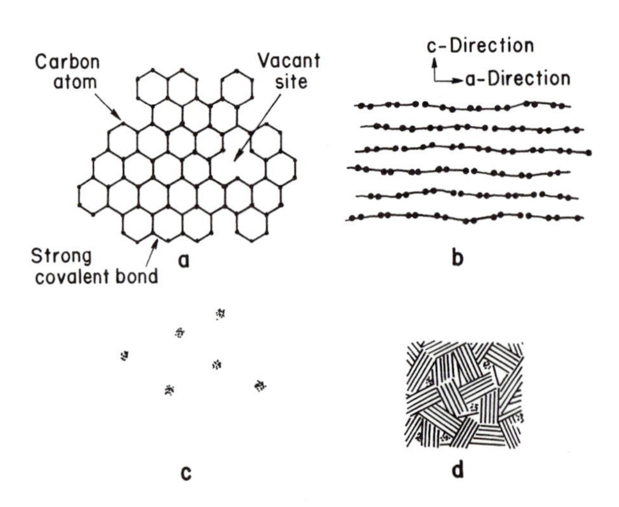

Figure 6-10. Schematic representation of poorly crystalline carbon. (a) Single layer plane; (b) parallel layers in a crystallite; (c) unassociated carbon; (d) an aggregate of crystallites, single layers, and unassociated carbon. From J. C. Bokros, "Deposition Structure and Properties of Pyrolitic Carbon," in *Chemistry and Physics of Carbon*, P. L. Walker (ed.), Marcel Dekker, New York, Vol. 5, pp. 70–81.

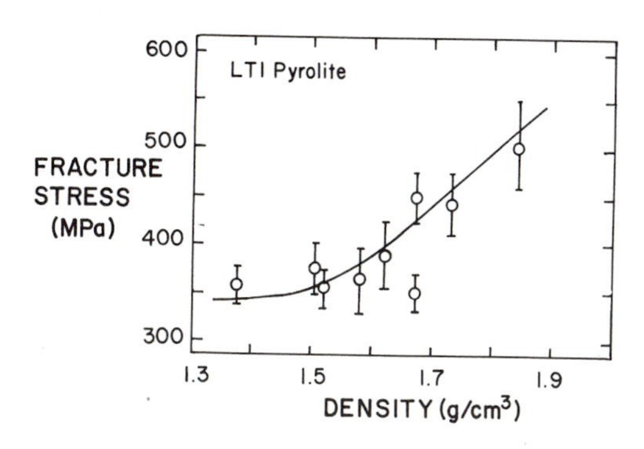

Figure 6-11. Fracture stress versus density for unalloyed LTI pyrolite carbons. From J. L. Kaae, "Structure and Mechanical Properties of Isotropic Pyrolytic Carbon Deposited below 1600°C," *J. Nucl. Mater.*, *38*, 42–50, 1971.

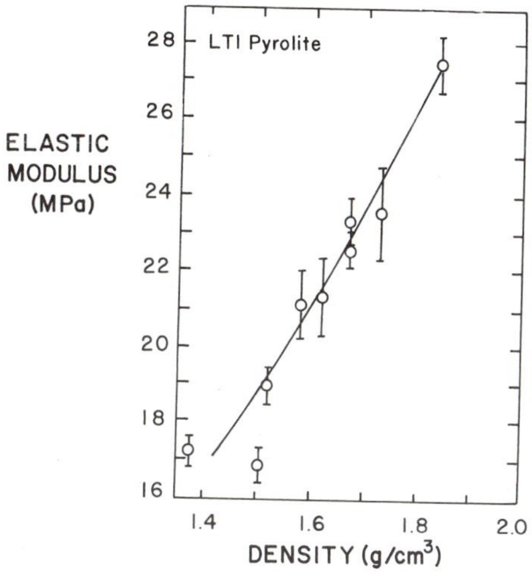

Figure 6-12. Elastic modulus versus density for unalloyed LTI pyrolite carbons. From J. L. Kaae, "Structure and Mechanical Properties of Isotropic Pyrolytic Carbon Deposited below 1600°C," *J. Nucl. Mater.*, *38*, 42–50, 1971.

Table 6-7. Properties of Various Types of Carbon

Property	Carbon type		
	Graphite	Glassy	Pyrolytic[a]
Density (g/cm^3)	1.5–1.9	1.5	1.5–2.0
Elastic modulus (GPa)	24	24	28
Compressive strength (MPa)	138	172	517 (575[a])
Toughness (mN/cm^3)[b]	6.3	0.6	4.8

[a] 1.0 wt% Si-alloyed pyrolytic carbon, Pyrolite® (Carbomedics, Austin, Tex.).
[b] $1\ mN/cm^3 = 1.45 \times 10^{-3}\ in\text{-}lb/in^3$.

Table 6-8. Mechanical Properties of Carbon Fiber-Reinforced Carbon[a]

Property	Fiber lay-up	
	Unidirectional	0–90° crossply
Flexural modulus (GPa)		
Longitudinal	140	60
Transverse	7	60
Flexural strength (MPa)		
Longitudinal	1200	500
Transverse	15	500
Interlaminar shear strength (MPa)	18	18

[a] From D. Adams and D. F. Williams, "Carbon Fiber-Reinforced Carbon as a Potential Implant Material," *J. Biomed. Mater. Res.*, *12*, 35–42, 1978.

to graphite and glassy carbon. This is again due to the lesser amount of flaws and unassociated carbons in the aggregate.

A composite carbon that is reinforced with carbon fiber has been considered for implants. The properties are highly anisotropic as given in Table 6-8. The density is in the range of 1.4–1.45 g/cm^3 with a porosity of 35–38%.

Carbons exhibit excellent compatibility with tissues. In particular, the compatibility with blood has made the pyrolytic carbon-deposited heart valves and blood vessel walls a widely accepted part of the surgical armamentarium.

6.6.3. Manufacturing of Implants

Pyrolytic carbons can be deposited onto finished implants from hydrocarbon gas in a *fluidized bed* at a controlled temperature and pressure as shown in Figure 6-13. The anisotropy, density, crystallite size, and structure of the deposited carbon can be controlled by temperature, composition of the fluidizing gas, bed geometry, and residence time (velocity) of the gas molecules in the bed. The microstructure of deposited carbon should be particularly controlled since the formation of growth features associated with uneven crystallization can result in a weaker material as shown in Figure 6-14. It is also possible to introduce various other elements into the fluidizing gas and codeposit them with carbon. Usually silicon (10–20 wt%) is codeposited (or *alloyed*) to increase hardness for applications requiring resistance to abrasion, such as heart valve discs.

Recently, success was achieved in depositing pyrolytic carbon onto the surfaces of blood vessel implants made of polymers. This is called ultra-low-temperature isotropic (ULTI) carbon instead of LTI (low-temperature isotropic) carbon. The deposited carbon is thin enough not to interfere with the flexibility of the grafts and yet exhibited excellent blood compatibility.

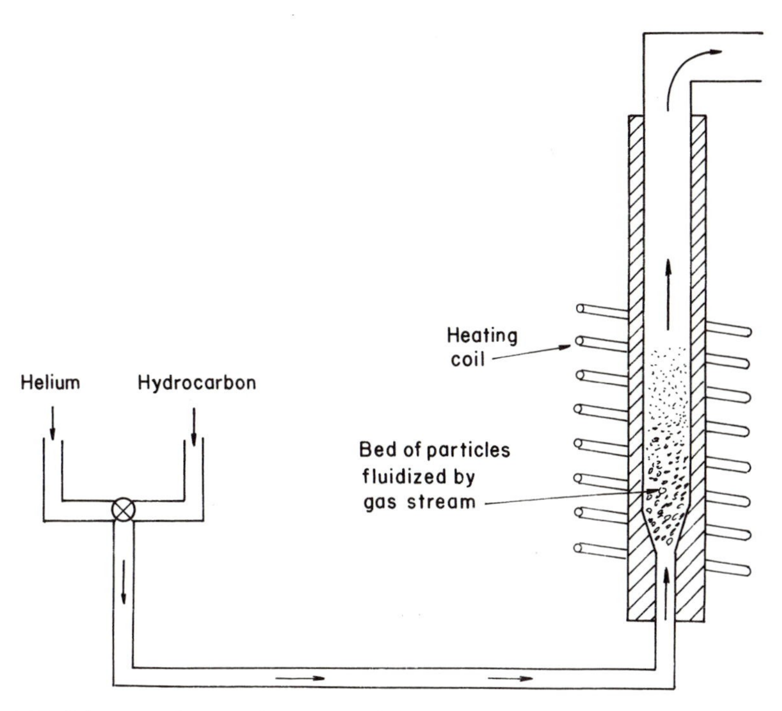

Figure 6-13. Schematic diagram showing particles being coated with carbon in a fluidized bed. From J. C. Bokros, "Deposition Structure and Properties of Pyrolitic Carbon," in *Chemistry and Physics of Carbon*, P. L. Walker (ed.), Marcel Dekker, New York, 1972, Vol. 5, pp. 70–81.

The vitreous or glassy carbon is made by controlled pyrolysis of polymers such as phenolformaldehyde, rayon (cellulose), and polyacrylonitrile at high temperature in a controlled environment. This process is particularly useful for making carbon fibers and textiles that can be used themselves or as components of composites.

6.7. DETERIORATION OF CERAMICS

It is of great interest to know whether the inert ceramics such as alumina undergo significant static or dynamic fatigue. In one study it was shown that above a critical stress level the fatigue strength of alumina is reduced by the presence of water. This is due to the delayed crack growth, which is accelerated by the water molecules. However, another study showed that a reduction in strength occurred if evidence of penetration by water was observed under a scanning electron microscope. No decrease in strength was observed for samples that showed no water marks on the fractured surface as shown in Figure 6-15. It was suggested that the presence of a minor amount of silica in one sample lot

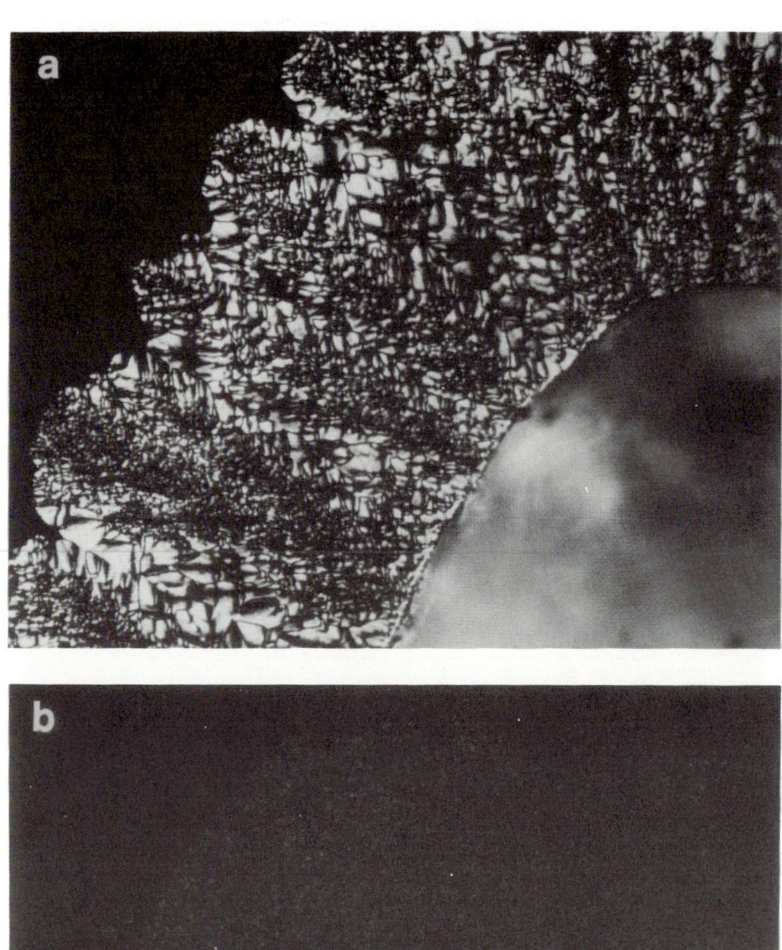

Figure 6-14. Microstructures of carbons deposited in a fluidized bed. (a) A granular carbon with distinct growth features. (b) An isotropic carbon without growth features. Both under polarized light, 240×. From J. C. Bokros, L. D. LaGrange, and G. J. Schoen, "Control of Structure of Carbon for Use in Bioengineering," in *Chemistry and Physics of Carbon*, P. L. Walker and P. A. Thrower (eds.), Marcel Dekker, New York, 1972, Vol. 9, pp. 103–171.

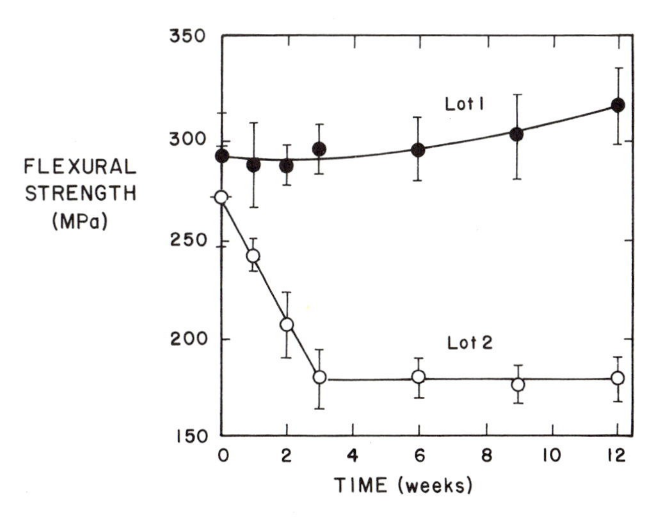

Figure 6-15. Flexural strength of dense alumina rods after aging under stress in Ringer's solution. Lot 1 and lot 2 are from different batches of production. From F. E. Krainess and W. J. Knapp, "Strength of a Dense Alumina Ceramic after Aging *in Vitro*," *J. Biomed. Mater. Res. 12*, 241–246, 1978.

may have contributed to the permeation of water molecules, which is detrimental to the strength. It is not clear whether the same static fatigue mechanism operates in single-crystal alumina or not. It is, however, reasonable to assume that the same static fatigue will occur if the ceramic contains flaws or impurities which will act as the source of crack initiation and growth under stress.

Study of the fatigue behavior of vapor-deposited pyrolytic carbon fibers (4000–5000 Å thick) onto stainless steel substrate showed that the film did not break unless the substrate underwent plastic deformation at 1.3×10^{-2} strain and up to one million cycles of loading. Therefore, the fatigue is closely related to the substrate as shown in Figure 6-16. Similar substrate–carbon adherence is the basis for the pyrolytic carbon-deposited polymer arterial grafts as mentioned earlier.

The fatigue life of ceramics can be predicted by assuming that the fatigue fracture is due to the slow growth of preexisting flaws. Generally the strength

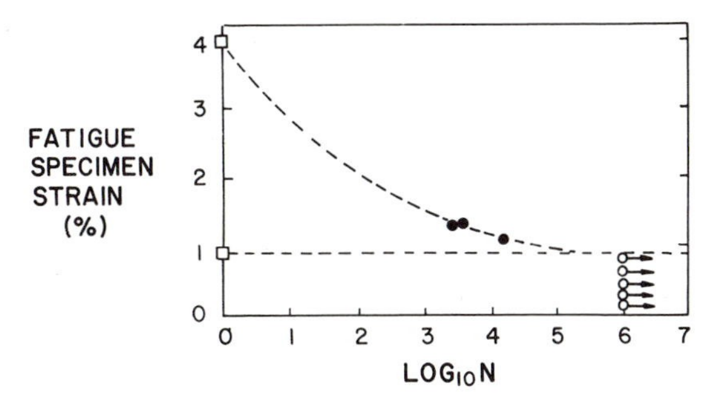

Figure 6-16. Strain versus number of cycles to failure. ○, absence of fatigue cracks in carbon film; ●, fracture of carbon film due to fatigue failure of substrates; □, data for substrate determined in single-cycle tensile test. From H. S. Shim and A. D. Haubold, "The Fatigue Behavior of Vapor Deposited Carbon Films," *Biomater. Med. Devices Artif. Organs, 8*, 333–344, 1980.

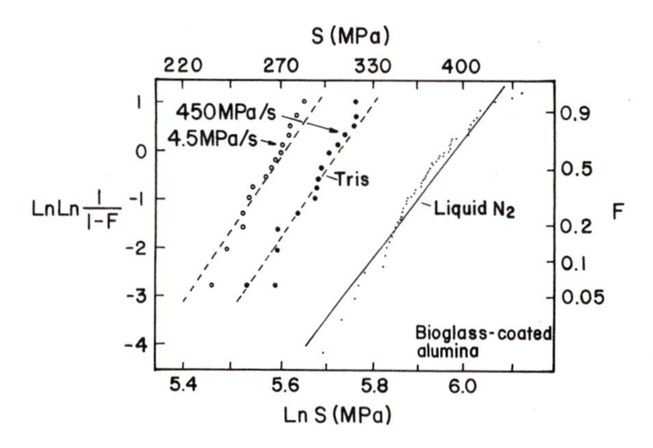

Figure 6-17. Plot of ln ln [1/(1 − *F*)] versus ln *s* for Bioglass-coated alumina in a tris hydroxyaminomethane buffer and liquid nitrogen. *F* is the probability of failure and *s* is strength. From J. E. Ritter, Jr., D. C. Greenspan, R. A. Palmer, and L. L. Hench, "Use of Fracture of an Alumina and Bioglass Coated Alumina," *J. Biomed. Mater. Res.*, *13*, 251–263, 1979.

distribution (s_i) of ceramics in an inert atmosphere can be correlated with the probability of failure *F*, by the following equation:

$$\ln \ln\left(\frac{1}{1-F}\right) = m \ln\left(\frac{s_i}{s_0}\right) \tag{6-2}$$

in which m and s_0 are constants. Figure 6-17 shows a good fit for Bioglass®-coated alumina.

A minimum service life (t_{min}) of a specimen can be predicted by means of a proof test wherein it is subjected to stresses that are greater than those expected in service. Proof tests also eliminate the weaker pieces. This minimum life can be predicted from the following equation:

$$t_{min} = B\sigma_P^{N-2}\sigma_a^{-N} \tag{6-3}$$

in which σ_P is the proof test stress, σ_a is the applied stress, and B and N are constants. Rearranging Eq. (6-3), we obtain

$$t_{min}\sigma_a^2 = B\left(\frac{\sigma_P}{\sigma_a}\right)^{N-2} \tag{6-4}$$

Figure 6-18 shows a plot of Eq. (6-4) for alumina on a logarithmic scale.

Example 6-4

Calculate the proof stress of an alumina sample if it is to last for 20 years at 100 MPa in air and in Ringer's solution.

Answer

$t_{min}\sigma_a^2 = 20 \text{ yr} \times 3.15 \times 10^7 \text{ sec/yr} \times (100 \text{ MPa})^2 = 63 \times 10^{11} \text{ sec(MPa)}^2$. $\log t_{min}\sigma_a^2 = 12.8$. This value is comparable to the dashed line for 50 years at 69 MPa in Figure 6-18. Therefore, $(\sigma_P/\sigma_a) = 2.0$ in air and 2.55 in Ringer's solution. So $\sigma_P = 200$ MPa in air, $\sigma_P = 255$ MPa in Ringer's solution. As one might expect, it is necessary to test the sample more rigorously in Ringer's solution than in air.

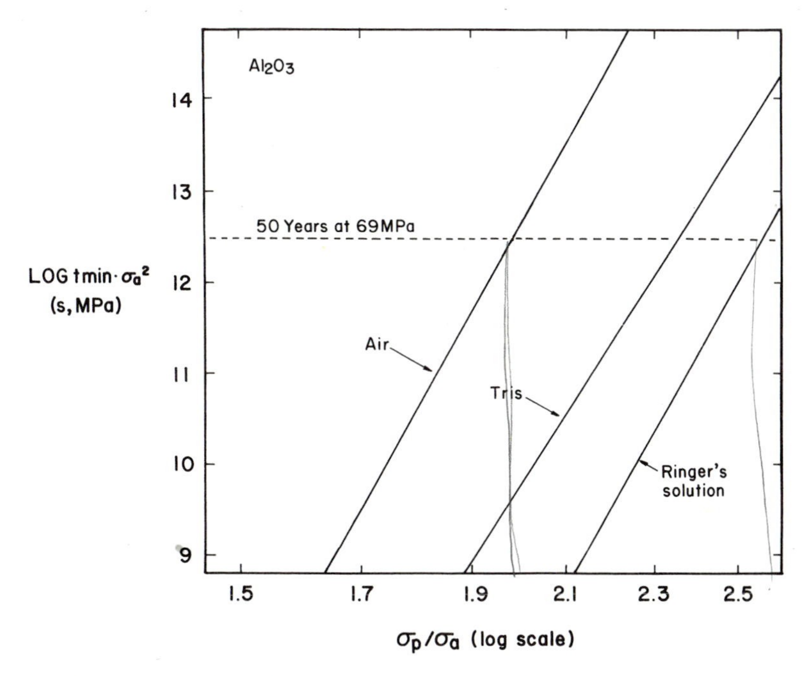

Figure 6-18. Plot of Eq. (6-4) for alumina after proof testing. $N = 43.85$, $m = 13.21$, and $s_0 = 55{,}728$ psi. From J. E. Ritter, Jr., D. C. Greenspan, R. A. Palmer, and L. L. Hench, "Use of Fracture of an Alumina and Bioglass Coated Alumina," *J. Biomed. Mater. Res.*, *13*, 251–263, 1979.

PROBLEMS

6-1. A ceramic is used to fabricate a hip joint. Assuming a simple ball and socket configuration with a surface contact area of 1.0 cm^2 and continuous static loading in a simulated condition similar to Figure 6-15 (extrapolate the data if necessary):

(a) How long will it last if the loading is 70 kg (mass), in water and in blood?

(b) Will the implant last a longer or shorter time with dynamic loading? Give reasons.

6-2. One is trying to coat the surface of an orthopedic implant with hydroxyapatite to enhance compatibility with tissues. List possible problems associated with this technique. Give two methods of applying the coating.

6-3. Discuss the advantages and disadvantages of coating an orthopedic implant with alumina versus pyrolytic carbon.

6-4. Consider the following mechanical properties.

Property	Bone	Vitreous carbon
Ultimate tensile strength (MPa)	100	120
Tensile modulus (GPa)	12–18	2.8

This comparison suggests that vitreous carbon would be an excellent material for bone replacement. What is wrong with this idea?

6-5. A bioengineer is designing a heart valve disc based on single-crystal alumina rather than LTI carbon. Discuss the advantages and disadvantages of both materials in this application.

DEFINITIONS

Alumina: Aluminum oxide (Al_2O_3), which is vary hard (Moh's hardness 9) and strong. Single crystals are called sapphire or ruby depending on color. Alumina is used to fabricate hip joint socket components or dental root implants.

Calcium phosphate: A family of calcium phosphate ceramics including hydroxyapatite and tricalcium phosphate, which are used to substitute or augment bony structures.

Electronegativity: Potential of an atom to attract electrons especially in the context of forming a chemical bond.

Glass-ceramics: A glass crystallized by heat treatment. Some of those have the ability to form chemical bonds with hard and soft tissues. Bioglass® and Ceravital® are well-known examples.

Hydroxyapatite: A calcium phosphate ceramic with a calcium-to-phosphorus ratio of 5:3 and nominal composition $Ca_{10}(PO_4)_6(OH)_2$. It has good mechanical properties and excellent biocompatibility. Hydroxyapatite is the mineral constituent of bone.

LTI carbon: A silicon-alloyed pyrolytic carbon deposited onto a substrate at low temperature with isotropic crystal morphology. Highly blood compatible and used for cardiovascular implant fabrication such as artificial heart valve.

Minimum radius ratio: The ratio of atomic radii computed by assuming the largest atom or ion that can be placed in a crystal's unit-cell structure without deforming the structure.

Moh's scale: A hardness scale in which 10 (diamond) is the hardest and 1 (talc) is the softest.

BIBLIOGRAPHY

J. C. Bokros, R. J. Arkins, H. S. Shim, A. D. Haubold, and N. K. Agarwal, "Carbon in Prosthetic Devices," in *Petroleum Derived Carbons* (M. L. Deviney and T. M. O'Grady, eds.), American Chemical Society, Washington, D.C., 1976.

K. de Groot (ed.), *Bioceramics of Calcium Phosphate*, CRC Press, Boca Raton, Fla., 1983.

R. M. Gill, *Carbon Fibres in Composite Materials*, Butterworths, London, 1972.

J. J. Gilman, "The Nature of Ceramics," in *Materials* (D. Flanagen et al., eds.), W. H. Freeman & Co., San Francisco, 1967.

G. W. Hastings and D. F. Williams (eds.), *Mechanical Properties of Biomaterials*, Part 3, J. Wiley and Sons, New York, pp. 207–274.

S. F. Hulbert and F. A. Young (eds.), *Use of Ceramics in Surgical Implants*, Gordon and Breach, New York, 1978.

S. F. Hulbert, F. A. Young, and D. D. Moyle (eds.), *J. Biomed. Mater. Res. Symp.*, No. 2, 1972.

H. Kawahara, M. Hirabayashi, and T. Shikita, "Single Crystal Alumina for Dental Implants and Bone Screws," *J. Biomed. Mater. Res.*, *14*, 597–606, 1980.

W. D. Kingery, H. K. Bowen, and D. R. Uhlmann, *Introduction to Ceramics*, 2nd ed., J. Wiley and Sons, New York, 1976.

D. McConell, *Apatite: Its Crystal Chemistry, Mineralogy, Utilization, and Biologic Occurrence*, Springer-Verlag, Berlin, 1973.

P. W. McMillan, *Glass-Ceramics*, 2nd ed., Academic Press, New York, 1979.

F. Norton, *Elements of Ceramics*, 2nd ed., Addison-Wesley Publ. Co., Reading, Mass., 1974.

CHAPTER 7

POLYMERIC IMPLANT MATERIALS

Polymers (*poly* = many, *mer* = unit) are made by linking small molecules (mers) through *primary covalent* bonding in the main chain backbone with C, N, O, Si, etc. One example is polyethylene, which is made from ethylene ($CH_2{=}CH_2$) where the carbon atoms share electrons with two other hydrogen and carbon atoms: $\text{-}CH_2\text{-}(CH_2\text{-}CH_2)_n\text{-}CH_2\text{-}$, in which n indicates the number of repeating units. Also note the repeating unit is $\text{-}CH_2CH_2\text{-}$, not $\text{-}CH_2\text{-}$.

In order to make a strong solid, the *repeating unit, n,* should be well over 1000, making the molecular weight of the polyethylene over 28,000 grams/mol. This is why the polymers are made of *giant molecules.* At low molecular weight the material behaves as a wax (paraffin wax used for household candles) and at still lower molecular weight as an oil and gas.

The main backbone chain can be of entirely different atoms, e.g., the polydimethyl siloxane (silicone rubber) $\text{-}Si(CH_3)_2[O\text{-}Si(CH_3)_2]_nO\text{-}$. The side group atoms can be changed; thus, if we substitute the hydrogen atoms in polyethylene with fluorine (F), the resulting material is well known as Teflon® (polytetrafluoroethylene).

7.1. POLYMERIZATION

In order to link the small molecules, one has to force them to lose their electrons by the chemical processes of condensation and addition. By controlling the reaction temperature, pressure, and time in the presence of catalyst(s), the degree to which repeating units are put together into chains can be manipulated.

7.1.1. Condensation or Step Reaction Polymerization

During condensation polymerization, a small molecule such as water will be condensed out by the chemical reaction:

$$\underset{\text{(amine)}}{R\text{-}NH_2} + \underset{\text{(carboxylic acid)}}{R'COOH} \Leftrightarrow \underset{\text{(amide)}}{R'CONHR} + \underset{\text{(condensation molecule)}}{H_2O} \qquad (7\text{-}1)$$

Table 7-1. Typical Condensation Polymers

Type	Interunit linkage
Polyester	$-\overset{\overset{\displaystyle O}{\|}}{C}-O-$
Polyamide	$-\overset{\overset{\displaystyle O}{\|}}{C}-\overset{\overset{\displaystyle H}{\|}}{N}-$
Polyurea	$-\overset{\overset{\displaystyle H}{\|}}{N}-\overset{\overset{\displaystyle O}{\|}}{C}-\overset{\overset{\displaystyle H}{\|}}{N}-$
Polyurethane	$-O-\overset{\overset{\displaystyle O}{\|}}{C}-\overset{\overset{\displaystyle H}{\|}}{N}-$
Polysiloxane	$-\overset{\overset{\displaystyle R}{\|}}{\underset{\underset{\displaystyle R}{\|}}{Si}}-O-$
Protein	$-\overset{\overset{\displaystyle O}{\|}}{C}-\overset{\overset{\displaystyle H}{\|}}{N}-$
Cellulose	$-C-O-C-$

This particular process is used to make polyamides (nylons). Nylon was the first commercial polymer, made in the 1930s.

Some typical condensation polymers and their interunit linkages are given in Table 7-1. One major drawback of condensation polymerization is the tendency for the reaction to cease before the chains grow to a sufficient length. This is due to the decreased mobility of the chains and reactant chemical species as polymerization progresses. This results in short chains. However, in the case of nylon the chains are polymerized to a sufficiently large extent before this occurs and the physical properties of the polymer are preserved.

Natural polymers, such as polysaccharides and proteins, are also made by condensation polymerization. The condensing molecule is always water.

7.1.2. Addition or Free Radical Polymerization

Addition polymerization can be achieved by rearranging the bonds within each monomer. Since each "mer" has to share at least two covalent electrons with other mers, the monomer has to have at least one double bond. For example, in case of ethylene:

$$n\,\underset{\underset{\displaystyle H}{|}}{\overset{\overset{\displaystyle H}{|}}{C}}=\underset{\underset{\displaystyle H}{|}}{\overset{\overset{\displaystyle H}{|}}{C}} \Rightarrow -\underset{\underset{\displaystyle H}{|}}{\overset{\overset{\displaystyle H}{|}}{C}}-(\underset{\underset{\displaystyle H}{|}}{\overset{\overset{\displaystyle H}{|}}{C}}-\underset{\underset{\displaystyle H}{|}}{\overset{\overset{\displaystyle H}{|}}{C}})_n-\underset{\underset{\displaystyle H}{|}}{\overset{\overset{\displaystyle H}{|}}{C}}- \tag{7-2}$$

The breaking of a double bond can be made with an *initiator.* This is usually a free radical such as benzoyl peroxide:

$$C_6H_5COO\text{-}OOCC_6H_5 \Rightarrow 2C_6H_5COO\cdot \Rightarrow \underset{(R\cdot)}{2C_6H_5\cdot} + 2CO_2 \tag{7-3}$$

The initiation can be activated by heat, ultraviolet light, and other chemicals.

The free radicals (initiators) can react with monomers:

$$R\cdot + CH_2{=}\overset{\overset{\displaystyle H}{|}}{\underset{\underset{\displaystyle X}{|}}{C}} \Rightarrow RCH_2{-}C\cdot \tag{7-4}$$

and this free radical can react with another monomer:

$$RCH_2{-}\overset{\overset{\displaystyle H}{|}}{\underset{\underset{\displaystyle X}{|}}{C}}\cdot + CH_2{=}CHX \Rightarrow RCH_2{-}CHX{-}CH_2{-}\overset{\overset{\displaystyle H}{|}}{\underset{\underset{\displaystyle X}{|}}{C}}\cdot \tag{7-5}$$

and the process can continue on.

This process is called *propagation* and can be written in short form (M = monomer):

$$\begin{aligned} R\cdot + M &\Rightarrow RM\cdot \\ RM\cdot + M &\Rightarrow RMM\cdot \end{aligned} \tag{7-6}$$

The propagation process can be *terminated* by combining two free radicals, by transfer, or by disproportionate processes, respectively, in the following equations:

$$RM_nM\cdot + R\cdot \text{ (or } RM\cdot) \Rightarrow RM_{n+1}R \text{ (or } RM_{n+2}R) \tag{7-7}$$

$$RM_nM\cdot + RH \Rightarrow RM_{n+1}H + R\cdot \tag{7-8}$$

$$RM_nM\cdot + \cdot MM_nR \Rightarrow RM_{n+1} + M_{n+1}R \tag{7-9}$$

An example of the disproportionate termination is

$$-CH_2\overset{\overset{\displaystyle H}{|}}{\underset{\underset{\displaystyle X}{|}}{C}}\cdot + \cdot\overset{\overset{\displaystyle H}{|}}{\underset{\underset{\displaystyle X}{|}}{C}}{-}CH_2{-} \Rightarrow -\overset{\overset{\displaystyle H}{|}}{\underset{\underset{\displaystyle X}{|}}{C}}H_2CH + \overset{\overset{\displaystyle H}{|}}{\underset{\underset{\displaystyle X}{|}}{C}}{=}CH{-}$$

Some of the commercially important monomers for addition polymers are given in Table 7-2.

There are three more types of initiating species for addition polymerization besides free radicals: cations, anions, and coordination (stereospecific) catalysts. Some monomers can use two or more of the initiation processes but others can use only one process as given in Table 7-2.

Table 7-2. Monomers for Addition Polymerization and Suitable Processes[a]

Monomer name	Chemical formula	Polymerization mechanism[b]			
		Radical	Cationic	Anionic	Coordin.
Acrylonitrile	$CH_2{=}CH{-}CN$	+	−	+	+
Ethylene	$CH_2{=}CH_2$	+	+	−	−
Methacrylate	$CH_2{=}CH{-}COOCH_3$	+	−	+	+
Methyl methacrylate	$CH_2{=}C(COOCH_3)CH_3$	+	−	+	+
Propylene	$CH_2{=}CHCH_3$	−	−	−	+
Styrene	$CH_2{=}CH{-}C_6H_5$	+	+	+	+
Vinyl chloride	$CH_2{=}CHCl$	+	−	−	+
Vinylidene chloride	$CH_2{=}CCl_2$	+	−	+	−

[a] From F. W. Billmeyer, *Textbook of Polymer Science*, 3rd ed., Wiley-Interscience, New York, 1984.
[b] +, high polymer formed, −, no reaction or oligomers only.

Example 7-1

Polyethylene crystal has a rhombohedral unit-cell structure as shown. Calculate its theoretical density.

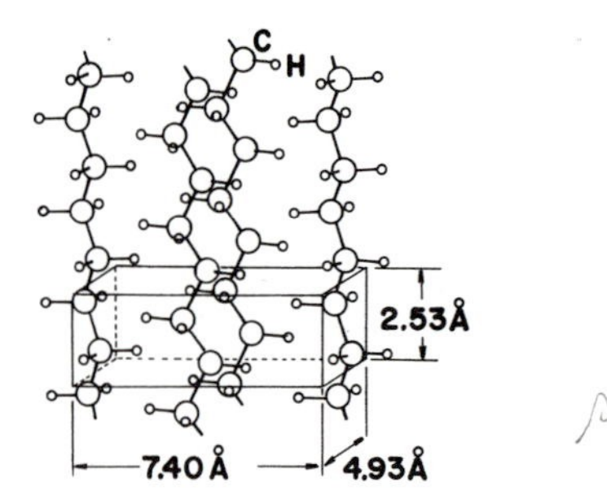

Answer

From the figure one can see that each unit cell has two mers of -(C_2H_4)-, therefore,

$$\rho = \frac{2 \text{ mers } (28 \text{ g}/6 \times 10^{23} \text{ mers})}{7.4 \times 4.93 \times 2.53 \times (10^{-8})^3}$$

$$= 1.01 \text{ g/cm}^3$$

7.2. EFFECT OF STRUCTURAL MODIFICATION AND TEMPERATURE ON PROPERTIES

The physical properties of polymers can be affected in many ways. In particular, the chemical composition and arrangement of chains will have a great effect on the final properties. By such means we can tailor the polymers to meet the end use.

7.2.1. Effect of Molecular Weight and Composition

The molecular weight and its distribution have a great effect on the properties of a polymer since its rigidity is primarily due to the immobilization or *entanglement* of the chains. This is because the chains are arranged like cooked spaghetti strands in a bowl. By increasing the molecular weight the polymer chains become longer and less mobile and a more rigid material results as shown in Figure 7-1. Equally important is that all chains should be equal in length since if there are short chains they will act as *plasticizers.* Plasticizers could lower the melting and glass transition temperatures, rigidity, i.e., modulus of elasticity, density, etc. since the small molecules will facilitate the movement of chains and interfere with (efficient) packing of long-chain molecules.

Another obvious way of changing properties is to change the *chemical composition* of the backbone or side chains. Substituting the backbone carbon of a polyethylene with divalent oxygen or sulfur will decrease the melting and glass transition temperatures since the chain becomes more flexible due to the increased rotational freedom:

$$-CH_2-CH_2-CH_2-CH_2-CH_2- \;\begin{cases} \nearrow \; -O-CH_2-CH_2-CH_2-CH_2-O- \\ \searrow \; -S-CH_2-CH_2-CH_2-CH_2-S- \end{cases} \qquad (7\text{-}10)$$

On the other hand, if the backbone chains can be made more rigid, then a stiffer polymer will result as in the case of polyester:

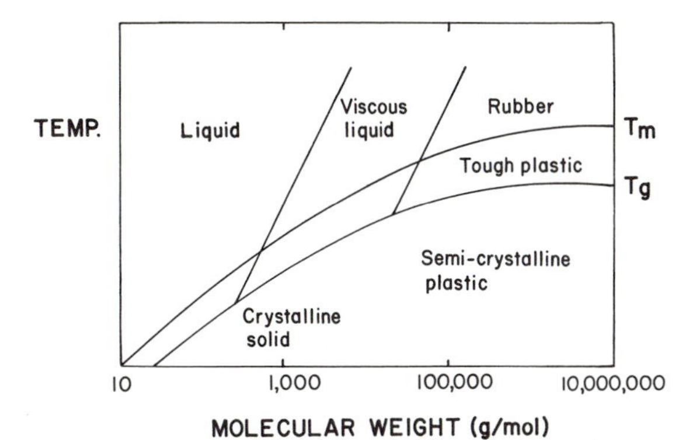

Figure 7-1. Approximate relations among molecular weight, T_g, T_m, and polymer properties.

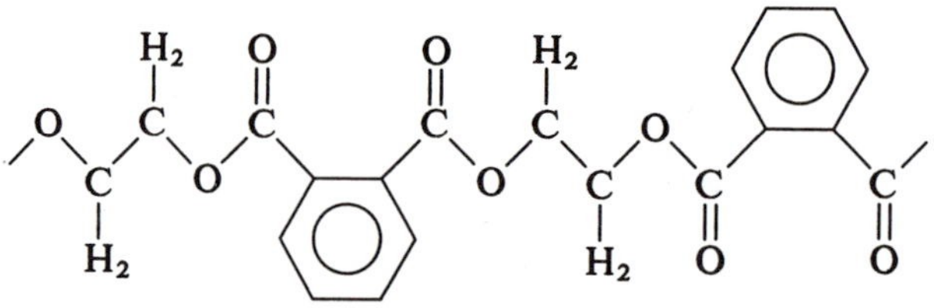

(7-11)

Polyethylene terephthalate (polyester, Dacron®)

7.2.2. Effect of the Side-Chain Substitution, Cross-Linking, and Branching

Increasing the size of side groups in linear polymers such as polyethylene will decrease the melting temperature because of the lesser perfection of molecular packing, i.e., decreased crystallinity. This effect is seen until the side group itself becomes large enough to hinder the movement of the main chains as shown in Table 7-3. Very long side groups can be thought of as being branches.

Cross-linking of the main chains (Figure 2-12) is in effect similar to the side-chain substitution with a small molecule, i.e., it lowers the melting temperature. This is because of the interference of the cross-linking, which causes lesser mobility of the chains resulting in further retardation of the crystallization rate. In fact, a large degree of cross-linking can prevent crystallization completely. However, when the cross-linking density increases for a rubber, the material becomes harder and the glass transition temperature also increases.

7.2.3. Effect of Temperature on Properties

Amorphous polymers undergo a substantial change in their properties as a function of temperature. The *glass transition temperature* T_g is a demarcation between the glassy region of behavior in which the polymer is relatively stiff and

Table 7-3. Effect of Side-Chain Substitution on Melting Temperature in Polyethylene

Side chain	T_m (°C)
$-H$	140
$-CH_3$	165
$-CH_2CH_3$	124
$-CH_2CH_2CH_3$	75
$-CH_2CH_2CH_2CH_3$	−55
$-CH_2CH(CH_3)CH_2CH_3$	196
$-CH_2-C(CH_3)_2-CH_2CH_3$	350

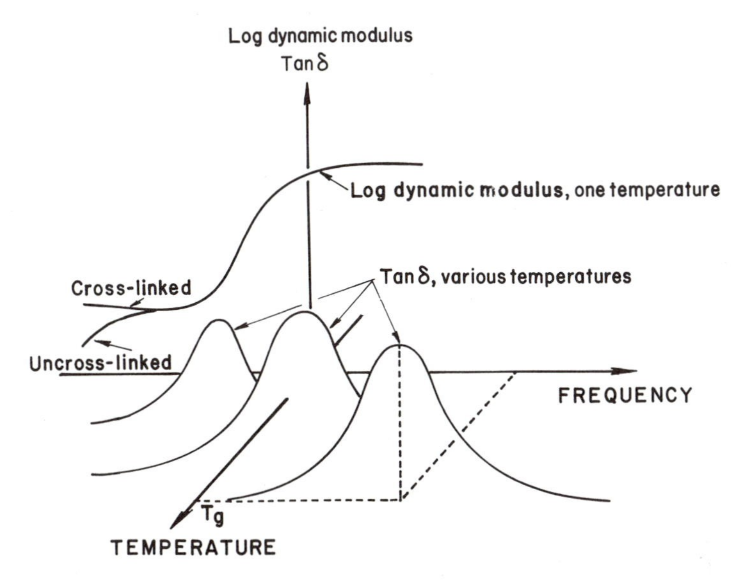

Figure 7-2. Glass transition temperature obtained from the peak in the loss tangent curve at a given frequency. The stiffness undergoes an abrupt change at the glass transition temperature. An un-cross-linked polymer exhibits continued mobility of chains resulting in a decrease of stiffness at low frequency corresponding to flow or creep at long times.

the rubbery region in which it is very compliant. T_g can also be defined as the temperature at which the slope of volume change versus temperature has a discontinuity in slope as discussed in Chapter 2. Since the polymers are viscoelastic, the value obtained in this measurement depends on how fast it is taken. An alternative definition of the glass transition temperature is the temperature of the peak in the viscoelastic loss tangent, as shown in Figure 7-2. Materials considered as hard polymers, such as PMMA, have glass transition temperatures well above room temperature. Rubbery materials have glass transition temperatures below room temperature; they become rigid at sufficiently low temperature but are flexible near room temperature.

As for cross-links, as discussed in Section 7.2.2, they serve the function of maintaining the solidity of the polymer when it is subjected to prolonged load. An uncross-linked polymer exhibits continued mobility of chains resulting in a decrease of stiffness at low frequency corresponding to flow or creep at long times, as shown in Figure 7-2. If the characteristic retardation times of this creep are less than the expected time of service of the implant, undesired deformation will occur.

7.3. POLYMERIC IMPLANT MATERIALS

Polymeric materials have a wide variety of applications for implantation since they can be easily fabricated into many forms: fibers, textiles, films, rods, and viscous liquids. Polymers bear a close resemblance to natural polymeric tissue components such as collagen. In some cases it is possible to achieve a bond between synthetic polymers and natural tissue polymers. An example is the bonding of heparin protein on the surface of polymers (silicone, urethane rubbers, etc.) for the prevention of blood clotting. Adhesive polymers can be used to close wounds or lute orthopedic implants in place.

Whenever possible, the ASTM standards for the specifications and tests should be utilized to get more uniform and better results when the material is implanted. Only a few plastics and rubbers are listed in the ASTM and these are indicated in the text whenever possible.

7.3.1. Polyamides (Nylons)

Polyamides are known as nylons and are designated by the number of carbon atoms in the repeating units. Nylons can be polymerized by step-reaction (or condensation) and ring-scission polymerization. They have excellent fiber-forming ability due to interchain hydrogen bonding and a high degree of crystallinity, which increases strength in the fiber direction.

The basic chemical structure of the repeating unit of polyamides can be written in two ways,

$$\text{-}[NH(CH_2)_xNHCO(CH_2)_yCO]_n\text{-} \tag{7-12}$$

and

$$\text{-}[NH(CH_2)_xCO]_n\text{-} \tag{7-13}$$

Equation (7-12) represents polymers made from diamine and diacids such as type 66 ($x = 6, y = 4$) and 610 ($x = 6, y = 8$). Polyamides made from ω-acids [caprolactams, Eq. (7-13)] are designated as nylon 6 ($x = 5$), 11 ($x = 10$), and 12 ($x = 11$). These types of polyamides can be produced by ring-scission polymerization, e.g.,

$$n \begin{array}{ccc} NH & — & C{=}O \\ | & & | \\ CH_2 & & CH_2 \\ | & & | \\ CH_2 & & CH_2 \\ & \diagdown \; \diagup & \\ & CH_2 & \end{array} \xrightarrow{\text{Heat}} —[NHCO(CH_2)_5]_n— \tag{7-14}$$

Caprolactam Nylon 6

The presence of -CONH- groups in polyamides attracts the chains strongly toward one another by hydrogen bonding as shown in Figure 7-3. Since the hydrogen bond plays a major role in determining properties, the number and distribution of -CONH- groups are important factors. For example, T_g can be

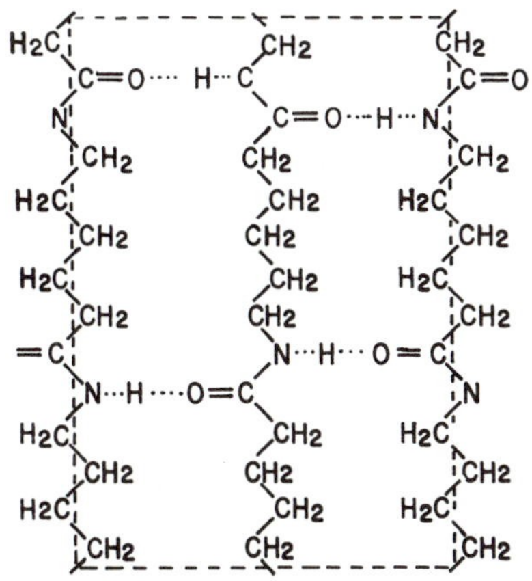

Figure 7-3. Hydrogen bonding in polyamide chains (nylon 6).

decreased by decreasing the number of -CONH- groups as given in Table 7-4. On the other hand, an increase in the number of -CONH- groups improves physical properties such as strength: Nylon 66 is stronger than nylon 610 and 6 is stronger than 11.

In addition to the higher nylons (610 and 11), there are aromatic polyamides termed aramids. One of them is poly(*p*-phenylene terephthalate) commonly known as Kevlar®, made by DuPont:

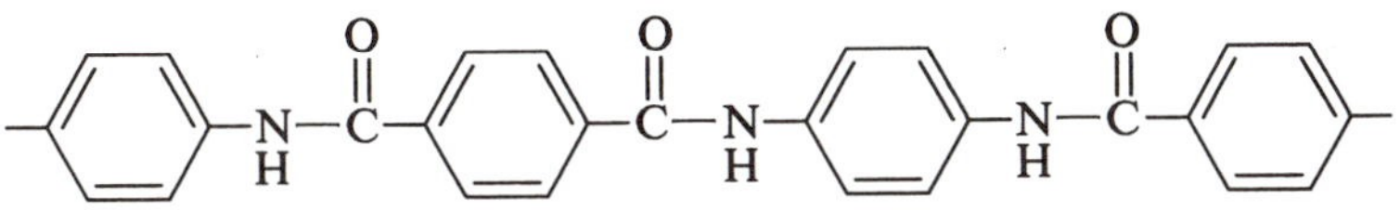

$$-C_6H_4-\underset{H}{N}-\overset{O}{\overset{\|}{C}}-C_6H_4-\overset{O}{\overset{\|}{C}}-\underset{H}{N}-C_6H_4-\underset{H}{N}-\overset{O}{\overset{\|}{C}}-C_6H_4- \qquad (7\text{-}15)$$

This material is readily made into fibers. The specific strength of such fibers is five times that of steel; consequently, it is most suitable to make composites.

The nylons are hygroscopic and lose their strength *in vivo* when implanted. The water molecules serve as *plasticizers* that attack the amorphous region. Proteolytic enzymes also aid hydrolysis by attacking the amide group. This is

Table 7-4. Properties of Polyamides

Property	Type					
	66	610	6	11	Aramid[a]	Kevlar[b]
Density (g/cm^3)	1.14	1.09	1.13	1.05	1.30	1.45
Tensile strength (MPa)	76	55	83	59	120	2700
Elongation (%)	90	100	300	120	<80	2.8
Modulus of elasticity (GPa)	2.8	1.8	2.1	1.2	>2.8	130
Softening temperature (°C)	265	220	215	185	275	—

[a] Molded parts, unfilled.
[b] Kevlar® 49 (DuPont) fibers.

probably due to the fact that the proteins also contain the amide group along their molecular chains which the proteolytic enzymes could attack.

7.3.2. Polyethylene

Polyethylene and polypropylene and their copolymers are called polyolefins. These are linear thermoplastics. Polyethylene is available commercially in three major grades: low and high density and ultrahigh molecular weight (UHMWPE). Polyethylene has the repeating unit structure:

$$-(\overset{\displaystyle H}{\underset{\displaystyle H}{\overset{|}{\underset{|}{C}}}}-\overset{\displaystyle H}{\underset{\displaystyle H}{\overset{|}{\underset{|}{C}}}})_n- \qquad (7\text{-}16)$$

which can be readily crystallized. In fact, it is almost impossible to produce noncrystalline polyethylene because of the small hydrogen side groups, which cause high mobility of chains.

The first polyethylene was synthesized by reacting ethylene gas at high pressure (100–300 MPa) in the presence of a catalyst (peroxide) to initiate polymerization. The process yields the *low*-density polyethylene. By using a *Ziegler catalyst* (stereospecific), high-density polyethylene can be produced at low pressure (10 MPa). Unlike the low-density variety, high-density polyethylene does not contain branches. The result is better packing of the chains, which increases density and crystallinity. The crystallinity is usually 50 to 70% and 70 to 80% for the low- and high-density polyethylene, respectively. Some important physical properties of polyethylenes are given in Table 7-5.

The *ultra-high-molecular-weight* polyethylene ($>2\times10^6$ g/mol) has been used extensively for orthopedic implant fabrications, especially for load-bearing surfaces such as total hip and knee joints. Recently, a new UHMWPE has been

Table 7-5. Properties of Polyethylene

Property	Low density	High density	UHMWPE[a]	Enhanced UHMWPE[b]
Molecular weight (g/mol)	$3\sim4\times10^3$	5×10^5	2×10^6	Same
Density (g/cm^3)	0.90–0.92	0.92–0.96	0.93–0.94	Same
Tensile strength (MPa)	7.6	23–40	27 min	Higher
Elongation (%)	150	400–500	200–250	Same
Modulus of elasticity (MPa)	96–260	410–1240	*c*	2200
Crystallinity (%)	50–70	70–80	*d*	*e*

[a] Data from ASTM F648, also 2% deformation after 90 min recovery subjected to 7 MPa for 24 hr (D621).
[b] Same as the conventional UHMWPE (ASTM F648). Data from "A New Enhanced UHMWPE for orthopaedic applications: A Technical Brief," DePuy, Warsaw, Ind., 1989.
[c] Close to 2200 MPa.
[d] Higher than high-density polyethylene.
[e] Equal or slightly higher than *d*.

introduced by DuPont with DePuy (Warsaw, Indiana). It is claimed to have a longer chain fold length than the conventional UHMWPE, increasing the crystallinity. Since the folded chains are crystalline, the amount of amorphous region is reduced, thus reducing the possibility of environmental attack (usually oxidation). This material also exhibits enhanced mechanical properties (higher hardness, modulus of elasticity, tensile yield strength). Moreover, creep properties are superior (that is, less creep), which is an important factor for designing joint implants. Other properties such as coefficient of friction and wear are improved marginally. This material has no known effective solvent at room temperature; therefore, only high temperature and pressure sintering may be used to produce desired products. Conventional extrusion or molding processes are difficult to use.

Polyethylene has been used as a solid or porous form. Biocompatibility tests for nonporous (F981) and porous polyethylene (F639 and 755) are given by ASTM standards.

7.3.3. Polypropylene

Polypropylene can be synthesized using a Ziegler-type stereospecific catalyst that controls the position of each side group as it is being polymerized to allow the formation of a regular chain structure from the asymmetric repeating unit:

$$-(\overset{\displaystyle H}{\underset{\displaystyle H}{\overset{|}{\underset{|}{C}}}}-\overset{\displaystyle CH_3}{\underset{\displaystyle H}{\overset{|}{\underset{|}{C}}}})_n- \tag{7-17}$$

Three types of structure can exist, depending on the position of the methyl (CH_3) group along the polymer chain. The random distribution of methyl groups in the *atactic* polymer prevents close packing of chains and results in amorphous polypropylene. In comparison, the *isotactic* and *syndiotactic* structures have a regular position of the methyl side groups in the same side and alternate side, respectively. They usually crystallize. However, the presence of the methyl side groups restricts the movement of the polymer chains, and crystallization rarely exceeds 50–70% for material with isotacticity over 95%. Table 7-6 lists typical properties of commercial polypropylenes which exhibit largely an atactic-type structure.

Table 7-6. Properties of Polypropylene

Property	Value
Density (g/cm^3)	0.90–0.91
Tensile strength (MPa)	28–36
Elongation (%)	400–900
Modulus of elasticity (GPa)	1.1–1.55
Softening temperature (°C)	150

Polypropylene has an exceptionally high flex life, and hence was tested in integrally molded hinges for finger joint prostheses. It also has excellent environmental stress-cracking resistance. The permeability of polypropylene to gases and water vapor is between that of low- and high-density polyethylene.

Example 7-2

Calculate the percent crystallinity of UHMWPE assuming the noncrystalline and 100% crystalline polyethylenes have densities of 0.85 and 1.01 g/cm^3, respectively.

Answer

From Table 7-5, the density of UHMWPE is 0.93–0.94; therefore,

$$\%\text{ crystallinity} = \frac{0.94 - 0.85}{1.01 - 0.85} = 0.56, \text{ or } \underline{56\%}$$

This value is lower than that of the low- and high-density polyethylenes as given in Table 7-5. The tabulated crystallinities were based on X-ray diffraction measurement technique instead of the density method.

7.3.4. Polyacrylates

These polymers are used extensively in medical applications as (hard) contact lenses, implantable ocular lenses, and as bone cement for joint prosthesis fixation. Dentures and maxillofacial prostheses are also made from acrylics because they have excellent physical and coloring properties and are easy to fabricate.

7.3.4.1. Structure and Properties of Acrylics and Hydrogels. The basic chemical structure of repeating units of acrylics can be represented by

$$-(CH_2-\underset{\substack{|\\ COOR_2}}{\overset{\substack{R_1\\ |}}{C}})_n- \tag{7-18}$$

The only difference between polymethyl acrylate (PMA) and polymethyl methacrylate (PMMA) is the R groups of formula (7-18). The R_1 and R_2 group for PMA are H and CH_3 and for PMMA they are both CH_3. These polymers are addition (or free radical) polymerized. These polymers can be obtained in liquid monomer or fully polymerized beads, sheets, rods, etc.

Because of the bulky side groups, these polymers are usually obtained in a clear amorphous state. PMMA has bulkier side groups than PMA and consequently PMMA has a higher tensile strength (60 MPa) and softening temperature (125°C) than PMA (7 MPa and 33°C) if the molecular weights are similar. PMMA has an excellent light transparency (92% transmission), high index of refraction (1.49), and excellent weathering properties. This material can be cast, molded, or machined with conventional tools. It has an excellent chemical resistivity and is highly biocompatible in pure form. The material is very hard and brittle in comparison with other polymers.

The first *hydrogel* polymer developed is polyhydroxyethyl methacrylate or poly-HEMA, which can absorb water more than 30% of its weight. This property makes it useful for soft lens applications. The chemical formula is similar to formula (7-18),

$$\begin{array}{c} CH_3 \\ | \\ -(CH_2-C)_n- \\ | \\ COOCH_2CH \end{array} \tag{7-19}$$

where the OH group is the hydrophilic group responsible for hydration of the polymer. Generally, hydrogels for contact lenses are made by polymerization of certain hydrophilic monomers with small amounts of cross-linking agent such as ethylene glycol dimethacrylate (EGDM),

$$\begin{array}{l} \quad\;\; CH_3 \\ \quad\;\;\; | \\ CH_2{=}C-COOCH_2 \\ \qquad\qquad\quad\;\; | \\ CH_2{=}C-COOCH_2 \\ \quad\;\;\; | \\ \quad\;\; CH_3 \end{array} \tag{7-20}$$

Another type of hydrogel was developed in the United States at about the same time as the poly-HEMA hydrogels were developed. The monomer of polyacrylamide hydrogel has the following chemical formula:

$$\begin{array}{c} H \\ | \\ CH_2{=}CCONH_2 \end{array} \tag{7-21}$$

The water content of the copolymer can be increased to over 60%, while the normal water content for poly-HEMA is 40%.

The hydrogels have a relatively low oxygen permeability in comparison with silicone rubber (see Table 7-7). However, the permeability can be increased with

Table 7-7. Oxygen Permeability Coefficient of Contact Lens Materials[a]

Polymer	$P_g \times 10^4$ (μl cm/cm^2 hr kPa)[b]	Comments
Polymethyl methacrylate	0.27	Hard contact lens
Polydimethylsiloxane	1750	Flexible
Polyhydroxyethyl methacrylate	24	39% H_2O, soft contact lens

[a] From M. F. Refojo, "Contact Lenses," in *Encyclopedia of Polymer Science and Technology*, N. M. Bikales (ed.), Interscience, New York, 1964, Vol. 1, pp. 117–197.
[b] At STP to convert μl cm/cm^2 hr kPa to μl cm/cm^2 hr mm Hg, divide by 7.5.

increased hydration (water content) or decreased (lens) thickness. Silicone rubber is not a hydrophilic material but its high oxygen permeability and transparency make it an attractive lens material. It is usually used after coating with hydrophilic hydrogels by grafting.

7.3.4.2. Bone Cement (PMMA). Bone cement has been used for clinical applications to secure a firm fixation of joint prostheses for hip and knee joints. Bone cement is primarily made of PMMA powder and monomer methyl methacrylate liquid as given in Table 7-8. Hydroquinone is added to prevent premature polymerization, which may occur under certain conditions, e.g., exposure to light, elevated temperatures, etc. *N, N*-dimethyl-*p*-toluidine is added to promote or accelerate (cold) curing of the finished compound. The term *cold curing* is used here to distinguish it from the high temperature and pressure (hot) molding technique used to make articles in dental laboratories. The liquid component is sterilized by membrane filtration. The solid component is a finely ground white powder (mixture of PMMA, methyl methacrylate–styrene copolymer, barium sulfate, and benzoyl peroxide).

When the powder and liquid are mixed together, the monomer liquid is polymerized by the free radical (addition) polymerization process. The activator dibenzoyl peroxide [see formula (7-3)], which is mixed with the powder, will react with a monomer to form a monomer radical, which will then attack another monomer to form a dimer radical. The process will continue until long-chain molecules are produced. The monomer liquid will wet the polymer powder particle surfaces and link them together after polymerization, as shown in Figure 7-4. The ASTM standard (F451) specifies the characteristics of the powder–liquid mixture and cured polymer after setting as given in Tables 7-9 and 7-10.

Polymerization during curing obviously increases the degree of polymerization, i.e., increases the molecular weight as given in Table 7-11. However, the molecular weight distribution does not change significantly after curing as shown in Figure 7-5. The properties of cured bone cement are compared with those of commercial acrylic resins in Table 7-12. These studies show that bone cement properties can be affected by the intrinsic and extrinsic factors listed in Table

Table 7-8. Composition of Bone Cement[a]

Liquid component (20 ml)	
Methyl methacrylate (monomer)	97.4 vol%
N,N-dimethyl-*p*-toluidine	2.6 vol%
Hydroquinone	75 ± 15 ppm
Solid powder component (40 g)	
Polymethyl methacrylate	15.0 wt%
Methyl methacrylate–styrene copolymer	75.0 wt%
Barium sulfate ($BaSO_4$), USP	10.0 wt%
Dibenzoyl peroxide	Not available

[a] Surgical Simplex® P Radiopaque Bone Cement (Howmedica, Inc., Rutherford, N.J.). See *Surgical Simplex P Bone Cement Technical Monograph*, Howmedica, Inc., 1977.

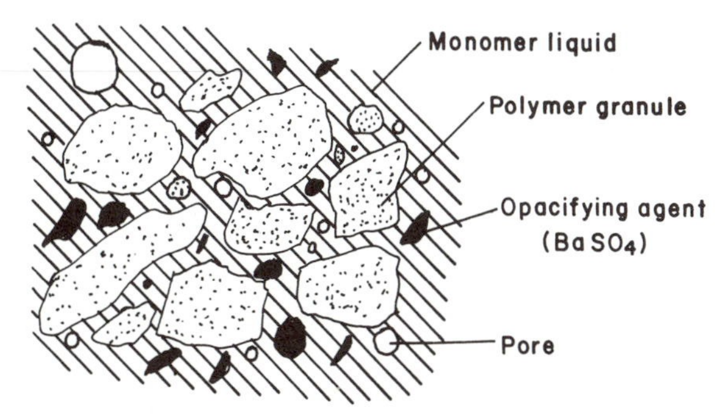

Figure 7-4. Two-dimensional representation of bone cement structure after curing. The monomer liquid will be polymerized and become solid.

Table 7-9. Requirements for Powder–Liquid Mixture

Maximum dough time	Setting time range	Maximum exotherm	Minimum intrusion
5.0 min	5–15 min	90°	2.0 mm

From ASTM F451.

7-13. The most important factor controlling the acrylic bone cement properties is the porosity developed during curing. Large pores (pores of several millimeters' diameter have been observed in clinical settings) are detrimental to the mechanical properties. Monomer vapors and air trapped during mixing are two reasons for the porosity. Obviously, one can reduce the porosity by exposure to vacuum and by centrifugation during mixing of monomer and powder. However, both techniques have some disadvantages such as depletion of monomer, difficulty of

Table 7-10. Requirements for Cured Polymer after Setting

Minimum compressive strength	Maximum indentation	Minimum recovery	Maximum water sorption	Maximum water solubility
70 MPa	0.14 mm	60%	0.7 mg/cm^2	0.05 mg/cm^2

From ASTM F451.

Table 7-11. Molecular Weight of Bone Cement[a]

Type of M.W. (g/mol)	Monomer	Powder	Cured
M_n (number average)	100	44,000	51,000
M_w (weight average)	100	198,000	242,000

[a] From S. S. Haas, G. M. Brauer, and G. Dickson, "A Characterization of PMMA Bone Cement," *J. Bone Joint Surg.*, *57A*, 280–291, 1975.

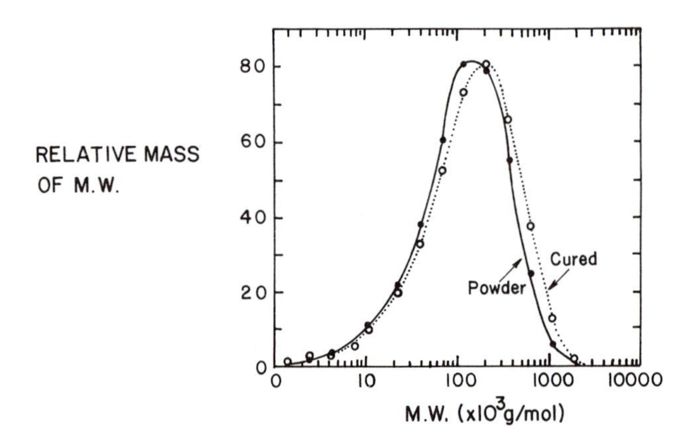

Figure 7-5. Molecular weight distribution of bone cement powder and after curing. See S. S. Haas, G. M. Brauer, and G. Dickson, "A Characterization of PMMA Bone Cement," *J. Bone Joint Surg.*, *57A*, 280–291, 1975.

Table 7-12. Physical Properties of Bone Cement Commercial Acrylic Resins

Property	Radiopaque bone cement[a]	Commercial acrylic resins[b]
Tensile strength (MPa)	28.9 ± 1.6	55-76
Compressive strength (MPa)	91.7 ± 2.5	76-131
Young's modulus (compressive loading, MPa)	2200 ± 60	2960-3280
Endurance limit[c]	0.3 uts[d]	0.3 uts
Density (g/cm^3)	1.10-1.23	1.18
Water sorption (%)	0.5	0.3-0.4
Shrinkage after setting (%)	2.75-5	—

[a] Data from S. S. Haas, G. M. Brauer, and G. Dickson, "A Characterization of PMMA Bone Cement," *J. Bone Joint Surg.*, *57A*, 280-291, 1975.
[b] Data from *Modern Plastics Encyclopedia*, Vol. 57, McGraw-Hill Book Co., New York, 1980, p. 533.
[c] Data from R. P. Kusy, "Characterization of Self-curing Acrylic Bone Cements," *J. Biomed. Mater. Res.*, *12*, 271-305, 1978.
[d] uts, ultimate tensile strength.

Table 7-13. Factors Affecting Bone Cement Properties

Intrinsic factors
- Composition of monomer and powder
- Powder particle size, shape, and distribution: degree of polymerization
- Liquid/powder ratio

Extrinsic factors
- Mixing environment: temperature, humidity, type of container
- Mixing technique: rate and number of beating with spatula
- Curing environment; temperature, humidity, pressure, contacting surface (tissue, air, water, etc.)

mixing while under vacuum, segregation of constituents by centrifuging, etc. in addition to the extra equipment needed.

7.3.5. Fluorocarbon Polymers

The best known fluorocarbon polymer is polytetrafluoroethylene (PTFE), commonly known as Teflon® (DuPont). Other polymers containing fluorine are polytrifluorochloroethylene, polyvinylfluoride, and fluorinated ethylene propylene. Only PTFE will be discussed here since the others have rather inferior chemical and physical properties and are rarely used for implant fabrication.

PTFE is made from tetrafluoroethylene under pressure with a peroxide catalyst in the presence of excess water for removal of heat. The repeating unit is similar to that of polyethylene, except that the hydrogen atoms are replaced by fluorine atoms:

$$-(\overset{\displaystyle F}{\underset{\displaystyle F}{\overset{|}{\underset{|}{C}}}}-\overset{\displaystyle F}{\underset{\displaystyle F}{\overset{|}{\underset{|}{C}}}})_n- \tag{7-22}$$

The polymer is highly crystalline (over 94% crystallinity) with an average molecular weight of $0.5\text{–}5 \times 10^6$ g/mol. This polymer has a very high density ($2.15\text{–}2.2\ \text{g/cm}^3$), low modulus of elasticity (0.5 GPa) and tensile strength (14 MPa). It also has a very low surface tension ($18.5\ \text{ergs/cm}^2$) and friction coefficient (0.1).

Standard specifications for the implantable PTFE are given by ASTM F754. PTFE also has an unusual property of being able to expand on a microscopic scale into a microporous material that is an excellent thermal insulator.

PTFE cannot be injection molded or melt extruded because of its very high melt viscosity and it cannot be plasticized. Usually the powders are sintered to above 327°C under pressure to produce implants.

7.3.6. Rubbers

Silicone, natural and synthetic rubbers, have been used for the fabrication of implants. Rubbers or elastomers are defined by ASTM as "a material which at room temperature can be stretched repeatedly to at least twice its original length and upon release of the stress, returns immediately with force to its approximate original length." Rubbers are stretchable because of the kinks of the individual chains such as seen in *cis*-1,4 polyisoprene as discussed in Section 2.4.

The *repeated* stretchability is due in part to the cross-links between chains, which hold the chains together. The amount of cross-linking for natural rubber controls the flexibility of the rubber: the addition of 2–3% sulfur results in a flexible rubber, while adding as much as 30% sulfur makes it a hard rubber.

Rubbers contain antioxidants to protect them against decomposition by oxidation, hence improving aging properties. Fillers such as carbon black or silica powders are also used to improve their physical properties.

Natural rubber is made mostly from the latex of the *Hevea brasiliensis* tree and the chemical formula is the same as that of *cis*-1,4 polyisoprene. Natural rubber was found to be compatible with blood in its pure form. Also, cross-linking by X rays and organic peroxides produces rubber with superior blood compatibility compared with rubbers made by the conventional sulfur vulcanization.

Synthetic rubbers were developed to substitute for natural rubber. The Natta and Ziegler types of stereospecific polymerization techniques have made this variety possible. The synthetic rubbers have been used rarely to make implants. One of these rubbers, neoprene (polychloroprene, $-CH_2-C(Cl)=CH-CH_2-$), is listed in Table 7-14 for comparison. The physical properties vary widely due to the wide variations in preparation recipes of these rubbers.

Silicone rubber, developed by Dow Corning, is one of the few polymers developed for medical use. The repeating unit is dimethyl siloxane, which is polymerized by a condensation polymerization. The reaction product is unstable and condenses, resulting in polymers:

$$n\,HO-\underset{\displaystyle CH_3}{\overset{\displaystyle CH_3}{\underset{|}{\overset{|}{Si}}}}-OH \Rightarrow -(\underset{\displaystyle CH_3}{\overset{\displaystyle CH_3}{\underset{|}{\overset{|}{Si}}}}-O)_n- + nH_2O \tag{7-23}$$

Low-molecular-weight polymers have low viscosity and can be cross-linked to make a higher-molecular-weight rubberlike material. Medical-grade silicone rubbers contain stannous octate as a catalyst and can be mixed with base polymer at the time of implant fabrication.

Silicone rubbers contain silica (SiO_2) powder as fillers to improve their mechanical properties. The density and stiffness of filled rubber increase with the volume fraction of filler. Filled rubber is actually a particle reinforced composite.

Polyurethanes are usually thermosetting polymers; they are widely used to coat implants. Polyurethane rubbers are produced by reacting a prepared prepolymer chain A with an aromatic diisocyanate to make very long chains possessing active isocyanate groups for cross-linking. The polyurethane rubber is quite strong

Table 7-14. Properties of Rubbers

Property	Natural	Neoprene	Silicone	Urethane
Tensile strength (MPa)	7–30	20	6–7	35
Elongation (%)	100–700	—	350–600	650
Hardness (Shore A Durometer)	30–90	40–95	—	65
Density (g/cm^3)	0.92	1.23	1.12–1.23	1.1–1.23

See also ASTM standards F604, F881 (silicone rubber), and F624 (urethane).

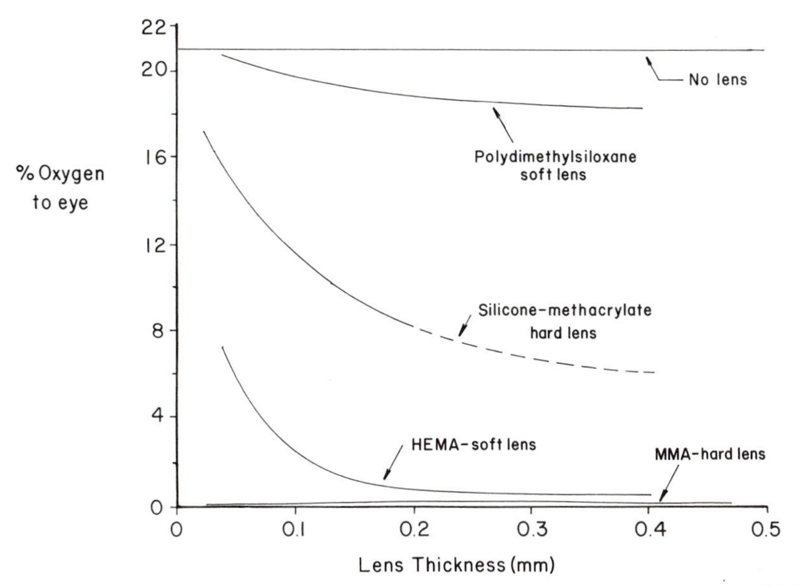

Figure 7-6. Oxygen permeability of contact lenses made of various materials versus thickness. From B. Arkles and P. Redinger, "Silicones in Biomedical Applications," in *Biocompatible Polymers, Metals and Composites*, M. Szycher (ed.), Technomic Publ., Westport, Conn., 1983, Chapter 32.

and has good resistance to oil and chemicals. Some physical properties of rubbers are summarized in Table 7-14 for quick reference.

Polydimethylsiloxane rubber can be (block) copolymerized with urethanes, carbonates, etc. in order to improve its properties. Depending on the applications, the copolymers can be tailor-made to meet a particular need such as contact lenses as shown in Figure 7-6. Oxygen permeability is one of the most important parameters for contact lens applications.

7.4. HIGH-STRENGTH THERMOPLASTICS

Recently, new polymeric materials have been developed to match the properties of light metals. These polymers have excellent mechanical, thermal, and chemical properties due to their stiffened main backbone chains. Polyacetals and polysulfones are being tested as implant materials, while polycarbonates have found applications in the heart/lung assist devices, food packaging, etc.

Polyacetals are produced by reacting formaldehyde as follows:

$$n\,\begin{matrix} H \\ | \\ C{=}O \\ | \\ H \end{matrix} \Longrightarrow \sim O{-}CH_2{-}O{-}CH_2{-}O{-}\left(CH_2{-}O\right)_n \sim \qquad (7\text{-}24)$$

The polyformaldehyde is also sometimes called polyoxymethylene (POM) and is known widely as Delrin® (DuPont). These polymers have a reasonably high molecular weight (>20,000 g/mol) and excellent mechanical properties. More importantly, they display an excellent resistance to most chemicals and to water over wide temperature ranges. *Polysulfones* were developed by Union Carbide in the 1960s. The chemical formula for a polysulfone is

$$-(-C_6H_4-C(CH_3)_2-C_6H_4-O-C_6H_4-SO_2-C_6H_4-O-)_n- \quad (7\text{-}25)$$

These polymers have a high thermal stability due to the bulky side groups (therefore, they are amorphous) and rigid main backbone chains. They are also highly stable to most chemicals but are not so stable in the presence of polar organic solvents such as ketones and chlorinated hydrocarbons.

Polycarbonates are tough, amorphous, and transparent polymers made by reacting bisphenol A and diphenyl carbonate:

$$-(-O-C_6H_4-C(CH_3)_2-C_6H_4-O-\overset{O}{\overset{\|}{C}}-)_n- \quad (7\text{-}26)$$

The best-known commercial polycarbonate is Lexan® (General Electric). It has excellent mechanical and thermal properties. Some physical properties of the high-strength thermoplastics are summarized in Table 7-15 for comparison.

Example 7-3

Silica flour (finely ground SiO_2, $\rho = 2.65\ g/cm^3$) is used as a filler for a polydimethyl siloxane.

(a) What volume fraction of SiO_2 is required to make a silastic rubber with density of $1.25\ g/cm^3$?

(b) What is the weight percent of SiO_2?

Table 7-15. Properties of Polyacetal, Polysulfones, and Polycarbonates

Property	Polyacetal (Delrin®)	Polysulfone (Udel®)	Polycarbonate (Lexan®)
Density (g/cm^3)	1.425	1.24	1.20
Tensile strength (MPa)	70	70	63
Elongation (%)	15–75	50–100	60–100
Tensile modulus (GPa)	3.65	2.52	2.45
Water adsorption (%, 24 hr)	0.25	0.3	0.3

Answer

(a)

$$\rho = \rho_1 V_1 + \rho_2 V_2 + \cdots$$

$$V_1 + V_2 = 1$$

$$1.25 = 2.65 V_1 + 0.90(1 - V_1)$$

$$= 1.75 V_1 + 0.90$$

$$V_1 = \underline{0.20 \text{ (20 vol\%)}}$$

(b) Since the weight of 1 cm^3 of rubber is 1.25 g and the volume fraction is 0.2 (or $0.2 \text{ cm}^3/1 \text{ cm}^3$), therefore,

$$W_1 = \frac{0.2 \times 2.65}{1.25} = \underline{0.42 \text{ (42 wt\%)}}$$

7.5. DETERIORATION OF POLYMERS

Polymers deteriorate due to chemical, thermal, and physical factors. These factors may act synergistically, hence accelerating the deterioration process. The deterioration affects the main backbone chain, side groups, cross-links, and their original molecular arrangement.

7.5.1. Chemical Effects

If a linear polymer is undergoing deterioration, the main chain will usually be randomly scissioned (cut). Sometimes depolymerization occurs, which differs from random chain scission. This process is the inverse of the chain termination of addition polymerization [see Eqs. (7-8) and (7-9)].

Cross-linking of a linear polymer may result in deterioration. An example of this is low-density polyethylene in which the cross-linking interferes with regular orderly arrangement of chains resulting in lower crystallinity, which decreases mechanical properties. On the other hand, if cross-linking is broken by the oxygen or ozone attack on (poly-)isoprene rubber, the rubber becomes brittle.

It is also undesirable to change the nature of bonds, as in the case of polyvinyl chloride:

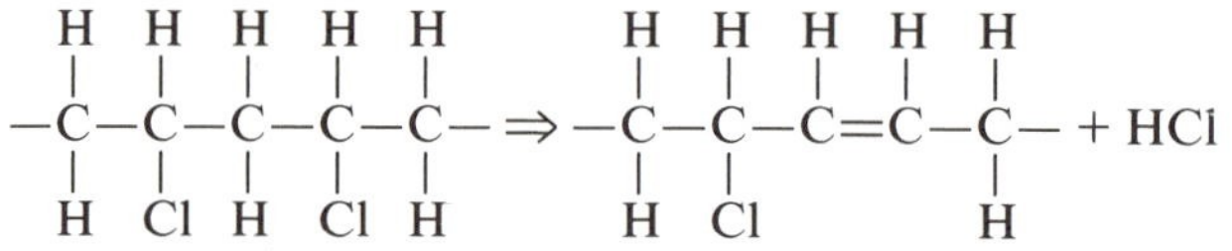

$$
\begin{array}{ccccccccccccccccccccccc}
 & \mathrm{H} & & \mathrm{H} & & \mathrm{H} & & \mathrm{H} & & \mathrm{H} & & & & \mathrm{H} & & \mathrm{H} & & \mathrm{H} & & \mathrm{H} & & \mathrm{H} & \\
 & | & & | & & | & & | & & | & & & & | & & | & & | & & | & & | & \\
- & \mathrm{C} & - & \mathrm{C} & - & \mathrm{C} & - & \mathrm{C} & - & \mathrm{C} & - & \Rightarrow & - & \mathrm{C} & - & \mathrm{C} & - & \mathrm{C} & = & \mathrm{C} & - & \mathrm{C} & - \;+\; \mathrm{HCl} \\
 & | & & | & & | & & | & & | & & & & | & & | & & & & & & | & \\
 & \mathrm{H} & & \mathrm{Cl} & & \mathrm{H} & & \mathrm{Cl} & & \mathrm{H} & & & & \mathrm{H} & & \mathrm{Cl} & & & & & & \mathrm{H} &
\end{array}
\tag{7-27}
$$

The by-products of degradation (HCl) can be irritable to tissues since they are, in this case, acids.

7.5.2. Sterilization Effects

Sterilization is essential for all implanted materials. Some methods of sterilization may result in polymer deterioration. In *dry heat sterilization* the temperature varies between 160 and 190°C. This is above the melting and softening temperature of many linear polymers like polyethylene and PMMA. In the case of polyamide (nylon), oxidation will occur at the dry sterilization temperature although it is below its melting temperature. The only polymers that can be safely dry sterilized are PTFE (Teflon®) and silicone rubber.

Steam sterilization (autoclaving) is performed under high steam pressure at relatively low temperature (120–135°C). However, if the polymer is subjected to attack by water vapor, this method cannot be employed. Polyvinyl chloride, polyacetals, polyethylenes (low-density variety), and polyamides (nylons) belong in this category.

Chemical agents such as ethylene and propylene oxide gases, and phenolic and hypochloride solutions are widely used for sterilizing polymers since they can be used at low temperatures. Chemical sterilization takes a longer time and is more costly. Sometimes chemical agents cause polymer deterioration even when sterilization takes place at room temperature. However, the time of exposure is relatively short (a few hours to overnight), and most polymeric implants can be sterilized with this method.

Radiation sterilization using the isotope cobalt 60 can also deteriorate polymers since at high dosage the polymer chains can be broken and recombined. In the case of polyethylene, at high dosage (above 10^6 Gy) it becomes a brittle, hard material. This is due to a combination of random chain scission and cross-linking.

7.5.3. Mechanochemical Effects

It is well known that cyclic or constant loading deteriorates polymers. This effect can be accelerated if the polymer is simultaneously subjected to chemical processes under mechanical loading. Thus, if the polymer is stored in water or in saline solution, its strength will decrease as shown in Figure 7-7. Another

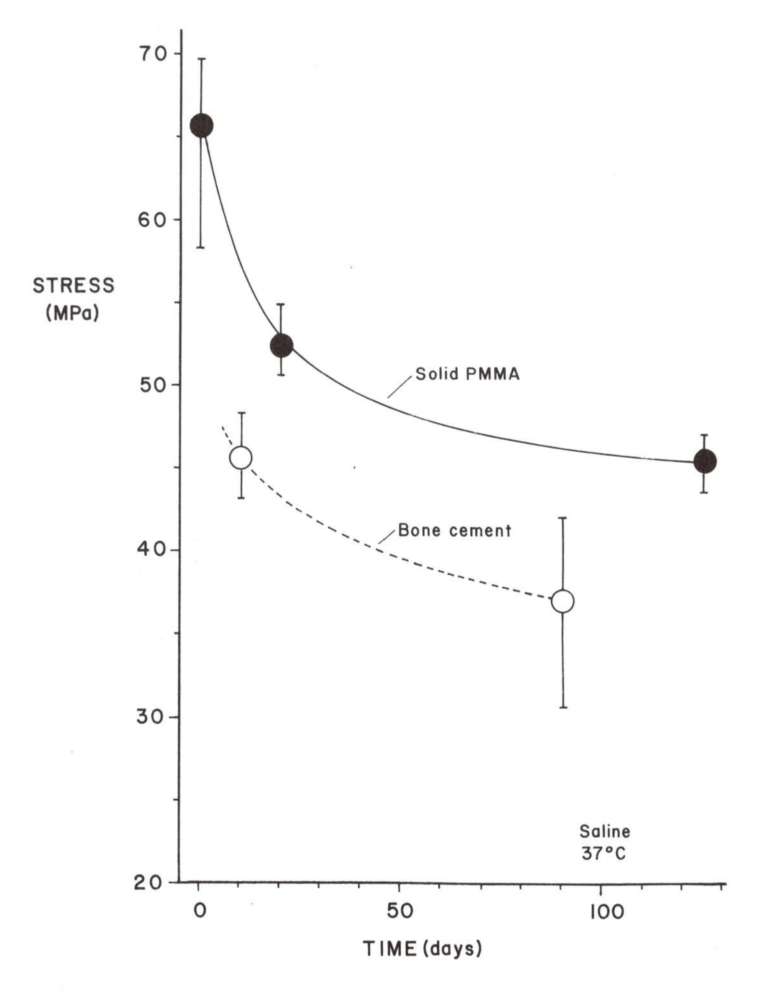

Figure 7-7. Ultimate tensile strength versus time for polymethyl methacrylate in saline solution at 37°C. Note the large decrease in tensile strengths for both solid and porous bone cement. Unpublished data of T. Parchinski, G. Cipoletti, and F. W. Cooke, Clemson University, 1977.

reason for the decrease in strength is the plasticizing effect of the water molecules at higher temperature. However, the plasticizing effect compensates for the deleterious effect of the saline solution under cyclic loading, so that there is no difference between samples stored in saline or in air as shown in Figure 7-8.

7.5.4. *In Vivo* Environmental Effects

Even though the materials implanted inside the body are not subjected to light, radiation, oxygen, ozone, and temperature variations, the body environment is very hostile, and all polymers start to deteriorate as soon as they are implanted.

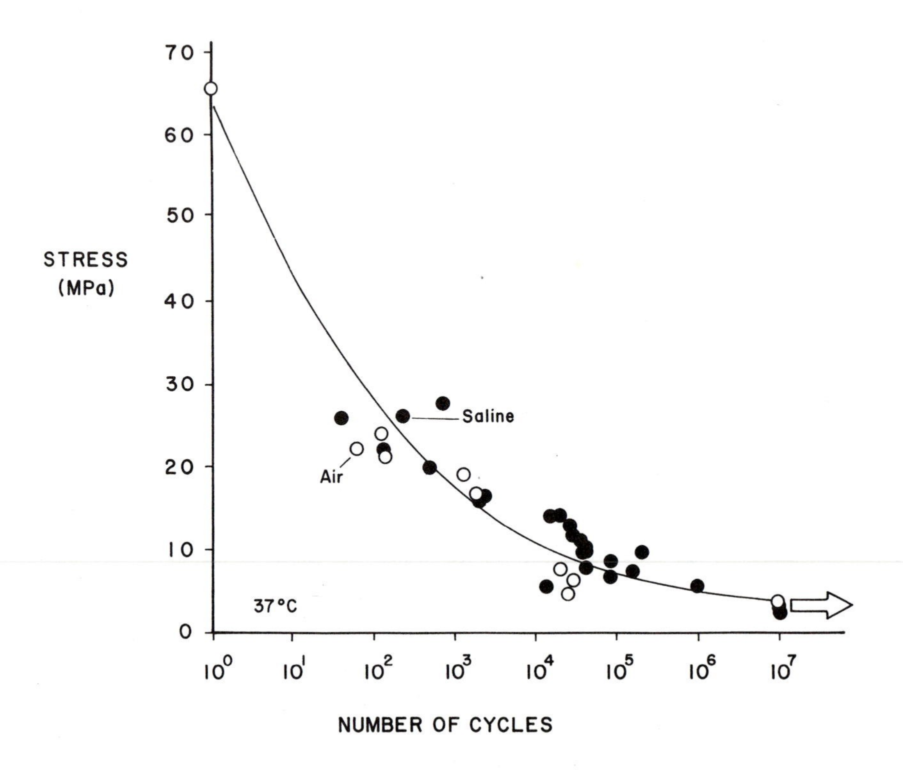

Figure 7-8. Fatigue test of solid polymethyl methacrylate. Note the S–N curve is the same for both untreated samples and samples soaked in saline solution at 37°C. After T. Parchinski and F. W. Cooke, Clemson University, 1977.

Table 7-16. Effect of Implantation on Polymers[a]

Polymers	Effects of implantation
Polyethylene	Low-density ones absorb some lipids and lose tensile strength. High-density ones are inert and no deterioration occurs.
Polypropylene	Generally no deterioration.
Polyvinyl chloride (rigid)	Tissue reaction, plasticizers may leach out and become brittle.
Polyethylene terephthalate	Susceptible to hydrolysis and loss of tensile (polyester) strength.
Polyamides (nylon)	Absorb water and irritate tissue, lose tensile strength rapidly.
Silicone rubber	No tissue reaction, very little deterioration.
Polytetrafluoroethylene	Solid specimens are inert. If it is fragmented into pieces, irritation will occur.
Polymethyl methacrylate	Rigid form: crazing, abrasion, and loss of strength by heat sterilization. Cement form: high heat generation, unreacted monomers during and after polymerization may damage tissues.

[a] From B. Bloch and G. W. Hastings, *Plastic Materials in Surgery*, 2nd ed., Charles C. Thomas, Springfield, Ill., 1972.

The most probable cause of deterioration is ionic attack (especially hydroxyl ion, OH^-) and dissolved oxygen. Enzymatic degradation may also play a significant role if the implant is made from natural polymeric materials like reconstituted collagen.

It is safe to predict that if a polymer deteriorates in physiological solution *in vitro,* the same will be true *in vivo.* Most hydrophilic polymers such as polyamides and polyvinyl alcohol will react with body water and undergo a rapid deterioration. The hydrophobic polymers like PTFE (Teflon®) and polypropylene are less prone to deteriorate *in vivo.*

The deterioration products may induce tissue reactions. In the case of *in vivo* deterioration, the original physical properties will be changed if the implant deteriorates. For example, polyolefins (polyethylene and polypropylene) will lose their flexibility and become brittle. For polyamides, the amorphous region is selectively attacked by water molecules, which act as plasticizers, making polyamides more flexible. Table 7-16 shows the effects of implantation on several polymeric materials.

PROBLEMS

7-1. Porous polyethylene (with interconnecting pores, μm diameter) is tested in the form of rods with a diameter of 3.4 mm following 2 months of exposure *in vitro* and *in vivo.* Typical force–elongation curves are obtained with a 1-cm gage length and as shown in the accompanying figure.

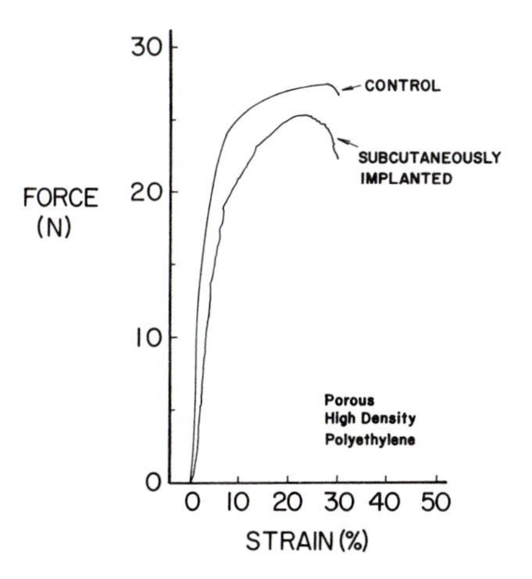

(a) Calculate the modulus of elasticity and tensile strength from both curves.

(b) Explain why the curve for the sample tested *in vivo* is not as smooth as the control sample, and why the strength is lower *in vivo.*

7-2. A sample of methacrylate ($CH_2{=}CHCOOCH_3$) is polymerized. The resulting polymer has a DP (degree of polymerization, n) of 1000. Draw the structure for the repeating unit of the polymer and calculate the polymer molecular weight.

7-3. An applied strain of 0.3 produces an immediate stress of 10 MPa in a piece of rubber, but after 42 days the stress is only 5 MPa.
(a) What is relaxation time assuming a single relaxation time?
(b) What is the stress after 90 days?

7-4. Polymers containing chemical bonds similar to those found in the body, e.g., amide and ester groups, are less biodegradable than polymers containing C–C and C–F bonds. Is this statement true?

7-5. The average bond energy of C–Cl for polyvinyl chloride is 340,000 J/mol. Can visible light ($\lambda = 4000$–7000 Å) have enough energy to break the bond? Recall $E = h\nu$ where h is Planck's constant and ν is vibration frequency.

7-6. The average and end-to-end distance (L) of a chain of an amorphous polymer can be expressed as $L = l\sqrt{m}$ where l is the interatomic distance (1.54 Å for C–C) and m is the number of bonds. If the average molecular weight of a polystyrene is 20,800 g/mol, what is the average end-to-end distance (L) of a chain?

7-7. Name the following polymers with the repeating unit given:

(a) $-CH_2CH_2-$
(b) $-CH_2-O-$
(c) $-CF_2-CF_2-$
(d) 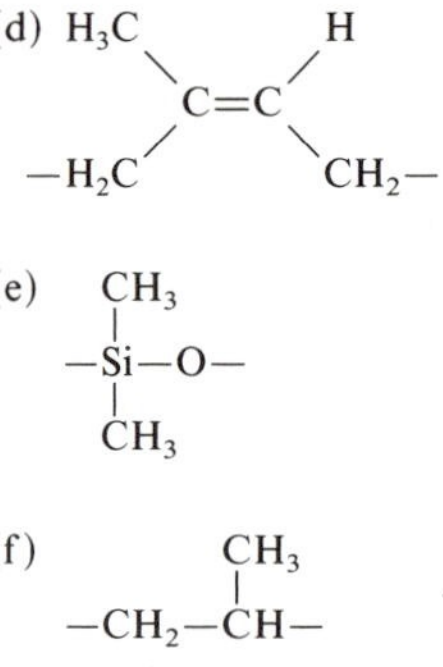 H_3C and H on C=C; $-H_2C$ and CH_2- (main chain: $-H_2C-C(CH_3)=CH-CH_2-$)

(e) $-Si(CH_3)_2-O-$

(f) $-CH_2-CH(CH_3)-$

(g) $-CH_2-CH(Cl)-$

(h) $-N(H)(CH_2)_5-C(=O)-$

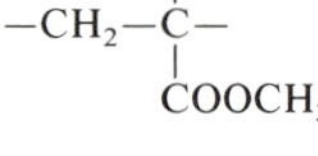

(i) $-CH_2-CH(CH)-$

(j) $-CH_2-C(CH_3)(COOCH_3)-$

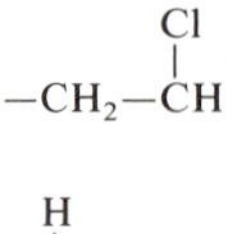

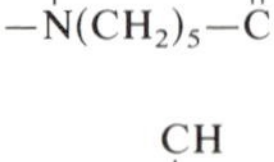

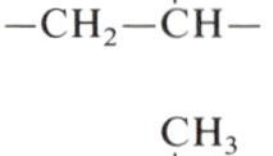

DEFINITIONS

Addition (or free radical) polymerization: Polymerization in which monomers are added to the growing chains, initiated by free radical agents.

Barium sulfate: Inert ceramic mixed in the bone cement as powder to make it radio-paque to X rays for better visualization of the implant on X-ray film.

Bone cement: Mixture of polymethyl methacrylate powder and methyl methacrylate monomer liquid to be used as a grouting material for the fixation of orthopedic implants.

Branching: Chains grown from the sides of the main backbone chains.

Condensation (step reaction) polymerization: Polymerization in which two or more chemicals are reacted to form a polymer by condensing out small molecules such as water and alcohol.

Delrin®: Polyacetal made by Union Carbide.

Filler: Materials added as a powder to a rubber to improve its mechanical properties.

Hydrogel: Polymer that can absorb water 30% or more of its weight.

Hydroquinone: Chemical inhibitor added to the bone cement liquid monomer to prevent accidental polymerization during storage.

Initiator: Chemical used to initiate the addition polymerization by becoming a free radical that in turn reacts with a monomer.

Kevlar®: Aromatic polyamides made by DuPont.

Lexan®: Polycarbonate made by General Electric.

N,N-dimethyl-p-toluidine: Chemical added to the powder portion of the bone cement for the acceleration of its polymerization.

Repeating unit: Basic molecular unit that can represent a polymer backbone chain. The average number of repeating unit is called the degree of polymerization.

Side group: Chemical group attached to the main backbone chain. It is usually shorter than the branches and exists before polymerization.

Simplex®: Acrylic bone cement made by Howmedica, inc.

Tacticity: Arrangement of side groups of a linear polymer chain, which can be atactic (random), isotactic (same side), and syndiotactic (alternating side).

Teflon®: Polytetrafluoroethylene made by DuPont.

Udel®: Polysulfone made by General Electric.

Vulcanization: Cross-linking of a (natural) rubber by adding sulfur.

Ziegler catalyst: Organometallic compounds that have the remarkable capacity of polymerizing a wide variety of monomers to linear and stereoregular polymers.

BIBLIOGRAPHY

B. Bloch and G. W. Hastings, *Plastic Materials in Surgery*, 2nd ed., Charles C. Thomas, Springfield, Ill., 1972.

S. D. Bruck, *Blood Compatible Synthetic Polymers: An Introduction*, Charles C. Thomas, Springfield, Ill., 1974.

S. D. Bruck, *Properties of Biomaterials in the Physiological Environment*, CRC Press, Boca Raton, Fla., 1980.

S. Dawids (ed.), *Polymers: Their Properties and Blood Compatibility*, Kluwer Academic, Dordrecht, 1989.

Guidelines for Blood-Material Interactions, Report of the National Heart, Lung, and Blood Institute Working Group, Devices and Technology Branch, NHLBI, NIH Publication No. 80-2185, Sept. 1980.

Guidelines for Physiochemical Characterization of Biomaterials, Report of the National Heart, Lung, and Blood Institute Work Group, Devices and Technology Branch, NHLBI, NIH Publication No. 80-2186, Sept. 1980.

A. S. Hoffman, "A Review of the Use of Radiation plus Chemical and Biochemical Processing Treatments to Prepare Novel Biomaterials," *Radiat. Phys. Chem.*, *18*, 323–342, 1981.

R. L. Kronenthal and Z. Oser (eds.), *Polymers in Medicine and Surgery*, Plenum Press, New York, 1975.

H. Lee and K. Neville, *Handbook of Biomedical Plastics*, Pasadena Technology Press, Pasadena, Calif., 1971.

R. I. Leininger, "Polymers as Surgical Implants," *CRC Crit. Rev. Bioeng.*, *2*, 333–360, 1972.

S. N. Levine (ed.), "Polymers and Tissue Adhesives," *Ann. N.Y. Acad. Sci.*, Part IV, *146*, 1968.

M. F. Refojo, "Contact Lenses," in *Encyclopedia of Chemical Technology*, 3rd ed., Vol. 16, J. Wiley and Sons, New York, 1979, pp. 720–742.

M. Szycher and W. J. Robinson (eds.), *Synthetic Biomedical Polymers, Concepts and Applications*, Technomic, Westport, Conn., 1980.

H.-G. Willert, G. H. Buchhorn, and P. Eyerer (eds.), *Ultra-High Molecular Weight Polyethylene as Biomaterial in Orthopedic Surgery*, Hogrefe & Huber, Toronto, New York, 1991.

CHAPTER 8

COMPOSITES AS BIOMATERIALS

Composite materials are those that contain two or more distinct constituent materials or phases, on a microscopic or macroscopic size scale. The term *composite* is usually reserved for those materials in which the distinct phases are separated on a scale larger than the atomic, and in which properties such as the elastic modulus are significantly altered in comparison with those of a homogeneous material. Accordingly, fiberglass and other reinforced plastics as well as bone are viewed as composite materials, but alloys such as brass, or metals such as steel with carbide particles are not. Natural biological materials tend to be composites; these are discussed in Chapter 9. Natural composites include bone, wood, dentin, cartilage, and skin. Natural foams include lung, cancellous bone, and wood. Natural composites often exhibit hierarchical structures in which particulate, porous, and fibrous structural features are seen on different microscales. In this chapter composite material fundamentals and applications in biomaterials are explored.

8.1. STRUCTURE

The properties of composite materials depend very much on *structure* (see Chapter 2), as they do in homogeneous materials. Composites differ in that considerable control can be exerted over the larger scale structure, and hence over the desired properties. In particular, the properties of a composite material depend on the *shape* of the inhomogeneities, on the *volume fraction* occupied by them, and on the *interface* among the constituents. The shape of the inhomogeneities in a composite material is classified as follows. The principal inclusion shape categories are the particle, with no long dimension; the fiber, with one long dimension; and the platelet or lamina, with two long dimensions, as shown in Figure 8-1. The inclusions may vary in size and shape within a

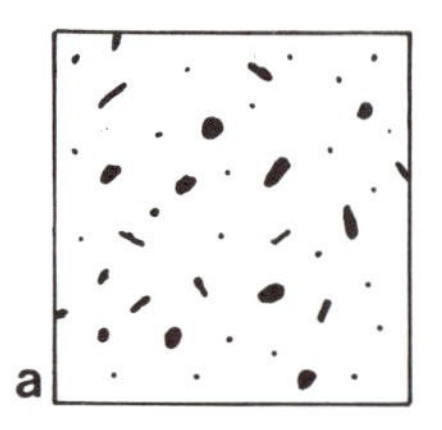

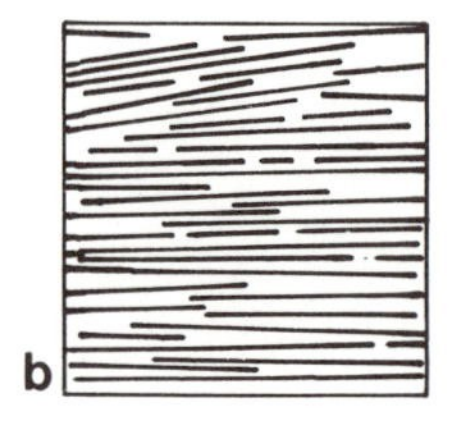

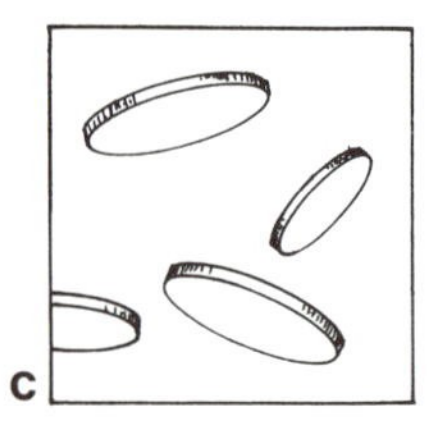

Figure 8-1. Morphology of basic composite inclusions. (a) Particle; (b) fiber; (c) platelet.

category. For example, particulate inclusions may be spherical, ellipsoidal, polyhedral, or irregular. Cellular solids are those in which the "inclusions" are voids, filled with air or liquid. In the context of biomaterials, it is necessary to distinguish the above cells, which are structural, from biological cells, which occur only in living organisms.

Several examples of composite material structures were presented in Chapter 2. The dental composite filling material shown in Figure 2-18 has a particulate structure. Figure 2-19 shows a fibrous material fracture surface, with fibers that have been pulled out. Figure 2-20 shows a section of a cross-ply laminate. Figure 2-21 shows several types of cellular solid.

8.2. MECHANICS OF COMPOSITES

In the context of mechanical properties, we may classify two-phase composite materials according to their microstructure. The inclusions within the matrix may be particles, fibers, or platelets. If either the inclusions or the matrix consists of air or liquid, the material is a cellular solid (foam). In each of the above types of structure, we may moreover make the distinction between random orientation and preferred orientation. The use of composite materials is motivated by the fact that they can provide more desirable material properties than those of homogeneous materials.

Mechanical properties in many composite materials depend on structure in a complex way, although for some structures, the prediction of properties is relatively simple. The simplest composite structures are the idealized Voigt and Reuss models, shown in Figure 8-2. The dark and light areas in these diagrams represent different constituent materials in the composite. In contrast to most composite structures, it is easy to calculate the stiffness of materials with the Voigt and Reuss structures.

Example 8-1

Determine Young's modulus of materials with the Voigt structure, assuming that Young's modulus is known for each constituent, and the volume fraction of each.

Answer

For the Voigt model, if tension is applied as in Figure 8-2, the inclusions (dark) and the matrix (light) deform together, with equal strain, so $\varepsilon_c = \varepsilon_i = \varepsilon_m$ in which c refers to the

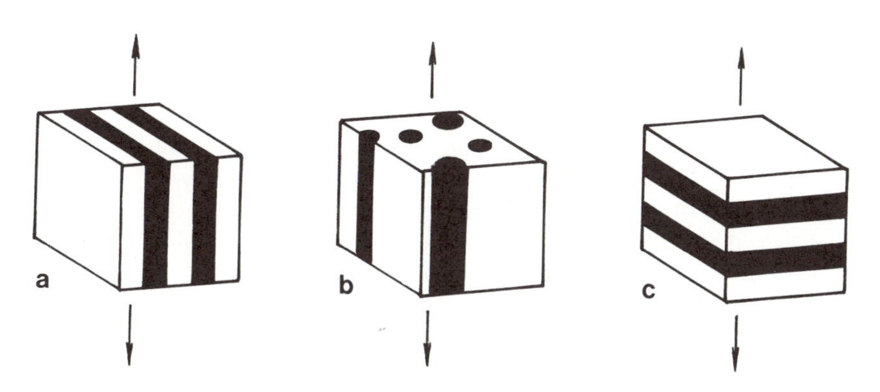

Figure 8-2. Tension force indicated by arrows, applied to Voigt (a, laminar; b, fibrous) and Reuss (c) composite models.

composite, i refers to the inclusions, and m refers to the matrix. Assume linearly elastic behavior so that $\sigma_c = E_c \varepsilon_c$ and similarly for the inclusions and matrix. The force F_c on the composite block is the sum of the forces on the inclusions and on the matrix. Therefore, $F_c = \sigma_c A_c = \sigma_i A_i + \sigma_m A_m = E_i \varepsilon_i A_i + E_m \varepsilon_m A_m$. Now divide both sides by the total cross-sectional area of the composite block, A_c, to get the stress in the composite. Then divide by the strain, which is equal in both constituents, and observe that for this geometry the volume fraction V_i is the same as the cross-sectional area fraction A_i / A_c.

Therefore, $E_c = E_f V_f + E_m V_m = E_f V_f + E_m (1 - V_f)$. This is the "rule of mixtures."

The stiffness of the Reuss model can be obtained in a similar manner (see Problem 8-1). This stiffness is quite different from that of the Voigt model. However, the Reuss laminate is identical to the Voigt laminate, except for a rotation with respect to the direction of load. Therefore, the stiffness of the laminate is *anisotropic*, i.e., dependent on direction. Anisotropy is characteristic of composite materials. The relationship between stress σ_{ij} and strain ε_{kl} in anisotropic materials is given by the tensorial form of Hooke's law:

$$\sigma_{ij} = \sum_{i=1}^{3} \sum_{j=1}^{3} C_{ijkl} \varepsilon_{kl} \tag{8-1}$$

Here C_{ijkl} is the elastic modulus tensor. It has $3^4 = 81$ elements; however, since the stress and strain are represented by symmetric matrices with six independent elements each, the number of independent modulus tensor elements is reduced to 36. An additional reduction to 21 is achieved by considering elastic materials for which a strain energy function exists. A *triclinic* crystal, which is the least symmetric crystal form, would be described by such a modulus tensor. The unit cell has three different oblique angles and three different side lengths. Triclinic modulus elements such as C_{1123} couple shearing deformations with normal stresses; this is undesirable in many applications. An *orthorhombic* crystal or an *orthotropic* composite has a unit cell with orthogonal angles. There are nine elastic moduli. The associated engineering constants are three Young's moduli, three

Poisson's ratio, and three shear moduli; there are no cross coupling constants. An example of such a composite is a unidirectional fibrous material with a rectangular pattern of fibers in the cross section. Bovine bone, which has a laminated structure, exhibits orthotropic symmetry, as does wood. In *hexagonal* symmetry, there are five independent elastic constants out of the nine remaining C elements. For directions in the transverse plane the elastic constants are the same, hence the alternate name *transverse isotropy*. A unidirectional fiber composite with a hexagonal or random fiber pattern has this symmetry, as does human Haversian bone. In *cubic* symmetry, there are three independent elastic constants, a Young's modulus E, a shear modulus G, and an independent Poisson's ratio ν. Cross-weave fabrics have cubic symmetry. Finally, an *isotropic* material has the same material properties in any direction. There are only two independent elastic constants. The others are related by equations such as

$$E = 2G(1 + \nu) \tag{8-2}$$

Random fibrous and random particulate composite materials are isotropic.

The properties of several composite material structures are shown in Table 8-1. V_i is the volume fraction (between zero and one) of inclusions, V_s is the volume fraction of solid in the case of foams, E is Young's modulus, and m refers to the matrix. As for strength, relationships are given only when they are relatively simple. The strength of composites depends not only on the strength of the constituents, but also on the stiffness and degree of ductility of the constituents. The Voigt relation for the stiffness is referred to as the rule of mixtures; related rules of mixtures are discussed elsewhere in this book. The Voigt and Reuss models provide upper and lower bounds, respectively, on the stiffness of a composite of arbitrary phase geometry. For composite materials

Table 8-1. Theoretical Properties of Composites[a]

Structure	Stiffness	Strength
Voigt model	$E = E_i V_i + E_m[1 - V_i]$	
Reuss model	$E = [V_i/E_i + (1 - V_i)/E_m]^{-1}$	
Isotropic: 3-D random orientation		
Particulate, dilute	$E = [5(E_i - E_m)V_i]/[3 + 2E_i/E_m] + E_m$	
Fibrous, dilute	$E = E_i V_i/6 + E_m$	
Platelet, dilute	$E = E_i V_i/2 + E_m$	
Foam, open cell	$E = E_s[V_s]^2$	
Crushing strength		$\sigma_{crush} = \sigma_{f,s} 0.65[V_s]^{3/2}$
Elastic collapse		$\sigma_{coll} = 0.05E_s[V_s]^2$
Anisotropic, oriented		
Unidirectional, fibrous	$E_{long} = E_i V_i + E_m[1 - V_i]$	$\sigma_{long} = \sigma_i V_i + \sigma_m[1 - V_i]$
	$E_{transv} = E_m[1 + 2nV_i/(1 - nV_i)]$ where $n = (E_f/E_m - 1)/(E_f/E_m + 2)$	

[a] Data from L. J. Gibson and M. F. Ashby, *Cellular Solids*, Pergamon, Oxford, 1988; and A. G. Agarwal and L. J. Broutman, *Analysis and Performance of Fiber Composites*, J. Wiley and Sons, New York, 1980.

that are isotropic, the more complex Hashin–Shtrickman relations provide tighter bounds on the moduli; both the Young's and shear moduli must be known for each constituent. The relations given in Table 8-1 for inclusions are valid for small volume fractions; the relations become much more complex in the case of large volume fractions. As for particles, they are assumed to be spherical and the matrix to have a Poisson's ratio of 0.5.

Observe that in isotropic systems, stiff platelet inclusions are the most effective in creating a stiff composite, followed by fibers; and the least effective geometry for stiff inclusions is the spherical particle. Even if the particles are perfectly rigid, their stiffening effect at low concentrations is modest:

$$E = E_m[1 + 5V_i/2] \quad (8\text{-}3)$$

Conversely, when the inclusions are more compliant than the matrix, spherical ones are the least harmful and platelet ones are the most harmful. Indeed, platelets in this case are suggestive of cracklike defects. Soft platelets, therefore, result not only in a compliant composite, but also a weak one. Soft spherical inclusions are used intentionally as crack stoppers to enhance the toughness of polymers such as polystyrene (high-impact polystyrene), with a small sacrifice in stiffness.

As for cellular solids, representative cellular solid structures are shown in Figure 2-21; measurement of density and porosity is described in Chapter 4, Section 4. The stiffness relationships in Table 8-1 for cellular solids are valid for all solid volume fractions; the strength relationships, only for relatively small density. The derivation of these relations is based on the concept of *bending* of the cell ribs and is presented in Gibson and Ashby (*Cellular Solids*, Pergamon, Oxford, 1988). Most man-made closed cell foams tend to have a concentration of material at the cell edges, so that they behave mechanically as open cell foams. The salient point in the relations for the mechanical properties of cellular solids is that the *relative density* dramatically influences the stiffness and the strength. As for the relationship between stress and strain, a representative stress–strain curve is shown in Figure 8-3. Observe that the physical mechanism for the

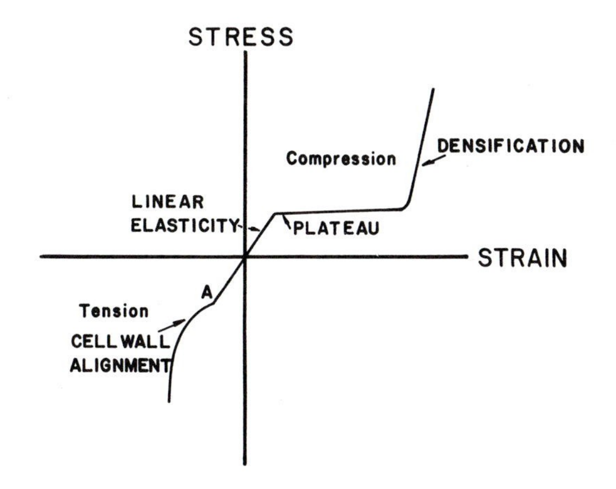

Figure 8-3. Representative stress–strain curve for a cellular solid. The plateau region for compression in the case of elastomeric foam (a rubbery polymer) represents elastic buckling; for an elastic–plastic foam (such as metallic foam) it represents plastic yield, and for an elastic–brittle foam (such as ceramic) it represents crushing. On the tension side, point A represents the transition between cell wall bending and cell wall alignment. In elastomeric foam the alignment occurs elastically, in elastic–plastic foam it occurs plastically, and an elastic–brittle foam fractures at A.

deformation mode beyond the elastic limit depends on the material from which the foam is made. Trabecular bone, for example, is a natural cellular solid, which tends to fail in compression by crushing. It is of interest to compare the predicted strength of an open cell foam from Table 8-1 with the observed dependence of strength on density for trabecular bone. Although many kinds of trabecular bone appear to behave as a normal open cell foam, there are different structures of trabecular bone and these may behave differently.

Anisotropic composites offer superior strength and stiffness in comparison with isotropic ones. Material properties in one direction are gained at the expense of properties in other directions. It is sensible, therefore, to use anisotropic composite materials only if the direction of application of the stress is known in advance.

Strength of composites depends on such particulars as the brittleness or ductility of the inclusions and the matrix. In fibrous composites, failure may occur by fiber breakage, buckling, or pullout; matrix cracking; or debonding of fiber from matrix. While unidirectional fiber composites can be made very strong in the longitudinal direction, they are weaker than the matrix alone when loaded transversely, as a result of stress concentration around the fibers. In many applications, short-fiber composites are used. While they are not as strong as those with continuous fibers, they can be formed economically by injection molding or by *in situ* polymerization. Choice of an optimal fiber length can result in improved toughness, due to the predominance of fiber pullout (Figure 2-19) as a fracture mechanism.

Example 8-2

Determine Young's modulus of trabecular bone of density 0.2 g/cm^3. Assume that the tissue behaves as an isotropic open cell foam. Observe that Young's modulus for human compact tibial bone is about 18 GPa and its tensile strength is about 140 MPa.

Answer

From Table 8-1, $E/E_s = [\rho/\rho_s]^2$ for open cell foams. So $E = 18$ GPa $[0.2/2]^2 = 180$ MPa. We have ignored the effect of tissue fluid or bone marrow in the pores. At low strain rates, it has been found that their effect on mechanical properties is negligible. However, at higher strain rates, the marrow and tissue fluid can contribute to viscoelastic behavior in trabecular bone and to its energy-absorbing capacity.

8.3. APPLICATIONS OF COMPOSITE BIOMATERIALS

Composite materials offer a variety of advantages in comparison with homogeneous materials. However, in the context of biomaterials, it is important that each constituent of the composite be biocompatible, and that the interface between constituents not be degraded by the body environment. Composites currently used in biomaterial applications include the following: dental filling composites; bone particle- or carbon fiber-reinforced methyl methacrylate bone cement and ultra-high-molecular-weight polyethylene; and porous surface orthopedic implants. Moreover, rubber used in catheters, rubber gloves, etc. is usually filled with very fine particles of silica to make the rubber stronger and tougher.

8.3.1. Dental Filling Composites

While metals such as silver amalgam and gold are commonly used in the restoration of posterior teeth, they are not considered desirable for anterior teeth for cosmetic reasons. Acrylic resins and silicate cements had been used for anterior teeth, but their poor material properties led to short service life and clinical failures. Dental composite resins have virtually replaced these materials and are very commonly used to restore posterior teeth as well as anterior teeth.

The composite resins consist of a polymer matrix and stiff inorganic inclusions. Representative structures are shown in Figures 2-18 and 8-4. Observe that the particles are very angular in shape. The inorganic inclusions confer a relatively high stiffness and high wear resistance on the material. Moreover, by virtue of their translucence and index of refraction similar to that of dental enamel, they are cosmetically acceptable. The inorganic inclusions are typically barium glass or silica (quartz, SiO_2). The matrix consists of BIS-GMA, an addition reaction product of bis(4-hydroxyphenol), dimethylmethane, and glycidyl methacrylate. Since the material is mixed, then placed in the prepared cavity to polymerize, the viscosity must be sufficiently low and the polymerization must

Figure 8-4. Microstructure of a dental composite. Miradapt® (Johnson & Johnson) 50% by volume filler: barium glass and colloidal silica.

Table 8-2. Composition and Shear Modulus of Dental Composites[a]

Name	Filler	Filler amount (wt%)	Particle size (μm)	G (GPa), 37°C
Adaptic	Quartz	78	13	5.3
Concise	Quartz	77	11	4.8
Nuva-fil	Barium glass	79	7	—
Isocap	Colloidal silica	33	0.05	—
Silar	Colloidal silica	50	0.04	2.3

[a] From Y. Papadogianis, D. B. Boyer, and R. S. Lakes, "Creep of Conventional and Microfilled Dental Composites," *J. Biomed. Mater. Res.*, *18*, 15–24, 1984.

be controllable. Low-viscosity liquids such as triethylene glycol dimethacrylate are used to lower the viscosity and inhibitors such as BHT (butylated trioxytoluene, or 2,4,6-tri-*tert*-butylphenol) are used to prevent premature polymerization. Polymerization can be initiated by a thermochemical initiator such as benzoyl peroxide, or by a photochemical initiator (benzoin alkyl ether) that generates free radicals when subjected to ultraviolet light from a lamp used by the dentist.

Compositions and stiffnesses of several representative commercially available dental composite resins are given in Table 8-2. In view of the greater density of the inorganic filler phase, a 77 wt% of filler corresponds to a volume percent of about 55%. Typical mechanical and physical properties of dental composite resins of about 50% filler by volume are shown in Table 8-3. As with other dental materials, the thermal expansion of these materials exceeds that of the tooth structure. The difference is thought to contribute to leakage of saliva, bacteria, etc. at the interface margins. Use of colloidal silica in the so-called "microfilled" composites allows these resins to be polished, so that less wear occurs and less plaque accumulates. It is more difficult, however, to make these with a high fraction of filler, since the tendency for high viscosity of the unpolymerized paste must be counteracted. An excessively high viscosity is problematical since it

Table 8-3. Typical Properties of Dental Composites[a]

Property	Value
Young's modulus E (GPa)	10–16
Poisson's ratio ν	0.24–0.30
Compressive strength (MPa)	170–260
Shear strength (MPa)	30–100
Porosity (vol%)	1.8–4.8
Polymerization contraction (%)	1.2–1.6
Thermal expansion α (10^{-6}/°C)	26–40
Thermal conductivity k [10^{-4} cal/sec/cm^2(°C/cm)]	25–33
Water sorption coeff. (mg/cm^2, 24 hr, r.m. temp.)	0.6–0.8

[a] From M. L. Cannon, "Composite Resins," in *Encyclopedia of Medical Devices and Instrumentation*, J. G. Webster, ed., J. Wiley and Sons, New York, 1988.

prevents the dentist from adequately packing the paste into the prepared cavity; the material will then fill in crevices less effectively. All of the dental composites exhibit creep. The stiffness changes by a factor of from 2.5 to 4 (depending on the particular material) over a time period from 10 seconds to 3 hours under steady load. This creep may result in indentation of the restoration, but wear seems to be a greater problem.

Dental composite resins have become established as restorative materials for both anterior and posterior teeth. The use of these materials is likely to increase as improved compositions are developed and in response to concern over long-term toxicity of silver–mercury amalgam fillings.

Example 8-3

Consider an isotropic composite in which the spherical particles are silica with a Young's modulus of 72 GPa and a polymer matrix with a Young's modulus of 1 GPa. Determine the modulus of the composite for an inclusion fraction of 33% by volume. Compare with the Reuss model.

Answer

Using the relation given in Table 8-1, $E = [5(72 - 1)0.33][3 + 2(72)/1]^{-1} + 1 = \underline{1.8\ \text{GPa}}$. The calculation is approximate in that 33% particle concentration is not really dilute. For comparison, the Reuss model (see Problem 8-1) gives $E = [0.33/72 + 0.67/1]^{-1} = \underline{1.5\ \text{GPa}}$. The stiffness of the particulate composite is not much greater than the Reuss lower bound; this is representative of spherical inclusions.

8.3.2. Porous Implants

Porous implants allow tissue ingrowth. The ingrowth is considered desirable in many contexts, since it allows a relatively permanent anchorage of the implant to the surrounding tissues; see Chapter 14 for more details. There are actually two composites to be considered in porous implants: the implant prior to ingrowth, in which the pores are filled with tissue fluid, which is ordinarily of no mechanical consequence; and the implant filled with tissue. In the case of the implant prior to ingrowth, it must be recognized that the stiffness and strength of the porous solid are much less than in the case of the solid from which it is derived, as described by the relationships in Table 8-1 (see also Figure 8-6).

Porous layers are used on bone-compatible implants to encourage bony ingrowth. The pore size of a cellular solid has no influence on its stiffness or strength (though it does influence the toughness), but pore size can be of considerable biological importance. Specifically, in orthopedic implants with pores larger than about 150 μm, bony ingrowth into the pores occurs and this is useful to anchor the implant. This minimum pore size is on the order of the diameter of osteons in normal Haversian bone. It was found experimentally that smaller pores less than 75 μm in size did not permit the ingrowth of bone tissue.

Figure 8-5. Knee prostheses with black carbon fiber-reinforced polyethylene tibial components.

Moreover, it was difficult to maintain fully viable osteons within pores in the 75–150 μm size range. Representative structure of such a porous surface layer is shown in Figure 8-6. Porous coatings are also under study for application in anchoring the artificial roots of dental implants to the underlying jawbone. When a porous material is implanted in bone, the pores become filled first with blood, which clots, then with osteoprogenitor mesenchymal cells, then, after about 4 weeks, bony trabeculae. The ingrown bone then becomes remodeled in response to mechanical stress. The bony ingrowth process depends on a degree of mechanical stability in the early stages. If too much motion occurs, the ingrown tissue will be collagenous scar tissue, not bone.

Figure 8-6. Structure of porous coating for bony ingrowth. The top scanning electron micrograph is a 5× magnification of the rectangular region in the bottom picture (200×, Ti5Al4V alloy). Note the irregular pore structure. S. H. Park and J. B. Park, unpublished data.

Porous materials used in soft tissue applications include polyurethane, polyimide, and polyester velours used in percutaneous devices. Porous reconstituted collagen has been used in artificial skin, and braided polypropylene has been used in artificial ligaments. As in the case of bone implants, the porosity encourages tissue ingrowth, which anchors the device. The healing and tissue response to a porous implant generally follows the sequence described elsewhere in this book. It must, however, be borne in mind that porous materials have a high ratio of surface area to volume. Consequently, the demands on the inertness and biocompatibility are likely to be greater for a porous material than a homogeneous one.

Ingrowth of tissue into implant pores is not always desirable: sponge (polyvinyl alcohol) implants used in early mammary augmentation surgery underwent ingrowth of fibrous tissue, and contracture and calcification of that tissue, resulting in hardened, calcified breasts. Current mammary implants make use of a ballonlike nonporous silicone rubber layer enclosing silicone oil or gel. A porous layer of polyester felt or velour attached to the balloon is provided at the back surface of the implant so that limited tissue ingrowth will anchor it to the chest wall and prevent it from migrating.

Porous blood vessel replacements encourage soft tissue to grow in, eventually forming a new lining, or neointima. This is another example of the biological role of porous materials as contrasted with the mechanical role. As discussed elsewhere in this book, a material in contact with the blood should be nonthrombogenic. The role of the neointima encouraged to grow into a replacement blood vessel is to act as a natural nonthrombogenic surface resembling the lining of the original blood vessel.

Porous materials are produced in a variety of ways such as by sintering of beads or wires in the case of bone-compatible surfaces. Vascular and soft tissue implants are produced by weaving or braiding fibers as well as by nonwoven "felting" methods. Protective foams for external use are usually produced by use of a "blowing agent," which is a chemical that evolves gas during the polymerization of the foam. An interesting approach to producing microporous materials is the replication of structures found in biological materials: the *replamineform* process. The rationale is that the unique structure of communicating pores is thought to offer advantages in the induction of tissue ingrowth. The skeletal structure of coral or echinoderms (such as sea urchins) is replicated by a casting process in metals and polymers; these have been tried in vascular and tracheal prostheses as well as in bone substitutes.

8.3.3. Fibrous and Particulate Composites in Orthopedic Implants

The rationale for incorporating stiff inclusions in a polymer matrix is to increase the stiffness, strength, fatigue life, and other properties. For that reason, carbon fibers have been incorporated in the high-density polyethylene used in total knee replacements. The reason for wishing to modify the standard UHMWPE used in these implants is that it should provide adequate wear resistance over 10 years' use. While this is sufficient for implantation in older patients, a longer wear-free lifetime is desirable in implants to be used in younger patients. Improvement in the resistance to creep of the polymeric component is also considered desirable, since excessive creep results in an indentation of the polymeric component after long-term use. Representative properties of carbon-reinforced UHMWPE are shown in Table 8-4. Enhancements of various properties by a factor of two are feasible.

Table 8-4. Properties of Carbon-Reinforced UHMWPE[a]

Fiber amount (%)	Density (g/cm^3)	Young's modulus (GPa)	Flexural strength (MPa)
0	0.94	0.71	14
10	0.99	1.01	20
15	1.00	1.4	23
20	1.03	1.5	25

[a] E. Sclippa and K. Piekarski, "Carbon Fiber Reinforced Polyethylene for Possible Orthopedic Usage," *J. Biomed. Mater. Res.*, 7, 59–70, 1973.

Fibers have also been incorporated into polymethyl methacrylate (PMMA) bone cement on an experimental basis. Significant improvements in the mechanical properties can be achieved. However, this approach has not found much acceptance since the fibers also increase the viscosity of the unpolymerized material. It is consequently difficult to form and shape the polymerizing cement during the surgical procedure. Metal wires have been used as macroscopic "fibers" to reinforce PMMA cement used in spinal stabilization surgery, but such wires are not useful in joint replacements owing to the limited space available.

Particle reinforcement has also been used to improve the properties of bone cement. For example, inclusion of bone particles in PMMA cement somewhat improves the stiffness and improves the fatigue life considerably, as shown in Figure 3-8. Moreover, the bone particles at the interface with the patient's bone are ultimately resorbed and are replaced by ingrown new bone tissue. This approach is in the experimental stages.

Example 8-4

How long should the fibers be in a short-fiber composite of graphite fibers in a polyethylene matrix, to be used for a knee replacement prosthesis?

Answer

To deal with the mechanical aspects of this question, we develop the "shear lag" model due to Cox. Consider the equilibrium conditions of a *segment* of a circular cross-section fiber of radius r and length dz acted upon by normal stress $\sigma_f|_{\text{end}}$ on the end (assumed constant in this model), a normal stress distribution σ_f in the fiber to be found, and shear stress τ on the lateral surfaces. The equilibrium equation is $\pi r^2 \sigma_f + \tau 2\pi r \, dz = \pi r^2(\sigma_f + d\sigma_f)$, so $2\tau \, dz = d\sigma_f$, so $d\sigma_f/dz = 2\tau/r$, so that

$$\sigma_f(z) = \sigma_f|_{\text{end}} + \frac{2}{r}\int_0^z \tau(z)\, dz$$

Further progress is facilitated by assuming that the matrix is rigid–perfectly plastic, so that the stress on the lateral surfaces of the fiber is equal to the matrix yield strength in shear τ_y. Such an assumption would approximate reality in the case of a ductile matrix subjected to a large load. Moreover, suppose that the fiber is sufficiently long that the stress in it is mostly due to shear from the lateral surfaces rather than tension at the ends. Then the above integral becomes simplified to $\sigma_f(z) = 2\tau_y z/r$, so that the fiber stress increases linearly with position on the fiber, up to half the fiber length L. The maximum fiber stress occurs at the fiber midpoint, $z = L/2$, so that

$$\frac{L}{2r} = \frac{\sigma_f^{\max}}{2\tau_y}$$

If we set the maximum fiber stress equal to its ultimate strength σ_f^{ult}, we obtain the *critical fiber length* L_c,

$$L_c = 2r\frac{\sigma_f^{ult}}{2\tau_y}$$

For fibers longer than this length, load is efficiently transferred to them, to achieve the maximum strength of the composite. Fibers about as long as the critical fiber length can either break or pull out of the matrix, leading to enhanced toughness. In applications such as the knee replacement, considerations such as the manufacture of the composite also are important in determining the fiber length.

PROBLEMS

8-1. For the Reuss model, show that the Young's modulus of the composite is given by $E_c = [E_f V_f + E_m V_m]^{-1}$. Hint: the stress is the same in each constituent and the elongations are additive (explain why).

8-2. Derive Eq. (8-3). Assume that the particulate inclusions are perfectly rigid.

8-3. Consider a bone plate made of a unidirectional fibrous composite. What fiber and matrix materials are suitable in view of the need for biocompatibility? Assume 50% fibers by volume and determine the Young's modulus in the longitudinal direction. Compare with a metal implant.

8-4. Consider an isotropic composite in which the fibers are silica needles with a Young's modulus of 72 GPa and a polymer matrix with a Young's modulus of 1 GPa. Determine the modulus of the composite for an inclusion fraction of 33% by volume. Compare with the Voigt model and with Example 8-3.

8-5. Calculate and plot the Voigt and Reuss bounds for a collagen-hydroxyapatite composite versus volume fraction hydroxyapatite. Plot on the same graph the Young's moduli for compact bone, dentin, and enamel. Use mechanical properties given elsewhere in this book.

8-6. Calculate the shear modulus of dental composites given in Table 8-3 using the relation $E = 2G(1 + \nu)$. Discuss the validity of this equation in the context of this problem.

SYMBOLS/DEFINITIONS

Greek Letters

ε:	Strain.
ρ:	Density.
σ:	Stress.
$\sigma_{f,s}$:	Fracture strength of a solid.

Latin Letters

C_{ijkl}:	Elastic modulus tensor.
E:	Young's modulus.
V:	Volume fraction of a constituent.

Terms

Anisotropic: Dependent on direction, referring to the material properties of composites.

Closed cell: A type of cellular solid in which a cell wall isolates the adjacent pores.

Composite: Composite materials are those that contain two or more distinct constituent materials or phases, on a microscopic or macroscopic size scale.

Cubic: A type of anisotropic symmetry in which the unit cells are cube-shaped. There are three independent elastic constants.

Hexagonal: A type of anisotropic symmetry in which the unit cells are hexagonally shaped. There are five independent elastic constants. Transverse isotropy is mechanically equivalent to hexagonal although the structure may be random in the transverse direction.

Inclusion: Embedded phase of a composite.

Isotropic: Independent of direction, referring to material properties.

Matrix: The portion of the composite in which inclusions are embedded. The matrix is usually less stiff than the inclusions.

Neointima: New lining of a blood vessel. It is stimulated to form by fabric-type blood vessel replacements.

Open cell: A type of cellular solid in which there is no barrier between adjacent pores.

Orthotropic: A type of anisotropic symmetry in which the unit cells are shaped like rectangular parallelepipeds. In crystallography, this is called orthorhombic. There are nine independent elastic constants.

Porous ingrowth: Growth of tissue into the pores of an implanted porous biomaterial. Such ingrowth may or may not be desirable.

Replamineform: Cellular solid made using a biological material as a mold.

Transverse isotropy: See hexagonal.

Triclinic: A type of anisotropic symmetry in which the unit cells are oblique parallelepipeds with unequal sides and angles. There are 21 independent elastic constants.

BIBLIOGRAPHY

M. F. Ashby, "The Mechanical Properties of Cellular Solids," *Metall. Trans.*, *14A*, 1755–1768, 1983.

R. M. Christensen, *Mechanics of Composite Materials*, J. Wiley and Sons, New York, 1979.

R. Craig, "Chemistry, Composition, and Properties of Composite Resins," in H. Horn (ed.), *Dental Clinics of North America*, W. B. Saunders, Co., Philadelphia, 1981.

L. J. Gibson and M. F. Ashby, "The Mechanics of Three Dimensional Cellular Materials," *Proc. R. Soc. London Ser. A*, *382*, 43–59, 1982.

Y. Papadogianis, D. B. Boyer, and R. S. Lakes, "Creep of Posterior Dental Composites," *J. Biomed. Mater. Res.*, *19*, 85–95, 1985.

M. Spector, M. Miller, and N. Beals, "Porous Materials," in *Encyclopedia of Medical Devices and Instrumentation*, J. G. Webster (ed.), J. Wiley and Sons, New York, 1988.

CHAPTER 9

STRUCTURE–PROPERTY RELATIONSHIPS OF BIOLOGICAL MATERIALS

The major difference between biological materials and biomaterials (implants) is *viability*. There are other equally important differences that distinguish living materials from artificial replacements. First, most biological materials are continuously bathed with body fluids. Exceptions are the specialized surface layers of skin, hair, nails, hooves, and the enamel of teeth. Second, most biological materials can be considered as *composites*.

Structurally, biological tissues consist of a vast network of intertwining fibers with polysaccharide ground substances immersed in a pool of ionic fluid. Attached to the fibers are cells which are the living tissues. The ground substances have definite structural organization and are not completely analogous to solute suspended in a solution. Physically, ground substances function as a glue, lubricant, and shock absorber in various tissues.

The structure and properties of a given biological material are dependent on the chemical and physical nature of the components present and their relative amounts. For example, neural tissues consist almost entirely of cells while bone is composed of organic materials and calcium phosphate minerals with minute quantities of cells and ground substances as a glue.

An understanding of the exact role played by a tissue and its interrelationship with the function of the entire living organism is essential if biomaterials are to be used intelligently. Thus, a person who wants to design an artificial blood vessel prosthesis has to understand not only the property-structure relationship of the blood vessel wall but also its systemic function. This is because the artery is not only a blood conduit, but also a component of a larger system, including a pump (heart) and an oxygenator (lung).

9.1. PROTEINS

Proteins are polyamides formed by step reaction polymerization between amino and carboxyl groups of amino acids,

$$\left(\begin{matrix} O \\ \| \\ C \end{matrix} - \begin{matrix} H \\ | \\ N \end{matrix} - \begin{matrix} H \\ | \\ C \\ | \\ R \end{matrix} \right)_n \tag{9-1}$$

where R is a side group. Depending on the side group, the molecular structure changes drastically. The simplest side group is hydrogen (H), which will form *glycine* (Gly). The geometry of glycine is shown in Figure 9-1a where the hypothetical flat sheet structure is shown. The structure has a repeating distance of 0.72 nm and the side groups (R) are crowded except in polyglycine, which has the smallest atom for the side group, H. If the side groups are larger, then the resulting structure is an α *helix* where the hydrogen bonds occur between different parts of the same chain and hold the helix together as shown in Figure 9-1b.

9.1.1. Collagen

One of the basic structural proteins is collagen, which has the general amino acid sequence of -Gly-Pro-Hyp-Gly-X- (X can be any other amino acid) arranged in a triple α helix. It has a high proportion of proline (Pro) and hydroxyproline (Hyp) as given in Table 9-1. Since the presence of hydroxyproline is unique in collagen (elastin contains a minute amount), the determination of collagen content in a given tissue is readily done by assaying the hydroxyproline.

Three left-handed-helical peptide chains are coiled together to give a right-handed coiled superhelix with periodicity of 2.86 nm. This triple superhelix is the molecular basis of *tropocollagen*, the precursor of collagen (see Figure 9-2).

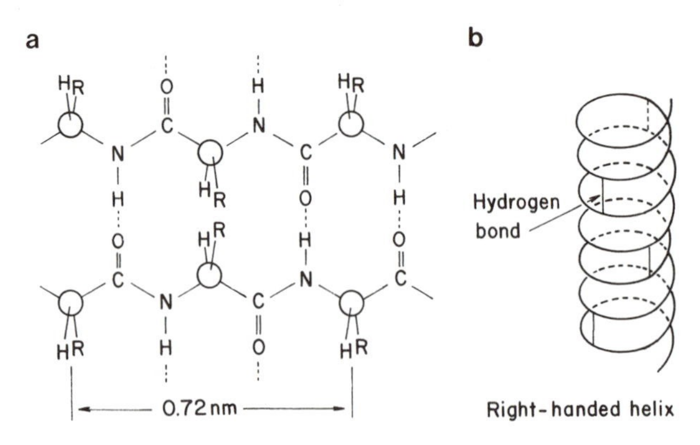

Figure 9-1. (a) Hypothetical flat sheet structure of a protein. (b) Helical arrangement of a protein chain.

Table 9-1. Amino Acid Content of Collagen[a]

Amino acids	Mol/100 mol amino acids
Gly	31.4-338
Pro	11.7-13.8
Hyp	9.4-10.2
Acid polar (Asp, Glu, Asn)	11.5-12.5
Basic polar (Lys, Arg, His)	8.5-8.9
Other	Residue

[a] From M. Chvapil, *Physiology of Connective Tissue*, Butterworths, London, 1967.

The three chains are held strongly to each other by H-bonds between glycine residues and between hydroxyl (OH) groups of hydroxyproline. In addition, there are cross-links via lysine among the (three) helices.

The primary factors stabilizing the collagen molecules are invariably related to the interactions among the α helices. These factors are H-bonding between the C=O and NH groups, ionic bonding between the side groups of polar amino acids, and the interchain cross-links between helices.

The collagen fibrils (20- to 40-nm diameter) form fiber bundles that have a diameter of 0.2–1.2 μm. Figure 9-3 shows scanning and transmission electron microscopic pictures of collagen fibrils in bone, tendon, and skin. Note the straightness of tendon collagen fibrils compared to the more wavy skin fibers.

The side groups of some amino acids are highly nonpolar in character and hence *hydrophobic*; therefore, chains with these amino acids avoid contact with water molecules and seek the greatest number of contacts with the nonpolar chains of amino acids. If we destroy the hydrophobic nature by an organic solvent solution (e.g., urea), the characteristic structure is lost resulting in microscopic changes such as shrinkage of collagen fibers. The same effect can be achieved by simply warming the collagen fibers. Another factor affecting the stability of the collagen is the incorporation of water molecules into the intra- and interchain structure. If the water content is lowered, the structural stability decreases. If the collagen is dehydrated completely (lyophilized), then the solubility also decreases (so-called *in vitro* aging of collagen).

It is known that acid mucopolysaccharides also affect the stability of collagen fibers by mutual interactions, forming mucopolysaccharide-protein complexes. It is believed that the water molecules affect the polar region of the chains making the dried collagen more disoriented than it is in the wet state.

9.1.2. Elastin

Elastin is another structural protein found in a relatively large amount in elastic tissues such as ligamentum nuchae (major supporting tissue in the head and neck of grazing animals), aortic wall, skin, etc. The chemical composition of elastin is somewhat different from that of collagen.

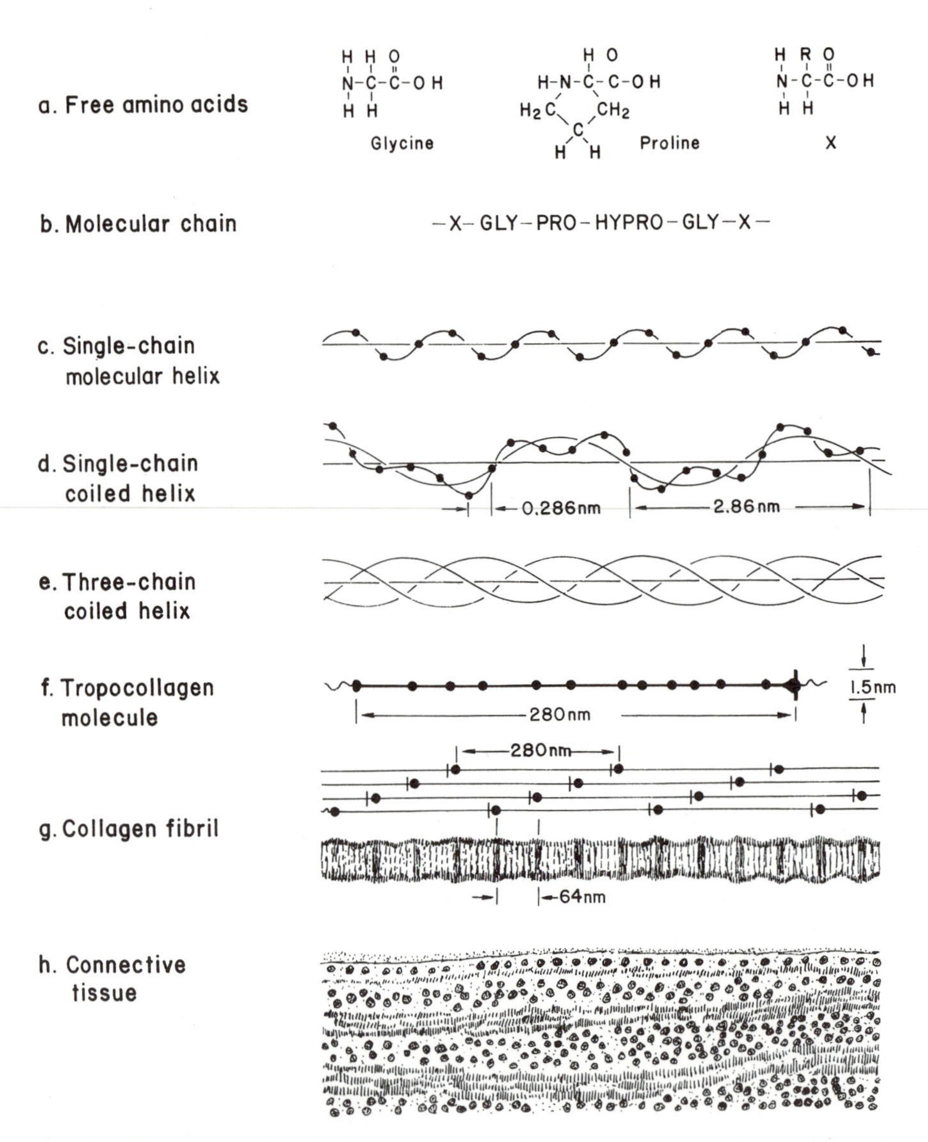

Figure 9-2. Diagram depicting the formation of collagen, which can be visualized as taking place in seven steps. The starting materials (a) are amino acids, of which two are shown and the side chain of any others is indicated by R in amino acid X. (b) The amino acids are linked together to form a molecular chain. (c) This then coils into a left-handed helix (d and e). Three such chains then intertwine in a triple-stranded helix, which constitutes the tropocollagen molecule (f). Many tropocollagen molecules become aligned in staggered fashion, overlapping by a quarter of their length to form a cross-striated collagen fibril (g). From J. Gross, "Collagen," *Sci. Am.*, *204*, 121–130, 1961.

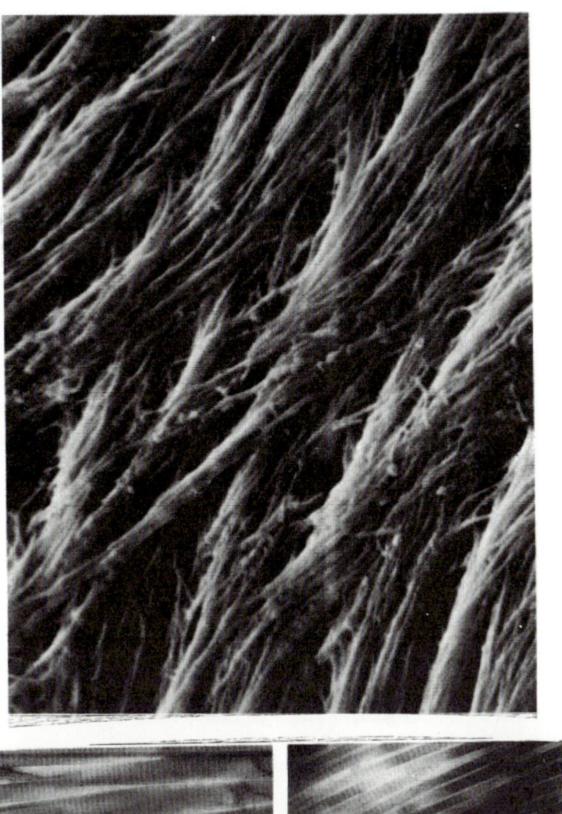

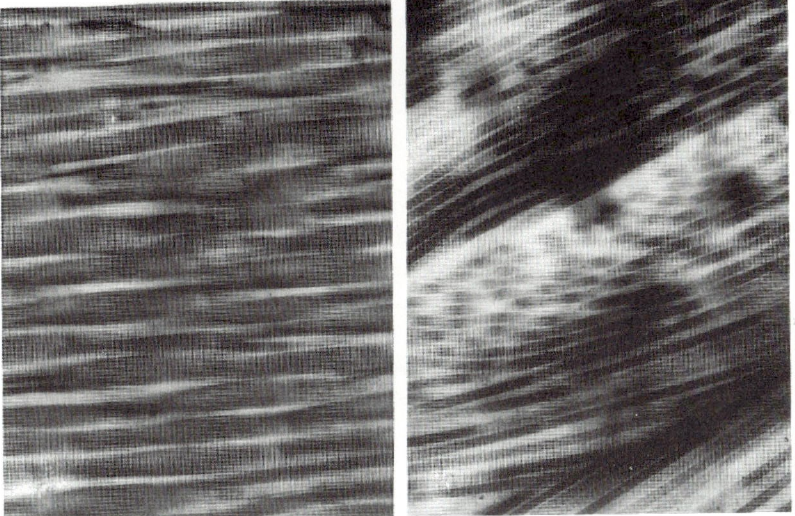

Figure 9-3. (Top) Scanning electron micrograph of the surface of adult rabbit bone matrix, showing how the collagen fibrils branch and interconnect in an intricate, woven pattern (×4800). (Bottom) Transmission electron micrographs of (left) parallel collagen fibrils in a tendon, and (right) meshwork of fibrils in skin (×24,000). a, From J. T. Tiffit, "The Organic Matrix of Bone Tissue," in *Fundamental and Clinical Bone Physiology,* M. R. Urist (ed.), J. B. Lippincott Co., Philadelphia, 1980; b, from Y. C. Fung, *Biomechanics: Mechanical Properties of Living Tissues,* Springer-Verlag, Berlin, 1981.

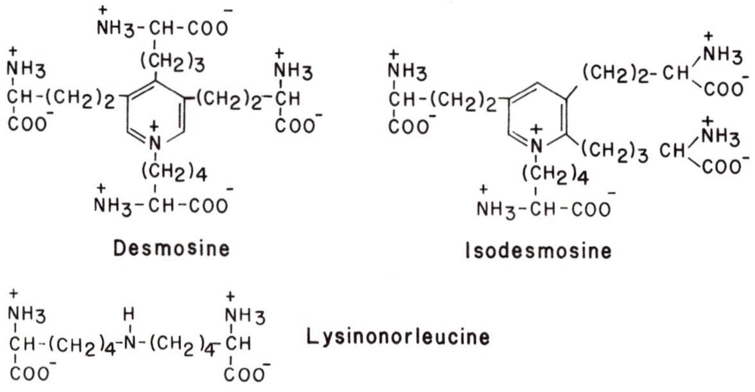

Desmosine

Isodesmosine

Lysinonorleucine

Figure 9-4. Structure of desmosine, isodesmosine, and lysinonorleucine.

Table 9-2. Amino Acid Content of Elastin

Amino acids	Residues/1000
Gly	324
Hyp	26
Cationic (Asp, Glu)	21
Anionic (His, Lys, Arg)	13
Nonpolar (Pro, Ala, Val, Met, Leu, Ile, Phe, Tyr)	595
Half-cystine	4

The high elasticity of elastin is due to the cross-linking of lysine residues via *desmosine*, *isodesmosine*, and *lysinonorleucine* shown in Figure 9-4. The formation of desmosine and isodesmosine is only possible with copper and lysyl oxidase enzyme present; hence, deficiency of copper in the diet may result in non-cross-linked elastin. This, in turn, will result in tissue that is viscous rather than normal rubberlike elastic tissue; this abnormality can lead to rupture of the aortic walls.

Elastin is very stable at high temperature in the presence of various chemicals owing to its very low content of polar side groups (hydroxyl and ionizable groups). The specific staining of elastin in tissue by lipophilic stains such as Weigert's resorcin–fuchsin occurs for the same reason. Elastin contains a high percentage of amino acids with aliphatic side chains such as *valine* (six times that of collagen). It also lacks all of the basic and acidic amino acids so that it has very few ionizable groups. The most abundant of these, glutamic acid, occurs only one-sixth as often as in collagen. *Aspartic acid*, *lysine*, and *histidine* are all below 2 residues per 1000 in mature elastin. The composition of elastin is given in Table 9-2.

9.2. POLYSACCHARIDES

Polysaccharides are polymers of simple sugars. They exist in tissues as a highly viscous material that interacts readily with proteins, including collagen, resulting in *glycosaminoglycans* (also known as *mucopolysaccharides*) or *proteoglycans.* These molecules readily bind both water and cations due to the large content of anionic side chains. They also exist at physiological concentrations not as viscous solids but as viscoelastic gels. All of these polysaccharides consist of disaccharide units polymerized into unbranched macromolecules as shown in Figure 9-5.

9.2.1. Hyaluronic Acid and Chondroitin

Hyaluronic acid is made of residues of *N*-acetylglucosamine and D-glucuronic acid, but it lacks the sulfate residues. The animal hyaluronic acid contains a protein component (0.33 wt% or more) and is believed to be chemically bound to at least one protein or peptide that cannot be removed. This, in turn, will result in proteoglycan molecules that may behave differently from the pure polysaccharides. Hyaluronic acid is found in the vitreous humor of the eye, synovial fluid, skin, umbilical cord, and aortic walls. Chondroitin is similar to hyaluronic acid in its structure and properties and is found in the cornea of the eyes.

9.2.2. Chondroitin Sulfate

This is the sulfated mucopolysaccharide which resists the hyaluronidase enzyme. It has three isomers as shown in Figure 9-5. Isomer A (chondroitin 4-sulfate) is found in cartilage, bones, and cornea, while isomer C (chondroitin

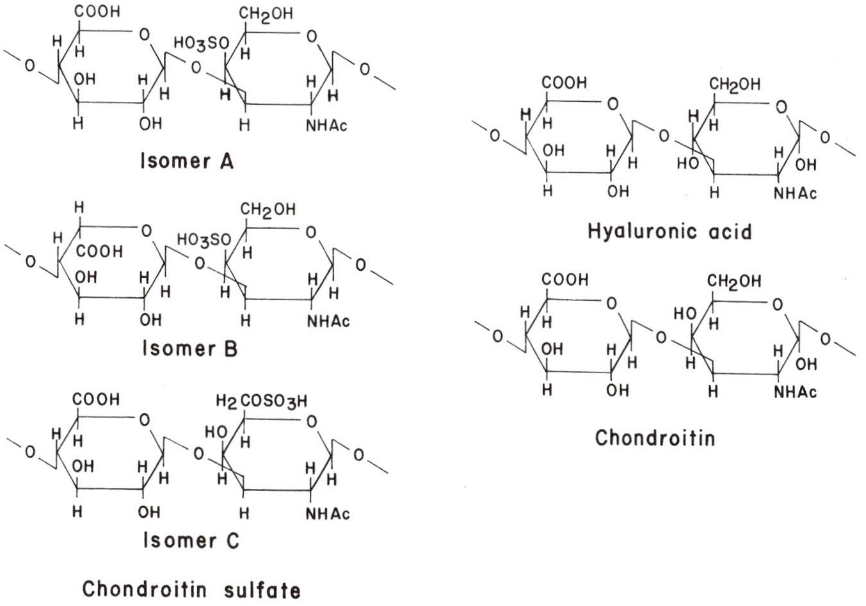

Isomer A

Isomer B

Isomer C

Chondroitin sulfate

Hyaluronic acid

Chondroitin

Figure 9-5. Structure of hyaluronic acid, chondroitin, and chondroitin sulfates.

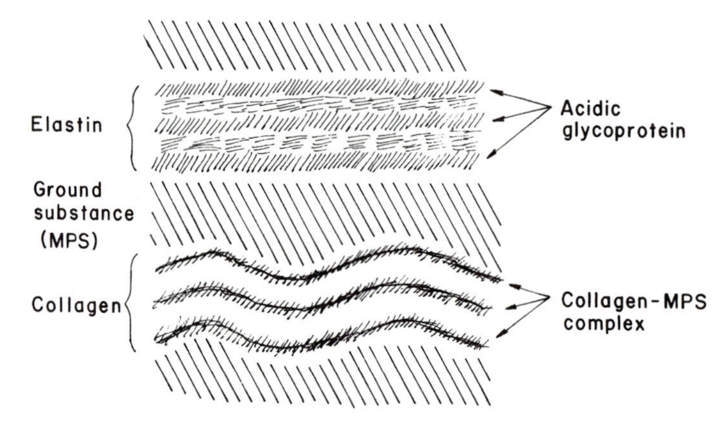

Figure 9-6. A schematic representation of mucopolysaccharide–protein molecules in connective tissues. Note the wavy nature of collagen fibers and straighter form of elastin.

6-sulfate) can be isolated from cartilage, umbilical cord, and tendon. Isomer B (dermatan sulfate) is found in skin and lungs and is resistant to testicular hyaluronidase.

The chondroitin sulfate chains in connective tissues are bound covalently to a polypeptide backbone through their reducing ends. Figure 9-6 shows a proposed macromolecular structure of protein-polysaccharides from which one can image the nature of the viscoelastic properties of the ground substance. These complexes of protein and mucopolysaccharides (ground substance) play an important role in the physical behavior of connective tissues as lubricating agents between tissues (e.g., joints) or between elastin and collagen microfibrils.

Example 9-1

Calculate the degree of polymerization of a chondroitin compound that has an average molecular weight of 100,000 g/mol.

Answer

Figure 9-5 (Ac is $COOCH_3$, acetyl) reveals that there are 14 carbon atoms, 14 oxygen atoms, 21 hydrogen atoms, and 1 nitrogen atom in a repeating unit; therefore, $14 \times 12 + 14 \times 16 + 21 \times 1 + 1 \times 14 = 427$.

$$\mathrm{DP} = 100{,}000/427 = \underline{234}$$

9.3. STRUCTURE–PROPERTY RELATIONSHIP OF TISSUES

Understanding the structure–property relationship of various tissues is important since one has to know what is being replaced by the artificial materials (biomaterials). Also, one may want to use natural tissues as biomaterials (e.g., porcine heart valves). The property measurements of any tissues are confronted with many of the following limitations and variations:

1. Limited sample size.
2. Original structure can undergo change during sample collection or preparation.
3. Inhomogeneity.
4. Complex nature of the tissues makes it difficult to obtain fundamental physical parameters.
5. Tissue cannot be frozen or homogenized without altering its structure or properties.
6. The *in vitro* and *in vivo* property measurements are sometimes difficult if not impossible to correlate.

The main objective of studying the property-structure relation of tissues is to design better-performing implants. Therefore, one should always ask, "What kind of physiological functions are being performed by the tissues or organs under study *in vivo* and how can one best assume their lost function?" Keeping this in mind, let us study the tissue structure-property relationships.

9.3.1. Mineralized Tissue (Bone and Teeth)

9.3.1.1. Composition and Structure. Bone and teeth are mineralized tissues whose primary function is "load-carrying." Teeth are in more extraordinary physiological circumstances since their function is carried out in direct contact with *ex vivo* substances, while functions of bone are carried out inside the body in conjunction with muscles and tendons. A schematic anatomical view of a long bone is shown in Figure 9-7.

Wet cortical bone is composed of 22 wt% organic matrix of which 90–96 wt% is collagen; 69 wt% mineral; and 9 wt% water as given in Table 9-3. The major subphase of the mineral consists of submicroscopic crystals of an apatite of calcium and phosphate, resembling hydroxyapatite in its crystal structure [$Ca_{10}(PO_4)_6(OH)_2$]. There are other mineral ions such as citrate ($C_6H_5O_7^{4-}$), carbonate (CO_3^{2-}), fluoride (F^-), and hydroxyl ions (OH^-) which may produce some other subtle differences in microstructural features of the bone. The apatite crystals are formed as slender needles, 20–40 nm in length by 1.5–3 nm in thickness, in the collagen fiber matrix. These mineral-containing fibrils are arranged into lamellar sheets (3–7 μm) that run helically with respect to the long axis of the cylindrical *osteons* (or sometimes called *Haversian systems*). The osteon is made up of 4 to 20 lamellae that are arranged in concentric rings around the Haversian canal. Osteons are typically from 150 to 250 μm in diameter. Between these osteons the interstitial systems are sharply divided by the *cementing line.* The metabolic substances can be transported by the intercommunicating pore systems known as *canaliculi, lacunae,* and *Volksmann's canals,* which are connected with the marrow cavity. These various interconnecting systems are filled with body fluids, and their volume can be as high as 18.9 ± 0.45% according to one estimate for compact beef bone. The external and internal surfaces of the

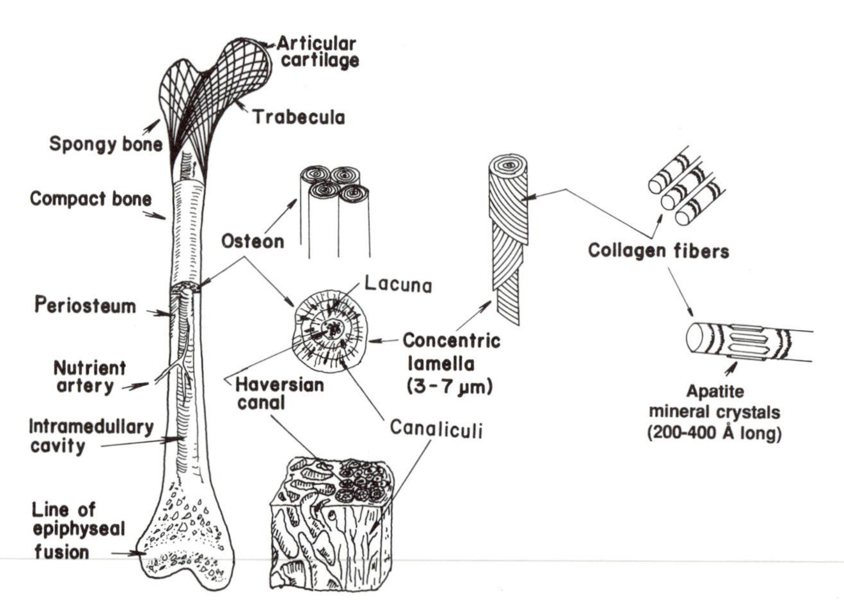

Figure 9-7. Organization of a typical bone.

bone are called the periosteum and endosteum, respectively, and both have osteogenic properties.

It is interesting to note that the mineral phase is not a completely discrete aggregation of the calcium phosphate mineral crystals. Rather it is made of a somewhat contiguous phase as evidenced in Figure 9-8 and by the fact that complete removal of the organic phase of the bone still gives a material with very good strength.

Long bones such as the femur contain cancellous (or spongy) and compact bone. The spongy bone consists of three-dimensional branches or bony trabeculae interspersed by the bone marrow. More spongy bone is present in the epiphyses of long bones while compact bone is the major form present in the diaphysis of the bone as shown in Figure 9-7.

There are two types of teeth, deciduous or primary and permanent of which the latter is more important for us from the biomaterials point of view. All teeth are made of two portions: the crown and the root, demarcated by the gingiva

Table 9-3. Composition of Bone[a]

Components	Amount (wt%)
Mineral (apatite)	69
Organic matrix	22
Collagen	(90–96% of organic matrix)
Others	(4–10% of organic matrix)
Water	9

[a] From J. T. Tiffit, "The Organic Matrix of Bone Tissue," in *Fundamental and Clinical Bone Physiology*, M. R. Urist (ed.), J. B. Lippincott Co., Philadelphia, 1980.

Figure 9-8. Scanning electron micrograph showing the mineral portion of an osteon's lamellae. The organic phase has been removed by ethylenediamine in a soxhlet apparatus. From K. Pierkarski, "Structure, Properties and Rheology of Bone," in *Orthopaedic Mechanics: Procedures and Devices*, D. N. Ghista and R. Roaf (eds.), Academic Press, New York, 1978.

(gum). The root is placed in a socket called the alveolus in the maxillary (upper) or mandibular (lower) bones. A sagittal cross section of a permanent tooth is shown in Figure 9-9 to illustrate various structural features. The enamel is the hardest substance found in the body and consists almost entirely of calcium phosphate salts (97%) in the form of large apatite crystals.

The dentin is another mineralized tissue whose distribution of organic matrix and mineral is similar to that of compact bone. Consequently, its physical

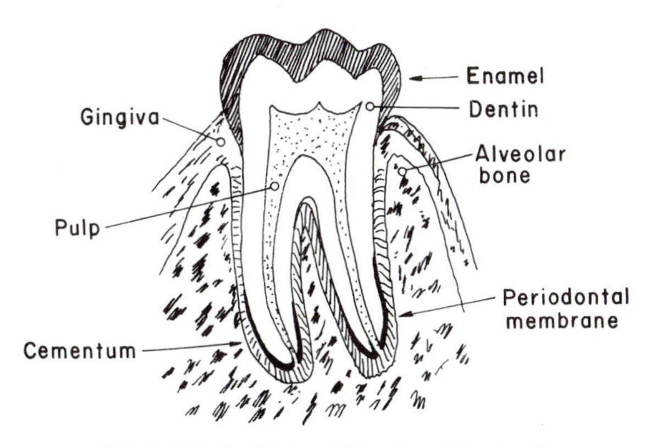

Figure 9-9. Sagittal section of a molar tooth.

Table 9-4. Physical Properties of Teeth

Tissue	Density (g/cm^3)	Modulus of elasticity (GPa)	Compressive strength (MPa)	Coefficient of thermal expansion (/°C)	Thermal conductivity (W/m K)
Enamel	2.2	48	241	11.4×10^{-6}	0.82
Dentin	1.9	13.8	138	8.3×10^{-6}	0.59

properties are also similar. The collagen matrix of the dentin might have a somewhat different molecular structure than normal bone: it is more cross-linked than that found in other tissues. Dentinal tubules (3–5 μm thick) radiate from the pulp cavity toward the periphery and penetrate every part of the dentin. The tubules contain collagen fibrils (2–4 μm thick) in the longitudinal direction and the interface is cemented by a protein–polysaccharide complex substance, as well as the processes of the odontoblasts, which are cells lining the pulp cavity.

Cementum covers most of the root of the tooth with a coarsely fibrillated bonelike substance that is devoid of canaliculi, Haversian systems, and blood vessels. The pulp occupies the cavity and contains thin collagenous fibers running in all directions and not aggregated into bundles. Ground substance, nerve cells, blood vessels, etc. are also contained in the pulp. The periodontal membrane anchors the root firmly into the alveolar bone and is made of mostly collagenous fibers plus glycoproteins (protein–polysaccharide complex).

Some of the physical properties of teeth are given in Table 9-4. As can be expected, the strength is the highest for enamel and dentin is between bone and enamel. The thermal expansion and conductivity are higher for enamel than for dentin.

Example 9-2

Calculate the volume percentage of each major component of a wet bone based on the weight percentage, i.e., 9, 69, and 22 wt% for water, mineral, and organic phase, respectively. Assume the densities of mineral and organic phase are 3.16 and 1.03 g/cm^3, respectively.

Answer

Based on 100 g of bone, the volume of each component can be calculated dividing the weight by its density:

Component	Wt%	Wt (g)	V (cm^3)	Vol%
Mineral	69	69	21.8	41.8
Organic	22	22	21.4	41.0
Water	9	9	9.0	17.2
Total	100	100	52.2	100.0

9.3.1.2. Mechanical Properties of Bone and Teeth. As with most other biological materials, the properties of bone and teeth depend substantially on the humidity, sign of load (compressive or tensile), rate of loading, and direction of the applied load with respect to the orientation of the microstructure. Therefore, one usually studies the effect of the above-mentioned factors and correlates the results with structural features. We will follow the same practice. Table 9-5 gives some general idea of the mechanical properties of various bones.

The *effect of drying* of the bone can be seen easily from Figure 9-10 where the dry sample shows slightly higher modulus of elasticity but lower toughness, fracture strength, and strain to failure. Thus, the wet bone in the laboratory, which behaves similarly to *in vivo* bone, can absorb more energy and elongate more before fracture.

The *effect of anisotropy* is expected since the osteons are longitudinally arranged along the long axis of the bone, and the load is mostly borne in that direction. The Young's modulus and the tensile and compressive strengths in the longitudinal direction are approximately 2 and 1.5 times higher than those in the radial or tangential directions, respectively. The tangential and radial directions differ little in their mechanical properties.

The *effect of the rate of loading* on the bone is shown in Figure 9-11. As can be seen, the Young's modulus, ultimate compressive and yield strength increase with increased rate of loading. However, the failure strain and the fracture toughness of the bone reach a maximum and then decrease. This implies that there is a critical rate of loading.

The *effect of mineral content* on the mechanical properties is given in Table 9-6. More highly mineralized bone has a higher modulus of elasticity and

Table 9-5. Properties of Bone[a]

Tissue	Direction of test	Modulus of elasticity (GPa)	Tensile strength (MPa)	Compressive strength (MPa)
Leg bones				
Femur	Longitudinal	17.2	121	167
Tibia	Longitudinal	18.1	140	159
Fibula	Longitudinal	18.6	146	123
Arm bones				
Humerus	Longitudinal	17.2	130	132
Radius	Longitudinal	18.6	149	114
Ulna	Longitudinal	18.0	148	117
Vertebrae				
Cervical	Longitudinal	0.23	3.1	10
Lumbar	Longitudinal	0.16	3.7	5
Spongy bone	Longitudinal	0.09	1.2	1.9
Skull	Tangential	—	25	—
	Radial	—	—	97

[a] From H. Yamada, *Strength of Biological Materials*, Williams and Wilkins, Baltimore, 1970.

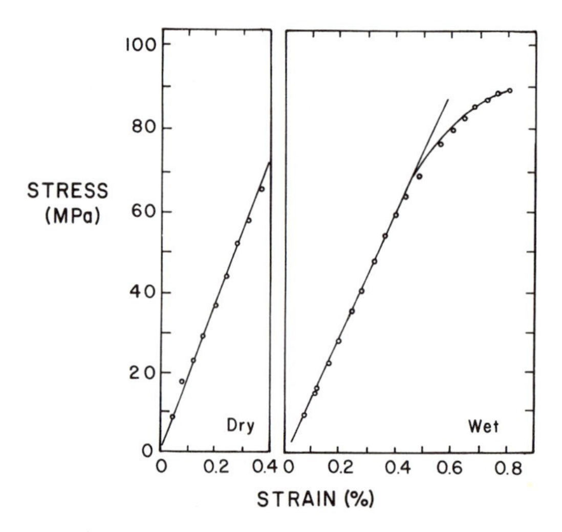

Figure 9-10. Effect of drying on the behavior of human compact bone. From F. G. Evans and M. Lebow, "The Strength of Human Compact Bone as Revealed by Engineering Technics," *Am. J. Surg.*, *83*, 326–331, 1952.

bending strength but lower toughness. This illustrates the importance of the organic phase in providing toughness and energy absorption capability in bone. Table 9-6 also illustrates the adaptation of bony tissue to different roles in biological systems. Antlers optimize toughness, while whale ear bones maximize stiffness; limb bones are intermediate.

The *effect of viscoelasticity* includes the loading rate effect described above. In viscoelastic materials, the stiffness depends on time under constant load; on rate of loading; and on frequency of (sinusoidal) loading. The viscoelastic properties of bone can be viewed conceptually in terms of mechanical models such as the one shown in Figure 9-12. The differential equation for the three-

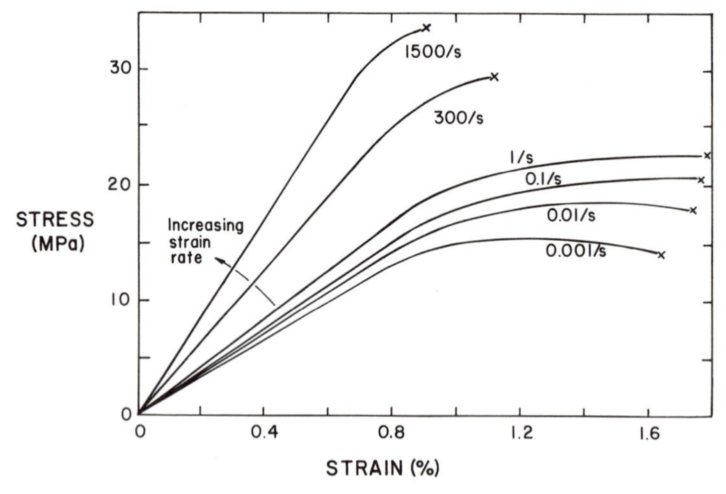

Figure 9-11. Stress as a function of strain and strain rate for human compact bone. From J. H. McElhaney, "Dynamic Response of Bone and Muscle Tissue," *J. Appl. Physiol.*, *21*, 1231–1236, 1966.

Table 9-6. Properties of Three Different Bones with Varying Mineral Contents[a]

Type of bone	Work of fracture (J/m^2)	Bending strength (MPa)	Young's modulus (GPa)	Mineral content (wt%)	Density (g/cm^3)
Deer antler	6190	179	7.4	59.3	1.86
Cow femur	1710	247	13.5	66.7	2.06
Whale tympanic bulla	200	33	31.3	86.4	2.47

[a] From J. D. Currey, "What Is Bone for? Property-Function Relationships in Bone," in *Mechanical Properties of Bone*, S. C. Cowin (ed.), ASME, AMD, New York, 1981, Vol. 45, pp. 13–26.

element model can be derived and be applied for various testing conditions (see Problem 9-5). In the case of bone, this is not particularly realistic unless only a small portion, less than a factor of 10, of the time/frequency range is considered. The measured effect of frequency of loading on the stiffness and dynamic mechanical damping of compact bone is shown in Figure 9-13. A simple spring-dashpot model such as that shown in Figure 9-12 would give substantial loss and modulus variation only over about a factor of 10 in time or frequency—a small part of the range shown in Figure 9-13. The stiffness increases with frequency (and hence also with rate of loading), while the damping (loss tangent) has a minimum at frequencies associated with walking, running, and other activities. Impact as well as ultrasonic waves used in medical diagnosis are absorbed in bone.

The *effect of density* (ρ in g/cm^3) and *strain rate* ($d\varepsilon/dt$ in sec^{-1}) on the compressive strength (σ_{ult} in MPa) of bone is summarized as

$$\sigma_{ult} = 68\rho^2(d\varepsilon/dt)^{0.06} \tag{9-2}$$

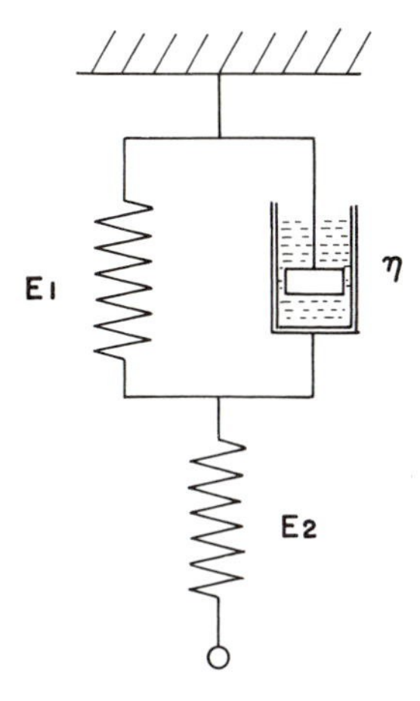

Figure 9-12. Three-element viscoelastic model of bone.

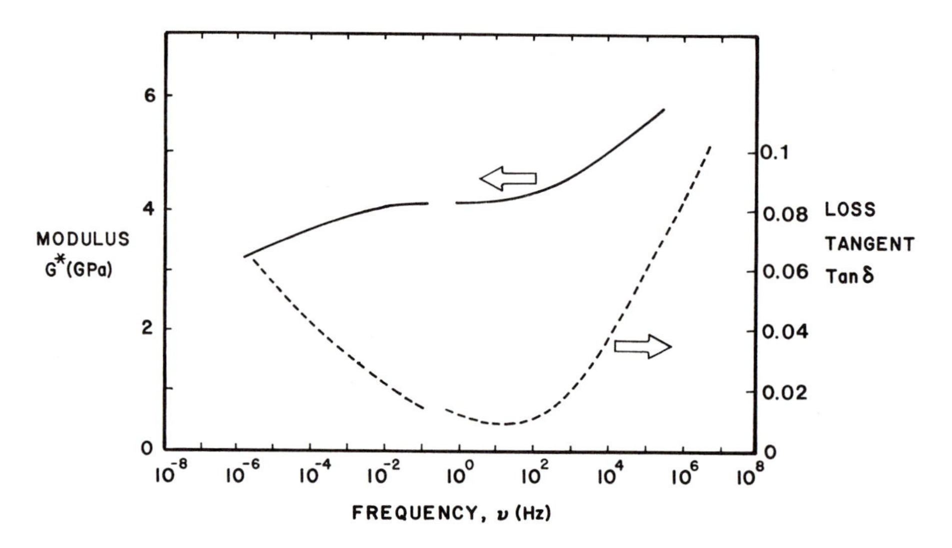

Figure 9-13. Loss tangent and effective shear modulus of compact bone (canine and human). More detailed description of the data can be found in R. S. Lakes, "Dynamic Study of Couple Stress Effects in Human Compact Bone," *J. Mech. Eng.*, *104*, 6–11, 1980.

This includes compact bone and, allowing some scatter in the comparison with experiment, trabecular bone as well. The strength of bone is seen to depend strongly on its density and weakly on strain rate as shown in Figure 9-14.

Example 9-3

Calculate the density of the mineral phase of dried cow femur (see Table 9-6) if the density of the organic phase and water is 1 g/cm³.

Answer

Using a simple mixture rule, neglecting water since the bone is dried:

$$\rho = \rho_1 V_1 + \rho_2 V_2 + \cdots + \rho_n V_n$$

$$V_1 + V_2 + \cdots + V_n = 1$$

From Example 9-2

$$V_1 = V_2 = 0.5$$

Therefore,

$$2.06 = 1 \times 0.5 + \rho_m \times 0.5$$

$$\rho_m = \underline{3.12 \text{ g/cm}^3}$$

This value is close to the value of hydroxyapatite mineral as given in Example 6-2.

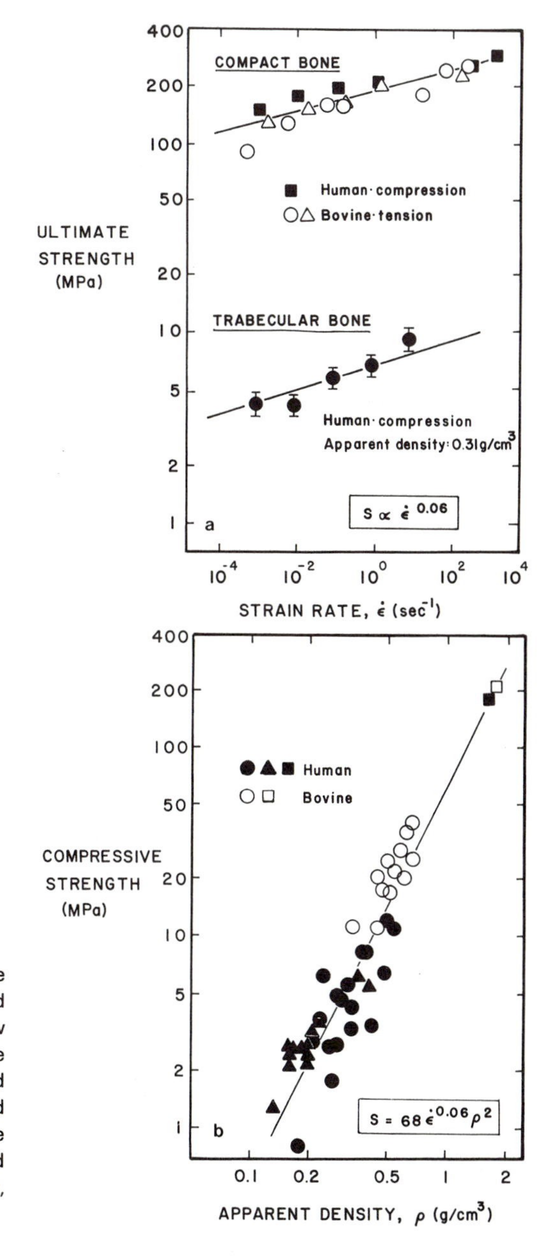

Figure 9-14. (A) Influence of strain rate on the ultimate strength of compact and trabecular bone tested without marrow *in vitro*. (B) Influence of density on the compressive strength of compact and trabecular bone. From D. R. Carter and W. C. Hayes, "Bone Compressive Strength: The Influence of Density and Strain Rate," *Science 194*, 1174–1176, 1976.

9.3.1.3. Modeling of Mechanical Properties of Bone. As mentioned earlier, bone is a composite material, and for that reason many researchers have proposed composite models based on two components, the mineral and the organic phase. In this section we present several simplified composite models. They are not intended as realistic descriptions of bone. If one assumes that the load is independently borne by the two components (collagen and mineral, hydroxyapatite) then the total load (P_t) is borne by mineral (P_m) and collagen (P_c):

$$P_t = P_m + P_c \tag{9-3}$$

Since $\sigma = P/A = E\varepsilon$, thus,

$$P_m = A_m \times E_m \times \varepsilon_m \tag{9-4}$$

where A, E, and ε are area, modulus, and strain, respectively. Suppose that the strain of collagen can be assumed to be equal to that of mineral, i.e.,

$$P_c = P_m \frac{A_c E_c}{A_m E_m} \tag{9-5}$$

therefore,

$$P_m = \frac{P_t A_m E_m}{A_m E_m + A_c E_c} \tag{9-6}$$

If we express Eq. (9-3) in terms of Young's modulus, it becomes

$$E_t = E_m V_m + E_c V_c \tag{9-7}$$

where V is volume fraction. This is the so-called "rule of mixtures" or Voigt model. It represents a rigorous upper bound on the stiffness of a composite. It also represents the actual stiffness of a composite with unidirectional, parallel fibers or laminae, for uniaxial stress along the fiber direction. It is only approximately correct for bone in the longitudinal direction. The same limitations apply to Eq. (9-3). If the fibers are arranged in the perpendicular direction, then one can derive the following equation:

$$\frac{1}{E_t} = \frac{V_m}{E_m} + \frac{V_c}{E_c} \tag{9-8}$$

Since not all collagenous fibers are exactly oriented in the same direction, one can propose another model:

$$\frac{1}{E_t} = \frac{x}{E_m V_m + E_c V_c} + (1 + x)\left(\frac{V_m}{E_m} + \frac{V_c}{E_c}\right) \tag{9-9}$$

where x is the fraction of bone that conforms to the parallel direction and $(1 - x)$ is the rest.

9.3.1.4. Electrical Properties of Bone. In the late 1950s it was shown that dry bone is piezoelectric in the classic sense, i.e., mechanical stress results in electric polarization, the indirect effect; and an applied electric field causes strain, the converse effect. (See the piezoelectric constitutive equations in Chapter 4.)

The piezoelectric properties of bone are of interest in view of their hypothesized role in bone remodeling. Wet collagen, however, does not exhibit piezoelectric response. Studies of the dielectric and piezoelectric properties of fully hydrated bone raise some doubt as to whether wet bone is piezoelectric at all at physiological frequencies. Piezoelectric effects occur in the kilohertz range, well above the range of physiologically significant frequencies. Both the dielectric properties and the piezoelectric properties of bone depend strongly on frequency. The magnitude of the piezoelectric sensitivity coefficients of bone depends on frequency, on direction of load, and on relative humidity. Values up to 0.7 pC/N have been observed in bone, to be compared with 0.7 and 2.3 pC/N for different directions in quartz, and 600 pC/N in some piezoelectric ceramics. It is, however, uncertain whether bone is piezoelectric in the classic sense at the relatively low frequencies that dominate in the normal loading of bone. It has been suggested that two different mechanisms are responsible for these effects: classical piezoelectricity resulting from the molecular asymmetry of collagen in dry bone, and fluid flow effects, possibly streaming potentials in wet bone. Streaming potentials are electrical signals resulting from flow of ionic electrolytes through channels in bone (canaliculi, Haversian canals) by the deformation of bones.

Bone exhibits additional electrical properties that are of interest. For example, the dielectric behavior (e.g., the dynamic complex permittivity, the real part of which governs the capacitance and the imaginary part of which governs the resistivity) controls the relationship between the applied electric field and the resulting electric polarization and current as discussed in Chapter 4. Dielectric permittivity of bone has been found to increase dramatically with increasing humidity and decreasing frequency. For bone under partial hydration conditions, the dielectric permittivity (which determines the capacitance) can exceed 1000 and the dielectric loss tangent (which determines the ratio of conductivity to capacitance) can exceed unity. Both the permittivity and the loss are greater if the electric field is aligned parallel to the bone axis. Bone under conditions of full hydration in saline behaves differently: the behavior of bovine femoral bone is essentially resistive, with very little relaxation. The resistivity is about 45–48 Ω-m for the longitudinal direction, and three to four times greater in the radial direction. These values are to be compared with a resistivity of 0.72 Ω-m for physiological saline alone. Since the resistivity of fully hydrated bone is about 100 times greater than that of bone under 98% relative humidity, it is suggested that at 98% humidity the larger pores are not fully filled with fluid.

Compact bone also exhibits a permanent electric polarization as well as pyroelectricity, which is a change of polarization with temperature. These phenomena are attributed to the polar structure of the collagen molecule; these molecules are oriented in bone. The orientation of permanent polarization has been mapped in various bones and has been correlated with developmental events.

Table 9-7. Electrical Properties of Bone

Property	Condition	Value	Ref.
Dielectric permittivity	Radial, 78% rh, 37°C, 1 Hz	10^5	*a*
Resistivity (Ω-m)	Longitudinal, 100% hydration, 0.1–30 sec	45	*b*
	Radial, 100% hydration, 0.1–30 sec	150	
Piezoelectric coefficients (pC/N)	75% rh, 23.5°C, 100 Hz	$d_{11} = 0.014$	*c*
		$d_{12} = 0.026$	
		$d_{13} = -0.032$	
		$d_{14} = 0.105$	
		$d_{15} = -0.013$	
		$d_{16} = -0.070$	

[a] Data from R. S. Lakes, R. A. Harper, and J. L. Katz, "Dielectric Relaxation in Cortical Bone," *J. Appl. Phys.*, *48*, 808–811, 1977.
[b] Data from D. A. Chakkalakal, M. W. Johnson, R. A. Harper, and J. L. Katz, "Dielectric Properties of Fluid-Saturated Bone," *IEEE Trans. Biomed. Eng.*, *BME-27*, 95–100, 1980.
[c] Data from A. J. Bur, "Measurements of the Dynamic Piezoelectric Properties of Bone as a Function of Temperature and Humidity," *J. Biomech.*, *9*, 495–507, 1976.

Electrical properties of bone are relevant not only as a hypothesized feedback mechanism for bone remodeling (see the following section), but also in the context of external electrical stimulation of bone to aid its healing and repair.

A short summary of the electrical properties of bone is given in Table 9-7.

9.3.2. Bone Remodeling

9.3.2.1. Phenomenology. The relationship between the mass and form of a bone to the forces applied to it was appreciated by Galileo, who is credited with being the first to understand the balance of forces in beam bending and with applying this understanding to the mechanical analysis of bone. Julius Wolff published his seminal 1892 monograph on bone remodeling; the observation that bone is reshaped in response to the forces acting on it is presently referred to as *Wolff's law.* Indirect evidence was presented in the drawing given in Figure 9-15 in which Wolff emphasized that the remodeling of cancellous bone structure followed mathematical rules corresponding to the principal stress trajectories. Many relevant observations regarding the phenomenology of bone remodeling have been suggested:

1. Remodeling is triggered not by principal stress but by "flexure."
2. Repetitive dynamic loads on bone trigger remodeling; static loads do not.
3. Dynamic flexure causes all affected bone surfaces to drift toward the concavity that arises during the act of dynamic flexure.

These rules are essentially qualitative and they do not deal with underlying causes. Additional aspects of bone remodeling may be found in the clinical

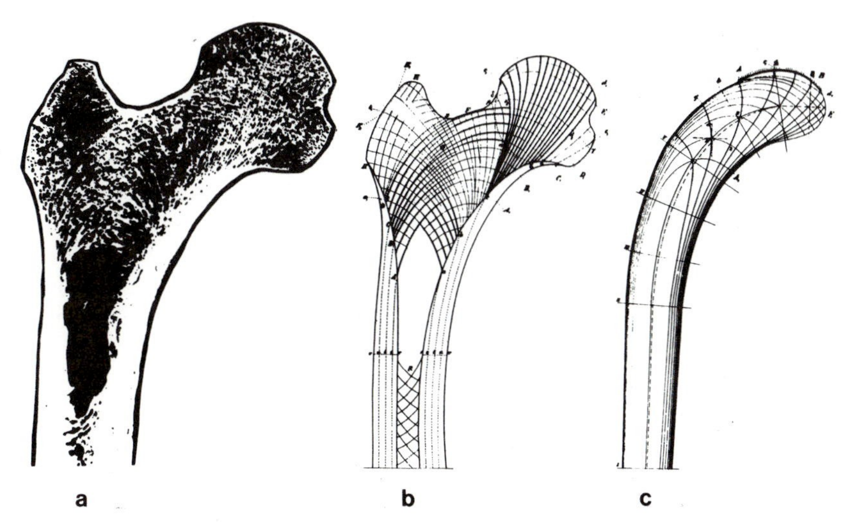

Figure 9-15. Frontal section through the proximal end of the femur (a); line drawing of the cancellous structure (b); Culmann's crane (c). From J. Wolff, *Das Gesetz der Transformation der Knochen*, A. Hirchwald, Berlin, 1892.

literature. For example, after complete removal of a metacarpal and its replacement with graft consisting of a strut of tibial bone, the graft becomes remodeled to resemble a real metacarpal; the graft continues to function after 52 years. In the standards of the Swiss Association for Internal Fixation, it is pointed out that severe osteoporosis can result from the use of two bone plates in the same region, as a result of the greatly reduced stress in the bone. It is suggested that as a result of bending stresses in the femur the medial and lateral aspects should be stiffer and stronger than the anterior and posterior aspects. Such a difference has actually been observed. Large cyclic stress causes more remodeling of bone than large static stress. Immobilization of humans causes loss of bone and excretion of calcium and phosphorus. Long spaceflights under zero gravity also cause loss of bone; hypergravity induced by centrifugation strengthens the bones of rats. Studies of stress-induced remodeling of living bone have been performed *in vivo* in pigs. In this study, strains were directly measured by strain gages before and after remodeling. Remodeling was induced by removing part of the pigs' ulna so that the radius bore all of the load. Initially, the peak strain in the ulna approximately doubled. New bone was added until, after 3 months, the peak strain was about the same as on the normal leg bones. *In vivo* experiments conducted in sheep have disclosed similar results. It is of interest to compare the response time noted in the above experiments with the rate of bone turnover in healthy humans. The life expectancy of an individual osteon in a normal 45-year-old man is 15 years and it will have taken 100 days to produce it.

Remodeling of Haversian bone seems to influence the quantity of bone but not its quality, i.e., Young's modulus, tensile strength, and composition are not substantially changed. However, the initial remodeling of primary bone to produce

Haversian bone results in a reduction in strength. As for the influence of the rate of loading on bone remodeling, there is good evidence to suggest that intermittent deformation can produce a marked adaptive response in bone, whereas static deformation has little effect. Experiments on rabbit tibiae bear this out. In the dental field, by contrast, it is accepted that static forces of long duration move teeth in the jawbone. In this connection, the direction (as well as the type) of stresses acting on the (alveolar) bone tissue should also be considered. The forces may be static but the actual loading on the alveolar bone would be dynamic owing to the mastication forces on the jawbones. Some have pointed out that the response of different bones in the same skeleton to mechanical loads must differ, otherwise lightly loaded bones such as the top of the human skull, or the auditory ossicles, would be resorbed.

Failure of bone remodeling to occur normally in certain disease states is of interest: for example, micropetrotic bone contains few if any viable osteocytes and usually contains a much larger number of microscopic cracks than adjacent living bone. This suggests that the osteocytes play a role in detecting and repairing the damage. In senile osteoporosis, bone tissue is removed by the body, often to such an extent that fractures occur during normal activities. Osteoporosis may be referred to as a remodeling error.

9.3.2.2. Feedback Mechanisms of Bone Remodeling. Bone remodeling appears to be governed by a feedback system in which the bone cells sense the state of strain in the bone matrix around them and either add or remove bone as needed to maintain the strain within normal limits (see Figure 13-1). The process or processes by which the cells are able to sense the strain and the important aspects of the strain field are unknown. It is suggested that bone is piezoelectric, i.e., that it generates electric fields in response to mechanical stress and it is thought that the piezoelectric effect is the part of the feedback loop by which the cells sense the strain field. This hypothesis obtained support from observations of osteogenesis in response to externally applied electric fields of the same order of magnitude as those generated naturally by stress via the piezoelectric effect. The study of bone bioelectricity has received impetus from observations that externally applied electric or electromagnetic fields stimulate bone growth. The electrical hypothesis, while favored by many, has not been proven. Indeed, other investigators have advanced competing hypotheses that involve other mechanisms by which the cells are informed of the state of stress around them. Such processes may include slip between lamellae impinging on osteocyte processes; slow slip at cement lines between osteons that would align the osteons to the stress; stress-induced fluid flow in channels such as canaliculi to stimulate or nourish the osteocytes; a direct hydrostatic pressure effect on osteocytes; or an effect of stress on the crystallization kinetics of the mineral phase.

Example 9-4

Calculate using a simple rule of mixture model the percentage of load borne by the mineral phase of a cow femur that is subjected to 500 N of load. The Young's moduli of mineral and collagen are about 17 and 0.1 GPa, respectively.

Answer

Since the area is proportional to the volume percentage of each component, hence from Example 9-2 and Eq. (9-6)

$$\frac{P_m}{P_i} = \frac{0.44 \times 17}{0.44 \times 17 + 0.40 \times 0.1} = \underline{0.9947}$$

so that 99.47% of the load is borne by the mineral phase. Actually the strength of demineralized bone is about 5–10% of that of whole bone. The rule of mixtures (Voigt model) represents an upper bound on the modulus of a composite. It is appropriate for a composite with fibers or laminae oriented in the direction of the applied load. The structure of bone is considerably more complex than these, so that the analysis is an approximation.

9.3.3. Collagen-Rich Tissues

Collagen-rich tissues function mostly in a load-bearing capacity. These tissues include skin, tendon, cartilage, etc. Special functions such as transparency for the lens of the eye and shaping of the ear, tip of nose, etc. can be carried out also by the collagenous tissues.

9.3.3.1. Composition and Structure. The collagen-rich tissues are mostly made up of collagen (over 75 wt% dry, see Table 9-8). The collagen is made of tropocollagen, a three-chain coiled superhelical molecule (Figure 9-2). The collagen fibrils aggregate to form fibers as shown in Figure 9-16. The fibrils and fibers are stabilized through intra- and intermolecular hydrogen bonding (C=O - - - HN; see Figure 9-1).

The physical properties of tissues vary according to the amount and structural variations of collagen fibers. Several different fiber arrangements are found in tissues: parallel fibers, crossed fibrillar arrays, and feltlike structures in which the fibers are more randomly arranged with some fibers going through the thickness of the tissue.

9.3.3.2. Physical Properties. The collagen-rich tissues can be thought of as a polymeric material in which the highly oriented crystalline collagen fibers are embedded in the ground substance of mucopolysaccharides and amorphous elastin (a rubberlike biopolymer). When the tissue is heated in the laboratory,

Table 9-8. Composition of Collagen-Rich Soft Tissues

Component	Composition (%)
Collagen	75 (dry), 30 (wet)
Mucopolysaccharides	20 (dry)
Elastin	<5 (dry)
Water	60–70

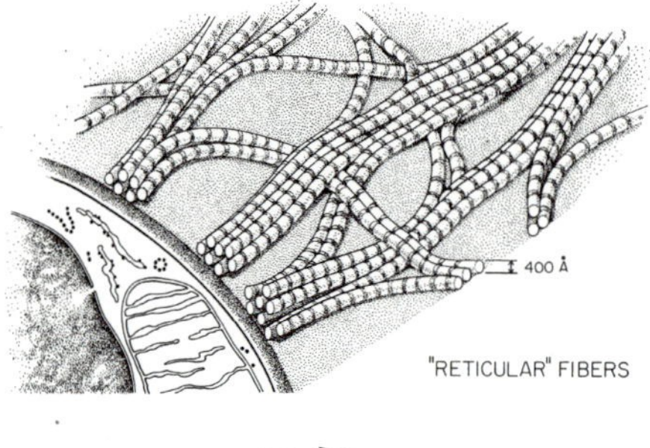

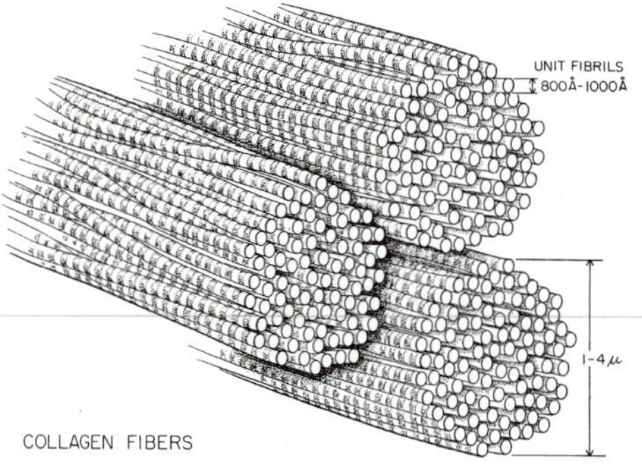

Figure 9-16. Diagram showing that the "reticular fibers" associated with the basal lamina of an epithelial cell (above) and the "collagen fibers" of the connective tissue in general (below) are both composed of unit fibrils of collagen. Those of the reticulum are somewhat smaller and interwoven in loose networks instead of in larger bundles. From W. Bloom and D. W. Fawcett, *A Textbook of Histology*, 9th ed., W. B. Saunders Co., Philadelphia, 1968.

its specific volume increases (density decreases, exhibiting glass transition temperature, $T_g \sim 40°C$ and shrinkage, $T_s \sim 56°C$). The shrinkage temperature is considered a denaturation point for collagen.

The stress-strain curves of a collagenous structure such as tendon exhibit nonlinear behavior as shown in Figure 9-17. Similar behavior is observed in synthetic fibers. The initial toe region represents alignment of fibers in the direction of stress, the steep rise in slope represents the majority of fibers stretched along their long axes. As for the later decrease in slope, individual fibers may be breaking prior to the final catastrophic failure. The highest slope of the stress-strain curve is about 1.0 GPa, which is close to the modulus of individual collagen

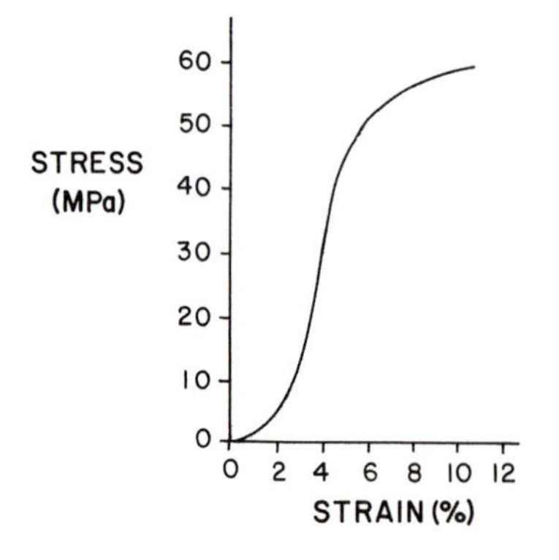

Figure 9-17. A typical stress–strain curve for tendon. From B. J. Rigby, N. Hiraci, J. D. Spikes, and H. Eyring, "The Mechanical Properties of Rat Tail Tendon," *J. Gen. Physiol.*, *43*, 265–283, 1959.

Table 9-9. Elastic Properties of Elastic and Collagen Fibers

Fibers	Modulus of elasticity (MPa)	Tensile strength (MPa)	Ultimate elongation (%)
Elastic	0.6	1	100
Collagen	1000	50–100	10

fibers. The tensile strength is, however, much lower than that of the individual fibers. Table 9-9 lists some mechanical properties of collagen and elastic fibers. The modulus of elasticity of collagen increases with the rate of loading due to the viscoelastic properties as in the case of bone discussed earlier.

Unlike tendon or ligament, skin has a feltlike structure consisting of continuous fibers that are randomly arranged in layers or lamellae. The skin tissues also show mechanical anisotropy as shown in Figure 9-18. This is the reason behind the Langer's lines shown in Figure 9-19. The Langer's lines are determined by puncturing small circular holes on the skin of a cadaver, and following the elongated holes (which become linelike) due to the anisotropy of the skin structure as shown in Figure 9-19. Surgeons use knowledge of this anisotropy in making incisions so that better wound healing is achieved.

Another feature of the stress–strain curve of the skin is its extensibility under a small load. This compliance is observed despite the high content of collagen, a relatively stiff material, in skin. The reason is that at small extension, the fibers are straightened and aligned rather than stretched. Upon further stretching the

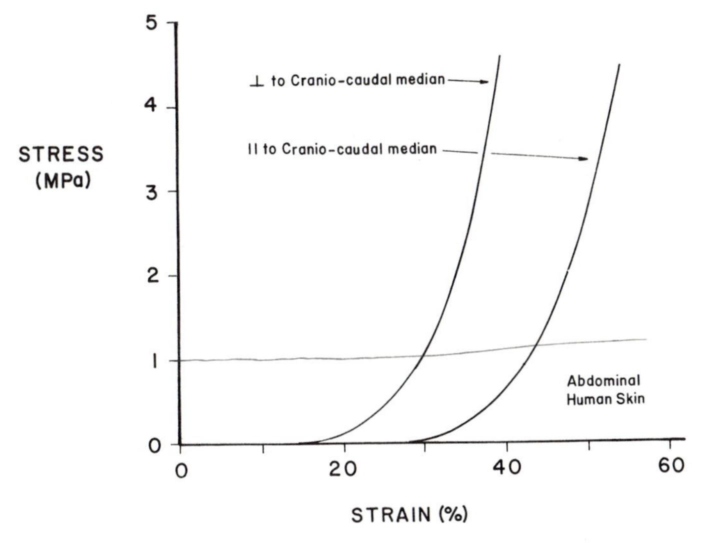

Figure 9-18. Stress–strain curves of human abdominal skin, which shows more extensibility in the direction of the main axis of the body than across the abdomen. From C. H. Daly, The Biomechanical Characteristics of Human Skin, Ph.D. thesis, University of Strathclyde, Scotland, 1966.

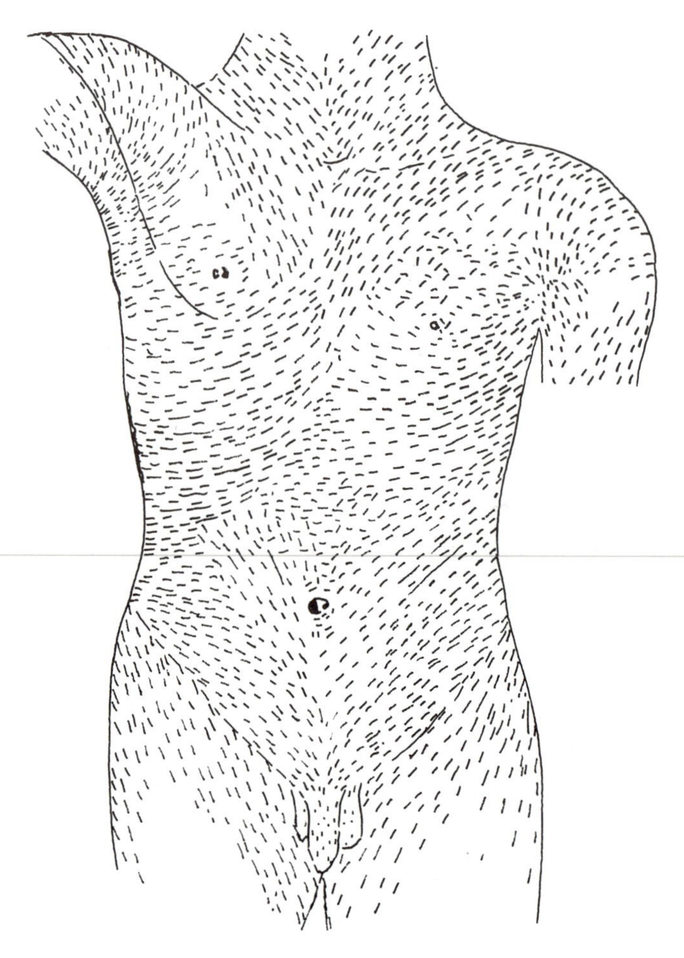

Figure 9-19. Distribution of Langer's lines produced by a conical stabbing instrument. From G. L. Wilkes, I. A. Brown, and R. H. Wildnauer, "The Biomechanical Properties of Skin," *CRC Crit. Rev. Bioeng.* *1*(4), 453–495, 1973.

fibrous lamellae align with respect to each other and resist further extension as shown in Figure 9-20. When the skin is highly stretched, the modulus of elasticity approaches that of tendon.

Cartilage is collagen-rich tissue that has two main physiological functions. One is the maintenance of shape (ear, tip of nose, and rings around the trachea), and the other is to provide bearing surfaces at joints. It contains very large and diffuse protein–polysaccharide molecules that form a gel in which the collagen-rich molecules are entangled (see Figure 9-6). They can affect the mechanical properties of the cartilage by hindering the movements through the interstices of the collagenous matrix network.

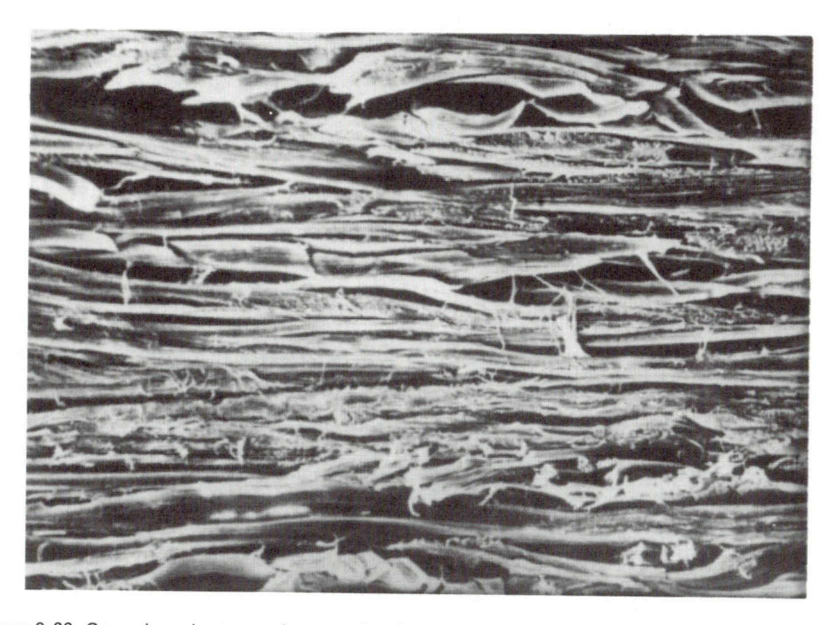

Figure 9-20. Scanning electron microscopic pictures of dermal skin before and after stretching. Stretching direction is horizontal. ×400. From G. H. Wilkes, I. A. Brown, and R. H. Wildnauer, "The Biomechanical Properties of Skin," *CRC Crit. Rev. Bioeng.* *1*(4), 453–495, 1973.

The joint cartilage has a very low coefficient of friction (<0.01). This is largely attributed to the squeeze-film effect between cartilage and synovial fluid. The synovial fluid can be squeezed out through highly fenestrated cartilage upon compressive loading while the reverse action will take place in tension. The lubricating function is carried out in conjunction with mucopolysaccharides, especially chondroitin sulfates as discussed in Section 9.2.2. The modulus of elasticity (10.3–20.7 MPa) and tensile strength (3.4 MPa) are quite low. However, wherever high stress is required, the cartilage is replaced by purely collagenous tissue.

Example 9-5

Estimate the wet% compositions of mucopolysaccharides (MPS) and elastin in Table 9-8 assuming the densities of collagen, MPS, and elastin are about 1 g/cm^3 and water content is about 65%.

Answer

Based on 100 g of wet tissue:

Composition	Weight (g)	Dry (g)
Collagen	30	35 g × 0.75 = 26.25
MPS	x	35 g × 0.20 = 7.00
Elastin	y	35 g × 0.05 = 1.75
Water	65	0

$x + y = 5$ g

Solving simultaneously yields:

$$x/y = 7/1.75, \qquad x = 4\text{ g and } y = 1\text{ g, i.e., MPS: } \underline{4\%}\text{, elastin: } \underline{1\%}$$

This indicates that a very small amount of elastin exists in the collagen-rich tissues.

9.3.4. Elastic Tissues

The elastic tissues are compliant and therefore undergo a large deformation in response to a small load. These tissues include blood vessels, ligamentum nuchae, muscles, etc.

9.3.4.1. Composition and Structure. Elastic tissues contain a relatively large amount of elastin, which is sometimes called "protein rubber." For example, ligamentum nuchae contains 80 wt% (dry) elastin. One of the most important elastic tissues is the blood vessel wall, which has three distinct layers when viewed in cross section (Figure 9-21): (1) *intima*, whose structural elements are oriented longitudinally; (2) *media*, which is the thickest layer of the wall and whose

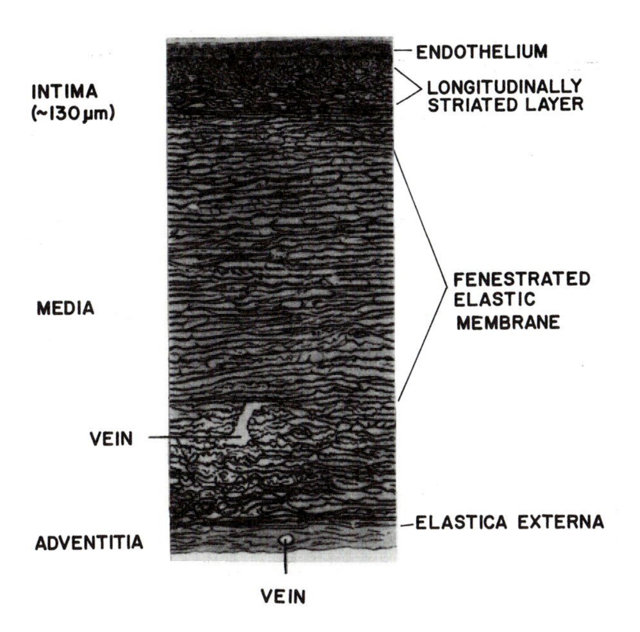

Figure 9-21. Structure of a blood vessel wall. From W. Bloom and D. W. Fawcett, *A Textbook of Histology*, 9th ed., W. B. Saunders Co., Philadelphia, 1968.

components are arranged circumferentially; and (3) *adventitia*, which connects the vessels firmly to surrounding tissue via fascia. The intima and media are fenestrated by the internal elastic membrane (*elastica interna*), which is predominant in arteries of medium size. Between the media and adventitia, a thinner external elastic membrane (*elastica externa*) can be found. The smooth muscle cells are found between adjacent elastic lamellae in helical array.

9.3.4.2. Properties of Elastic Tissues. The mechanical properties of elastic tissues such as elastin are similar to those of rubber. Chains in elastin are cross-linked by desmosine, isodesmosine, and lysinonorleucine as mentioned before. Figure 9-22 shows a stress–strain curve of bovine ligamentum nuchae at low extension. As can be seen, the modulus of elasticity and the amount of stored energy lost upon releasing the load are quite low. This is characteristic of elastic tissues whose primary function is the restoration of a deformed state to the original shape with minimum energy loss. Therefore, the relative amount of elastin varies along the blood vessel walls, the largest of which is the arch of the aorta, which functions as a "secondary pump" where the expelled blood from the heart is temporarily stored. The relative amount of elastin decreases with decreasing size of the vessel as shown in Figure 9-23.

As in the case of the skin, the anisotropy of mechanical properties is prominent in the longitudinal and circumferential direction of the blood vessel as shown in Figure 9-24. Note that the composition of the vessel walls changes along the length of the wall, and consequently, their physical properties also

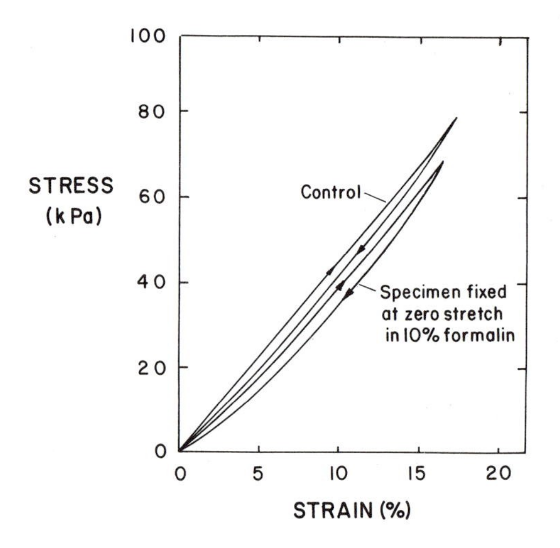

Figure 9-22. The stress–strain curve of elastin. The material is the ligamentum nuchae of cattle, which contains a small amount of collagen that was denatured by heating at 100°C for an hour. Such heating does not change the mechanical properties of elastin. The specimen is cylindrical with a rectangular cross section. Loading is uniaxial. The curve labeled "control" refers to native elastin. The curve labeled "10% formalin" refers to specimen fixed in formalin solution for a week without initial strain. From Y. C. Fung, *Biomechanics: Mechanical Properties of Living Tissues,* Springer-Verlag, Berlin, 1981.

change (see Figure 9-23). Another complicating factor describing mechanical properties is the existence of involuntary smooth muscle that is associated with arterial blood pressure regulation.

Table 9-10 illustrates that the mean pressure of the various blood vessels and the approximate tension developed at normal pressure are related as given

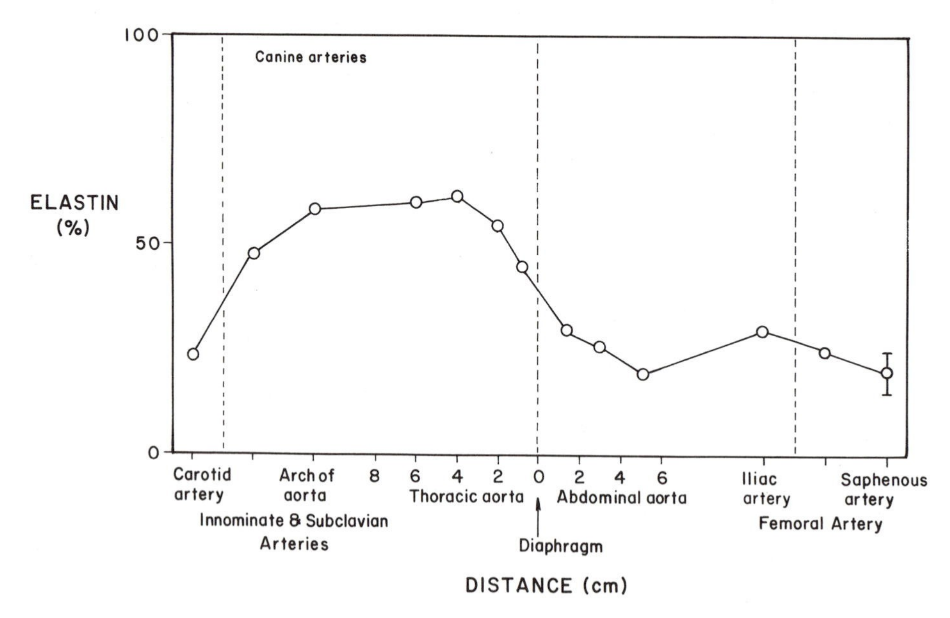

Figure 9-23. Variation of elastin percent per combined elastin and collagen content along the major arterial tree. From A. C. Burton, *Physiology and Biophysics of Circulation,* Year Book Medical Publ., Inc., Chicago, 1965.

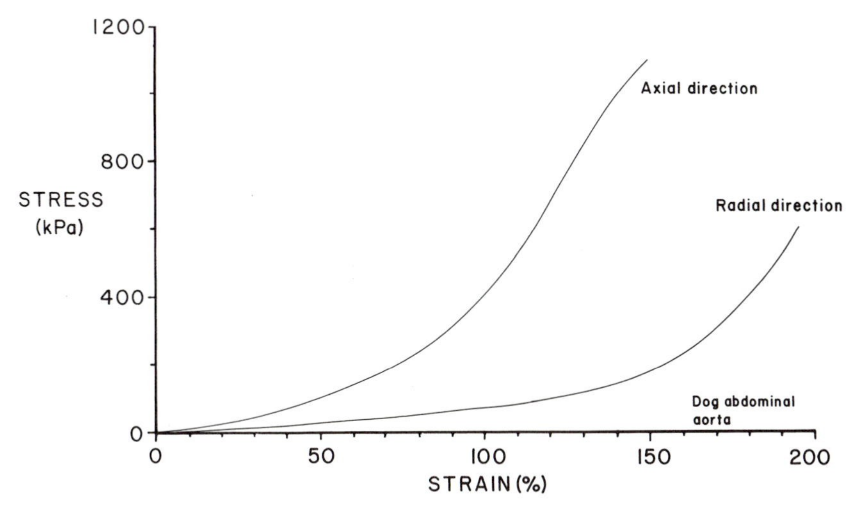

Figure 9-24. Stress–strain curves of human artery in the longitudinal and circumferential direction of the vessel (J. B. Park, unpublished data).

in the Laplace equation,

$$T = P_i \times r \tag{9-10}$$

where T is wall tension, P_i is internal pressure, and r is the radius of the vessel. This equation is usually applied to a uniform, thin, isotropic tube in the absence of longitudinal tension. It is obvious that none of these assumptions can be met strictly by the blood vessels.

Muscle is another elastic tissue. It is not discussed in detail in this book because of the "active" nature of the tissue the importance of which is more obvious in conjunction with physiological function. As passive tissue the stress-strain curve shows nonlinear, viscoelastic behavior (Figure 9-25). Table 9-11 gives mechanical properties of some of the nonmineralized tissues for comparison.

Table 9-10. Wall Tension and Pressure Relationship of Various Sizes of Blood Vessels[a]

Type of vessel	Mean pressure (mm Hg)	Internal pressure (dynes/cm^2)	Radius	Wall tension (dynes/cm)
Aorta, large artery	100	1.5×10^5	1.3 cm	170,000
Small artery	90	1.2×10^5	0.5 cm	60,000
Arteriole	60	0.8×10^5	62–150 μm	500–1200
Capillaries	30	0.4×10^5	4 μm	16
Venules	20	0.26×10^5	10 μm	26
Veins	15	0.2×10^5	200 μm	400
Vena cava	10	0.13×10^5	1.6 cm	21,000

[a] From A. C. Burton, *Physiology and Biophysics of Circulation*, Year Book Medical Publ., Inc., Chicago, 1965.

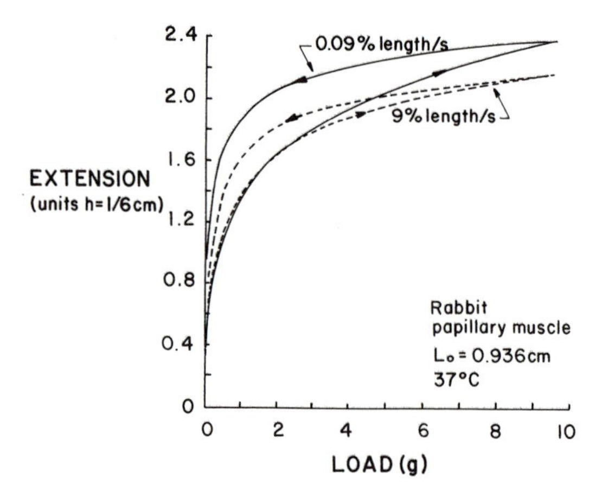

Figure 9-25. The length–tension curve of a resting papillary muscle from the right ventricle of the rabbit. Hysteresis curves at strain rates 0.09, 0.9, and 9% length/sec. Length at 9 mg = 0.936 cm at 37°C. From Y. C. Fung, *Biomechanics: Mechanical Properties of Living Tissues,* Springer-Verlag, Berlin, 1981.

Table 9-11. Mechanical Properties of Some of the Nonmineralized Human Tissues

Tissues	Tensile strength (MPa)	Ultimate elongation (%)
Skin	7.6	78
Tendon	53	9.4
Elastic cartilage	3	30
Heart valves (aortic)		
Radial	0.45	15.3
Circumferential	2.6	10.0
Aorta		
Transverse	1.1	77
Longitudinal	0.07	81
Cardiac muscle	0.11	63.8

Another important consideration for the physical properties of tissues is that any disturbance of the specimen may change its constituents, especially unbound water. For example, when a soft tissue specimen such as a segment of aorta is stretched in the laboratory, the unbound "free water" is squeezed out of the specimen since the laboratory preparation is an open system. The role of water with regard to the mechanical properties of tissues is not yet fully understood.

Example 9-6

From Figure 9-18, answer:

(a) How much stress will be developed if the abdominal skin is stretched 30% in the parallel and in the perpendicular direction to the cephalocaudal direction of the body?
(b) What are the strains if the skin is stressed to 1 MPa?
(c) What are the moduli of elasticity in the two principal directions?

Answer

(a) From Figure 9-18, the stresses in the perpendicular and parallel to the cephalocaudal direction of the main body are about 1 and 0.01 MPa, respectively.

(b) 30% and 43% strain will be developed in the perpendicular and parallel to the cephalocaudal direction of the body, respectively.
(c) Perpendicular direction: $E = (5.2 - 0)/(0.41 - 0.31) = \underline{52\ \text{MPa}}$. Parallel direction: $E = (4.6 - 0)/(0.54 - 0.44) = \underline{46\ \text{MPa}}$.

9.3.4.3. Further Considerations of the Mechanical Properties of Soft Tissues. As with other viscoelastic materials such as polymers, soft tissues also exhibit similar behavior that can be represented with traditional multielement Voigt and Maxwell models in series or in parallel fashion. Although this type of analysis gives a very useful handle on the mechanical behavior of tissues, it does not explicitly explain the relative importance of constitutive components in tissues. One might, therefore, try to understand the overall behavior of tissues by modeling the structure, as the example given in Figure 9-26. This model shows that the blood vessel is made up of three major components: smooth muscle fibers arranged in helical fashion with very short pitch; elastic fibers of elastin; and collagen forming a crimped network structure that can be extended at high load.

Some researchers have explored the soft tissue behavior based on enzymolysis experiments in which a particular component is removed by an appropriate enzyme. Typical stress–strain curves obtained after removal of each major component are given in Figure 9-27. One can see that the ligamentum nuchae shows rubberlike elasticity up to 50% elongation. However, if one removes the collagen component from the tissue by an enzyme (collagenase) or autoclaving, it behaves entirely like an elastomer up to 100% elongation. Conversely, if one removes the elastin component from the same tissue, then the remaining tissue behaves as collagenous tissue except the curve is shifted toward high extension, which used to be taken up by the elastin. The removal of ground substance did not alter the basic stress–strain behavior. From these experiments one can deduce a simple model as shown in Figure 9-28.

In this model the elastic fibers (elastin) are stretched followed by the much stronger collagen fibers when two components are pulled together. With a little more imagination, one can envision the distribution of fiber length and thus the response can be smoother than the two-fiber system as shown in Figure 9-28. From this kind of study relating the microstructure to the macrobehavior, one

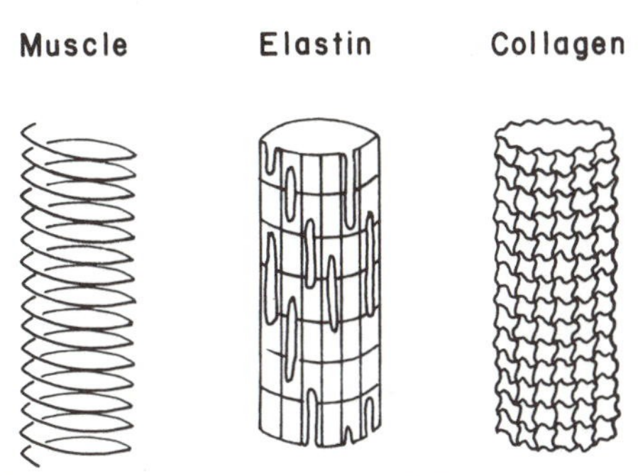

Figure 9-26. A sketch of the structure of the vein. From T. Azuma and M. Hasegawa, "Distensibility of the Vein: From the Architectural Point of View," *Biorheology, 10,* 469–479, 1973.

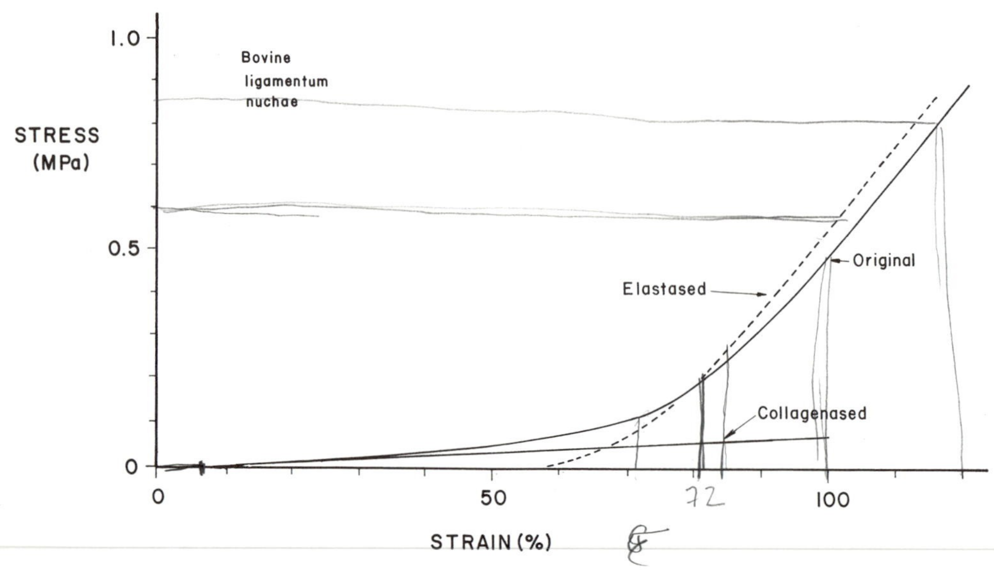

Figure 9-27. Stress–strain curves of bovine ligamentum nuchae after elastase and collagenase treatments. From A. S. Hoffman, L. A. Grande, P. Gibson, J. B. Park, C. H. Daly, and R. Ross, "Preliminary Studies on the Mechanochemical–Structure Relationships in Connective Tissues Using Enzymolysis Techniques," in *Perspectives of Biomedical Engineering*, R. M. Kenedi (ed.), University Park Press, Baltimore, 1972, pp. 173–176.

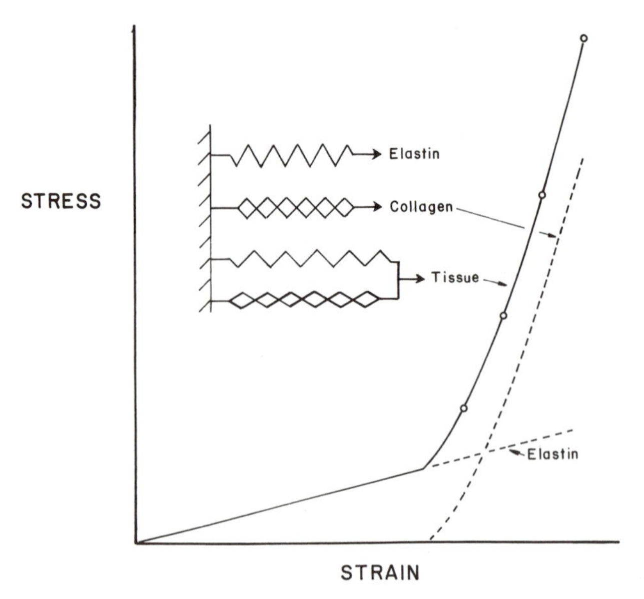

Figure 9-28. A schematic representation of structure–property relationship of connective tissues. The collagen is represented by a loosely knitted fabric.

can understand fully the nature of interaction between components in a connective tissue.

Example 9-7

Calculate the modulus of elasticity of the bovine ligamentum nuchae (Figure 9-27).

Answer

There are two distinct regions demarcated at about 60–70% strain:
Initial region

$$E = \frac{\sigma}{\varepsilon} = \frac{0.06\ \text{MPa}}{100\%} = \underline{0.06\ \text{MPa}}$$

Secondary region

$$E = \frac{0.85 - 0\ \text{MPa}}{120\% - 72\%} = \underline{1.77\ \text{MPa}}$$

These two values are lower and higher than the single value given in Table 9-9 for the elastic fibers.

PROBLEMS

9-1. Calculate x of Eq. (9-9) for bone with Young's modulus of 17 GPa by assuming $V_m = V_c$, $E_m = 100$ GPa, $E_c = 0.1$ GPa.

9-2. The force versus displacement curve shown below was obtained by tensile testing of canine skin. The skin specimen was cut by using a stamping machine that has a width of 4 mm. The thickness of the skin is about 3 mm and the length of the sample between the grips is 20 mm.

(a) What is the tensile strength and fracture strain of the skin?
(b) What is the modulus of elasticity in the initial and secondary regions?
(c) What is the toughness of the skin?

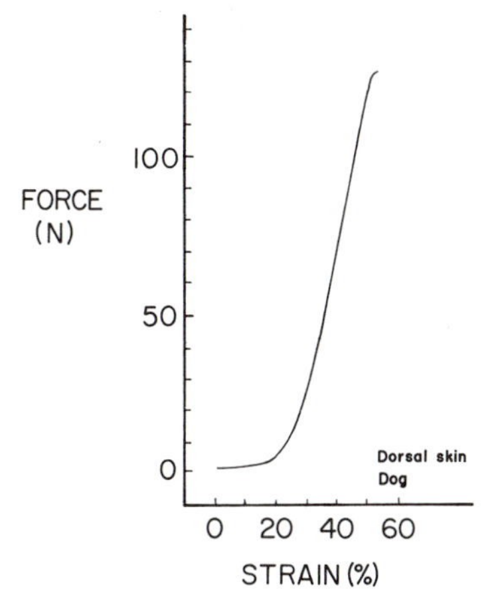

9-3. Ligamentum nuchae is made of elastin and collagen (others do not contribute toward mechanical properties). The relative amounts excluding water are 70, 25, and 5 wt% for elastin, collagen, and others, respectively. Assume they are homogeneously distributed.

(a) Calculate the percent contribution of the elastin toward the total strength assuming elastic behavior.
(b) Compare the result with Figure 9-26.

9-4. If a bone contains 18.9% porosity, what would be its modulus of elasticity in the absence of pores. Compare with the modulus of hydroxyapatite.

9-5. The viscoelastic properties of compact bone have been described by using the three-element model shown in Figure 9-12.

(a) Derive a differential equation for the model.
(b) Solve part a for γ (strain) for stress relaxation tests.
(c) Sketch the relaxation on a log scale similar to the one used in Figure 9-13.
(d) What are the shortcomings of the model?

SYMBOLS/DEFINITIONS

Greek Letters

ε	Strain.
ρ	Density.
σ	Stress.

Latin Letters

A	Area.
C	Coulomb, a measure of electric charge. 1 coulomb = 1 ampere × 1 second.
E	Young's modulus, a measure of stiffness.
P	Load (force) or pressure (force per area).
r	Radius.
T	Wall tension.
V	Volume fraction.

Terms

Aspartic acid (*Asp*):	One of the essential amino acids.
Canaliculi:	Small channels (~0.3 μm diameter) radiating from the lacunae in bone tissue.
Cementum:	Calcified tissue of mesodermal origin covering the root of a tooth.
Chondroitin:	One of the polysaccharides commonly found in the cornea of the eye.

Chondroitin sulfate: Sulfated mucopolysaccharides commonly found in cartilages, bones, cornea, tendon, lung, and skin.

Crown: A crown-shaped structure, especially the exposed or enamel-covered portion of a tooth. Largely (97%) made of hydroxyapatite mineral.

Dentin: The chief substance of the tooth, forming the body, neck, and roots, being covered by enamel on the exposed part of the tooth and by cementum on the root. It is similar in composition and properties to compact bone.

Desmosine: One of the cross-linking chemicals in elastin.

Elastin: One of the proteins in connective tissue. It is highly stable at high temperatures and in chemicals. It also has rubberlike properties, and hence nicknamed "tissue rubber."

Glutamic acid (Glu): One of the essential amino acids much more commonly occurring in collagen than in elastin.

Glycine (Gly): One of the amino acids having the simplest structure.

Haversian system: Same as osteon.

Histidine (His): One of the amino acids.

Hyaluronic acid: One of the polysaccharides commonly found in synovial fluid, aortic wall, etc.

Hydroxyapatite: Mineral component of bone and teeth. It is a type of calcium phosphate, with composition $Ca_{10}(PO_4)_6(OH)_2$.

Hydroxyproline (Hyp): One of the amino acids commonly occurring in collagen molecules.

Isodesmosine: Isomer of desmosine.

Lacuna: A pore ($\sim 10 \times 15 \times 25\ \mu m$) in Haversian bone; lacunae often contain osteocytes (bone cells).

Lysine (Lys): One of the amino acids from which hydrogen bonding takes place stabilizing collagen chains.

Lysinonorleucine: One of the cross-linking chemicals in elastin.

Mixture rule: Properties of a material made of many materials depend linearly on the amount of each material contributed.

Osteons: Large fiberlike structure ($150 \sim 250\ \mu m$ in diameter) in compact bone. Concentric layers or lamellae surround a central channel or Haversian canal which contains a small blood vessel. Each lamella contains smaller fibers. Osteons, also called Haversian systems, are separated by cement lines.

Piezoelectricity: Electric polarization resulting from mechanical stress on a material; conversely, deformation resulting from an imposed electric field.

Polysaccharides: Polymerized sugar molecules found in tissues as lubricant (synovial fluid), cement (between osteons, tooth root attachment), or complexed with proteins such as glycoproteins or mucopolysaccharides.

Proline (Pro): One of the amino acids commonly occurring in collagen molecules.

Pulp: Richly vascularized and innervated connected tissue inside a tooth.

Streaming potential: Electric potential generated in the (solid) wall of a channel when charged particles are flowing and polarization takes place in the wall.

Tropocollagen: Precursor of collagen, right-handed superhelical coil structure, which, in turn, is made of three left-handed helical peptide chains.

Valine (Val): One of the essential amino acids more commonly occurring in elastin than in collagen.

Volksmann's canal: Vascular channels in compact bone. They are not surrounded by concentric lamellae as are the Haversian canals.

Wolff's law: Remodeling of bone takes place in response to mechanical stimulation so that the new structure becomes better adapted to the load.

BIBLIOGRAPHY

R. Barker, *Organic Chemistry of Biological Compounds*, Chapters 4 and 5, Prentice-Hall, Inc., Englewood Cliffs, N.J., 1971.

J. Black, *Biological Performance of Materials*, Marcel Dekker, New York, 1981.

S. C. Cowin (ed.), *Mechanical Properties of Bone*, American Society of Mechanical Engineers, New York, 1981.

J. D. Curry, *Mechanical Adaptations of Bone*, Princeton University Press, Princeton, N.J., 1984.

H. R. Elden (ed.), *Biophysical Properties of the Skin*, J. Wiley and Sons, New York, 1971.

R. S. Lakes, "Properties of Bone and Teeth," in *Encyclopedia of Medical Devices and Instrumentation*, J. G. Webster (ed.), J. Wiley and Sons, New York, pp. 501–512.

M. R. Urist (ed.), *Fundamental and Clinical Bone Physiology*, J. B. Lippincott Co., Philadelphia, 1980.

A. Viidik, "Functional Properties of Collagenous Tissues," *Int. Rev. Connect. Tissue Res.*, *6*, 127–215, 1973.

S. A. Wainwright, W. D. Biggs, J. D. Currey, and J. M. Gosline, *Mechanical Design in Organisms*, Edward Arnold, London, 1976.

CHAPTER 10

TISSUE RESPONSE TO IMPLANTS

In order to implant a material, the surgeon has to injure the tissue first. Also the injured or diseased tissues should be removed to some extent in the process of implantation. The success of the entire operation depends on the kind and degree of tissue response. This is an important aspect of biocompatibility. Biocompatibility entails mechanical, chemical, pharmacological, and surface compatibility as mentioned in Chapter 1. The tissue response toward injury may vary widely according to the site, species, contamination, etc. However, the inflammation and the cellular response to the wound for both intentional and accidental injuries are the same regardless of the sites.

10.1. NORMAL WOUND HEALING PROCESS

10.1.1. Inflammation

Whenever tissues are injured or destroyed, the adjacent cells respond to repair them. An immediate response to any injury is the inflammatory reaction. Soon after the injury, constriction of capillaries occurs (stopping blood leakage); then dilatation occurs. Simultaneously there is a greatly increased activity in the endothelial cells lining the capillaries. The capillaries become covered by adjacent leukocytes, erythrocytes, and platelets. Concurrently with vasodilatation, leakage of plasma from capillaries occurs. The leaked fluid combined with the migrating leukocytes and dead tissues will constitute *exudate.* Once enough cells (see Table 10-1 for definitions of types of cells) are accumulated by lysis, the exudate becomes pus. It is important to know that pus can sometimes occur in nonbacterial (aseptic) inflammation.

At the time of damage to the capillaries, the local lymphatics are also damaged since they are more fragile than the capillaries. However, the leakage of fluids

Table 10-1. Definitions of Cells Appearing in this Chapter

Types of cells	Description
Chrondroblast:	An immature collagen (cartilage)-producing cell.
Endothelial:	A cell lining the cavities of the heart and the blood and lymph vessels.
Erythrocyte:	A formed element of the blood containing hemoglobin (red blood cell).
Fibroblast:	A common fixed cell of connective tissue that elaborates the precursors of the extracellular fibrous and amorphous components.
Giant cell	
Foreign body giant cell:	A large cell derived from a macrophage in the presence of a foreign body.
Multinucleated giant cell:	A large cell having many nuclei.
Granulocyte:	Any blood cell containing specific granules; included are neutrophils, basophils, and eosinophils.
Leukocytes:	A colorless blood corpuscle capable of ameboid movement, protects body from microorganisms and can be of five types: lymphocytes, monocytes, neutrophils, eosinophils, and basophils.
Macrophage:	Large phagocytic mononuclear cell. Free macrophage is an ameboid phagocyte and present at the site of inflammation.
Mesenchymal:	Undifferentiated cell having similar role as fibroblasts but often smaller and can develop into new cell types by certain stimuli.
Mononuclear:	Any cell having one nucleus.
Osteoblast:	An immature bone-producing cell.
Phagocyte:	Any cell that destroys microorganisms or harmful cells.
Platelet:	A small circular or oval disk-shaped cell (3-μm diameter) precursor of a blood clot.

from capillaries will provide *fibrinogen* and other *formed elements of the blood*, which will quickly plug the damaged lymphatics, thus localizing the inflammatory reaction.

All of the reactions mentioned above, vasodilatation of capillaries, leakage of fluid into the extravascular space, and plugging of lymphatics, will provide the classic inflammatory signs: redness, swelling, and heat which can lead to local pain.

When the tissue injury is extensive or the wound contains either irritants or bacteria, the inflammation may lead to extensive tissue destruction. The tissue destruction is done by collagenase, which is a proteolytic enzyme capable of digesting collagen. The collagenase is released from *granulocytes*, which in turn are lysed by the lower pH at the wound site. Local pH can drop from the normal values of 7.4–7.6 to below 5.2 at the injured site. If there is no drainage for the necrotic debris, lysed granulocytes, formed blood elements, etc., then the site becomes a severely destructive inflammation resulting in a necrotic abscess.

If the severely destructive inflammation persists and no healing process occurs within 3-5 days, a *chronic* inflammatory process commences. This is marked by the presence of mononuclear cells called *macrophages*, which can coalesce to form multinuclear giant cells (Figure 10-1). The macrophages are phagocytic and remove foreign materials or bacteria. Sometimes the mononuclear cells evolve

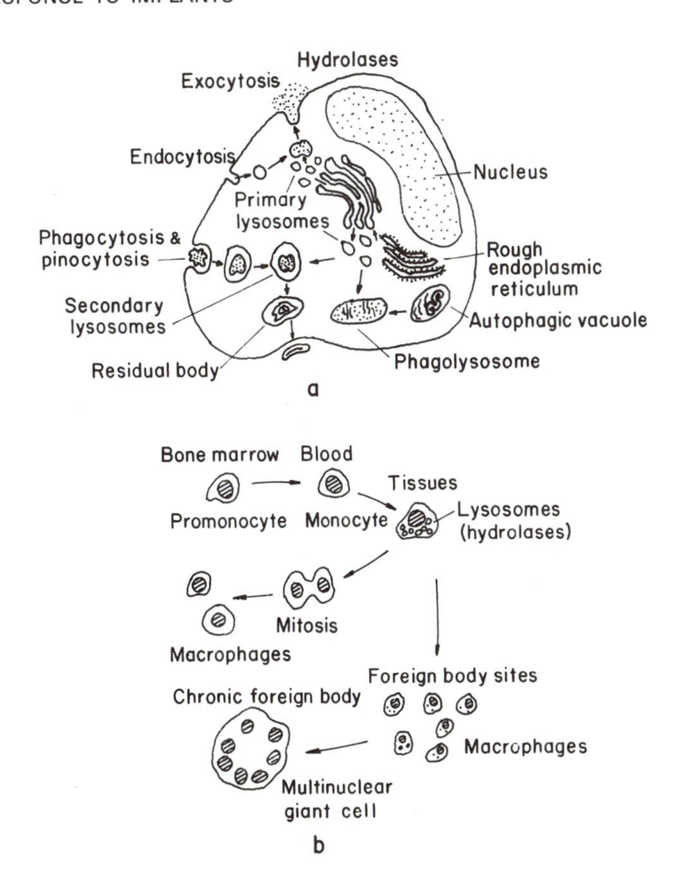

Figure 10-1. (a) The activated macrophage and (b) development of the multinuclear foreign body giant cell. From J. Black, *Biological Performance of Materials*, Marcel Dekker, New York, 1981.

into histiocytes, which regenerate collagen. This regenerated collagen is used to unite the wound or to wall-off unremovable foreign materials by encapsulation.

In chronic inflammatory reaction, lymphocytes occur as clumps or foci. These cells are a primary source of immunogenic agents which become active if foreign proteins are not removed by the body's primary defense.

10.1.2. Cellular Response to Repair

Soon after injury the mesenchymal cells evolve into migratory fibroblasts that move into the injured site while the necrotic debris, blood clots, etc. are removed by the granulocytes and macrophages. The inflammatory exudate contains fibrinogen, which is converted into fibrin by enzymes released through blood and tissue cells (see Section 10.3). The fibrin scaffolds the injured site. The migrating fibroblasts use the fibrin scaffold as a framework onto which the collagen is deposited. New capillaries are formed following the migration of fibroblasts and the fibrin scaffold is removed by the fibrinolytic enzymes activated by the endothelial cells. The endothelial cells together with the fibroblasts liberate collagenase, which limits the collagen content of the wound.

After 2 to 4 weeks of fibroblastic activities the wound undergoes remodeling during which the glycoprotein and polysaccharide content of the scar tissue decreases and the number of synthesizing fibroblasts also decreases. A new balance of collagen synthesis and dissolution is reached and the maturation phase of the wound begins. The time required for the wound healing process varies with various tissues although the basic steps described here can be applied in all connective tissue wound healing processes.

The healing of soft tissues and especially the healing of skin wounds has been studied intensively since this is germane to all surgery. The degree of healing can be determined by histochemical or physical parameters. A combined method will give a better understanding of the overall healing process. Figure 10-2 shows a schematic diagram of sequential events of cellular response of soft tissues after injury. The wound strength is not proportional to the amount of collagen deposited in the injured site as shown in Figure 10-3. This indicates that there is a latent period for the collagen molecules (procollagen is deposited by fibroblasts) to polymerize. It may take additional time to align the fibers in the direction of stress and cross-link fibrils to increase the physical strength closer to that of normal tissue. This collagen restructuring process requires more than 6 months to complete although the wound strength never reaches the original value. The wound strength can be affected by many variables, e.g., severe malnutrition

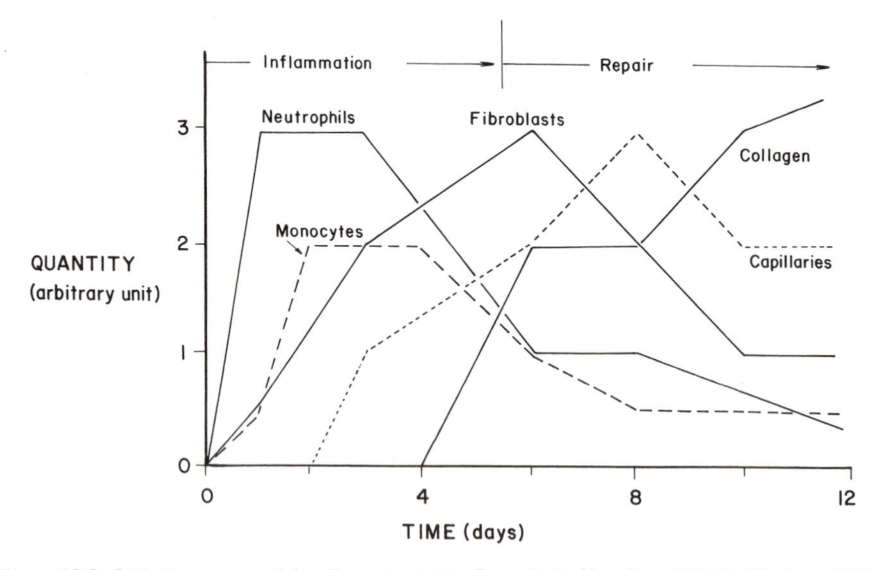

Figure 10-2. Soft tissue wound healing sequence. From L. L. Hench and E. C. Ethridge, "Biomaterial—The Interfacial Problem," *Adv. Biomed. Eng., 5,* 35–150, 1975.

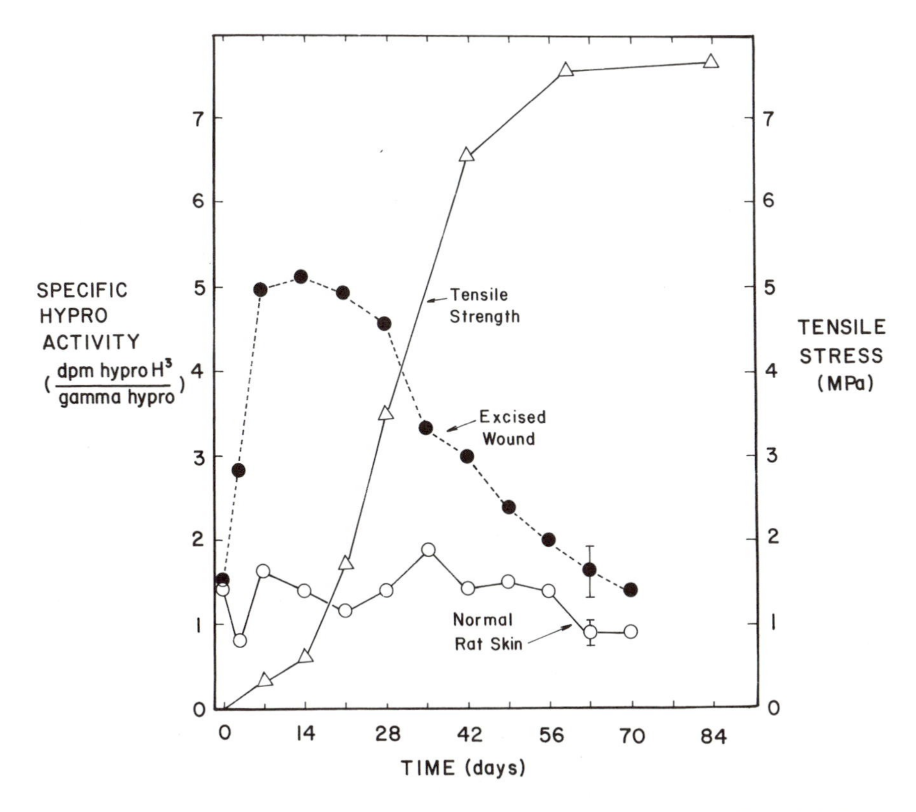

Figure 10-3. Tensile strength and rate of collagen synthesis of rat skin wounds. From E. E. Peacock, Jr., and W. Van Winkle, Jr., *Surgery and Biology of Wound Repair*, W. B. Saunders Co., Philadelphia, 1970.

resulting in protein depletion, temperature, presence of other wounds, and oxygen tension. Other factors such as drugs, hormones, irradiation, and electrical stimulation all affect the normal wound healing process.

The healing of bone fracture is *regenerative* rather than simple repair. The only other tissue that truly regenerates in humans is liver. However, the extent of regeneration is limited in humans. The cellular events following fracture of bone are illustrated in Figure 10-4. When a bone is fractured, many blood vessels (including those in the adjacent soft tissues) hemorrhage and form a blood clot around the fracture site. Shortly after fracture, the *fibroblasts* in the outer layer of periosteum and the osteogenic cells in the inner layer of periosteum migrate and proliferate toward the injured site. These cells lay down a fibrous collagen matrix called *callus*. *Osteoblasts* evolved from the osteogenic cells near the bone surfaces start to calcify the callus into trabeculae, which are the structural elements of spongy bone. The osteogenic cells migrating farther away from an established blood supply become chondroblasts, which lay down cartilage. Thus, after about 2 to 4 weeks the periosteal callus is made of three parts as shown in Figure 10-5.

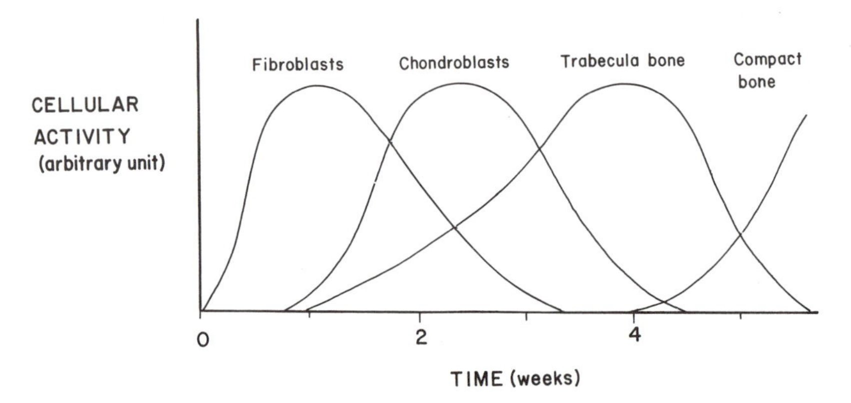

Figure 10-4. Sequence of events followed by bone fracture. From L. L. Hench and E. C. Ethridge, "Biomaterial—The Interfacial Problem," *Adv. Biomed. Eng., 5,* 35–150, 1975.

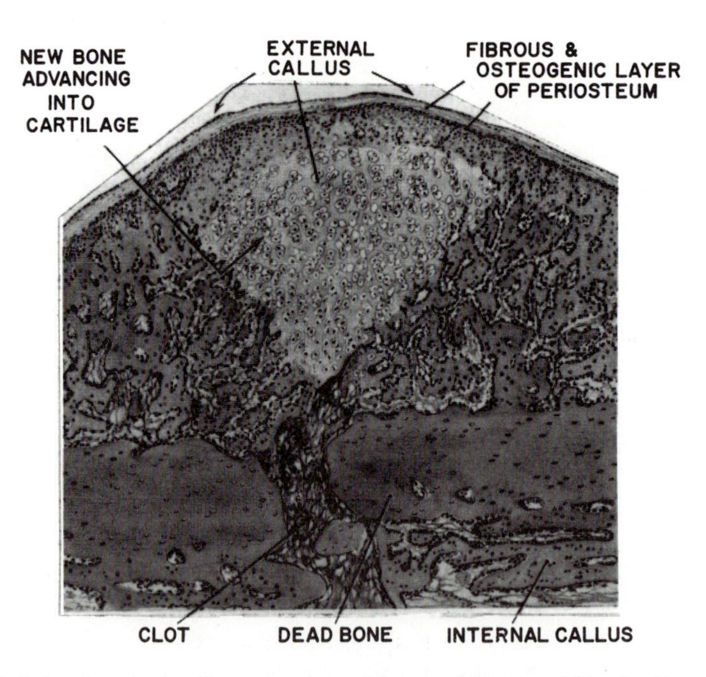

Figure 10-5. A drawing of a longitudinal section of fractured rib of a rabbit after 2 weeks, H & E stain. From A. H. Ham and W. R. Harris, "Repair and Transplantation of Bone," in *The Biochemistry and Physiology of Bone,* G. Bourne (ed.), Vol. 3, 2nd ed., Academic Press, New York, 1971, pp. 337–399.

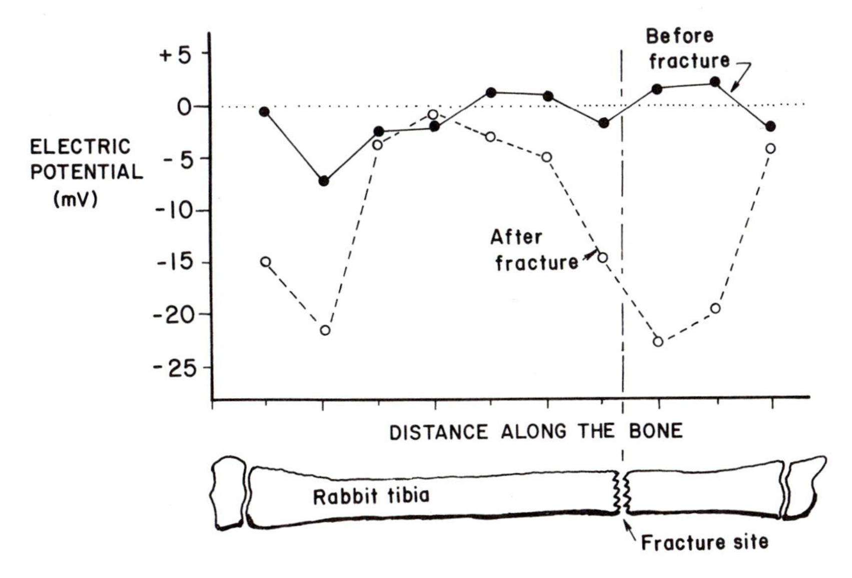

Figure 10-6. Skin surface of rabbit limb before and after fracture. Note that the fracture site increased electronegative potential. From Z. B. Friedenberg and C. T. Brighton, "Bioelectric Potentials in Bone," *J. Bone Joint Surg.*, *48A*, 915–923, 1966.

Simultaneous with the external callus formation, a similar repair process occurs in the marrow cavity. Since there is an abundant supply of blood the cavity turns into callus rather fast and becomes fibrous or spongy bone.

New trabeculae develop in the fracture site by *appositional growth* and the spongy bone turns into compact bone. This maturation process begins after about 4 weeks.

Some other interesting observations have been made on the healing of bone fractures in relation to the synthesis of polysaccharide on collagen. It is believed that the amount of collagen and polysaccharides is closely related to the cellular events following fracture. When the amount of collagen starts to increase, it marks the onset of the remodeling process. This occurs after about 1 week. Another interesting observation is the electrical potential (or biopotential) measured in the long bone before and after fracture as shown in Figure 10-6. The large *electronegativity* in the vicinity of fracture marks the presence of increased cellular activities in the tissues. Thus, there is a maximum negative potential in the epiphysis in the normal bone since this zone is more active (growth plate is in the epiphysis).

Example 10-1

The healing process of wounds in the skin has often been investigated since such healing is relevant to every surgery. In one study electrical stimulation was used to accelerate the healing of wounds in rabbit skin as shown in the following figure. The mean current flow was 21 μA and the mean current density was 8.4 $\mu A/cm^2$. After 7 days, the load to fracture

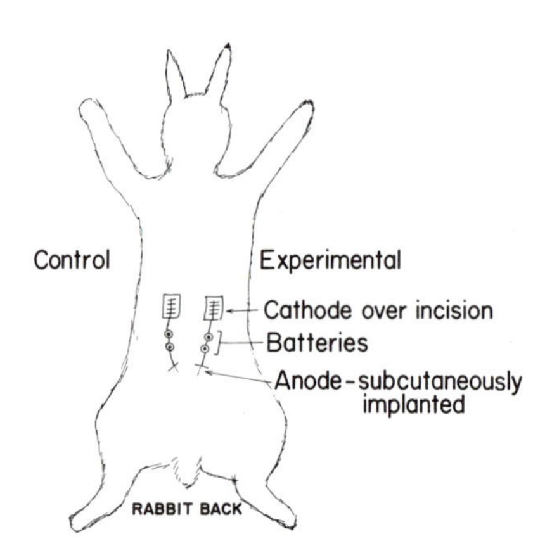

of the skin (removed from the dead animals) on the control samples was 797 g and on the stimulated experimental side it was 1224 g on the average (J.J. Konikoff, *Ann. Biomed. Eng.*, *4*, 1, 1976).

(a) Calculate the percent increase of strength by stimulation.
(b) The width of the testing sample was 1.6 cm. Assuming 1.8 mm thickness of skin, calculate the tensile stress for both control and experimental sample.
(c) Compared with the strength of normal skin (about 8 MPa), what percentages of the control and experimental skin wound strengths were recovered?
(d) Compare the results of (c) with the result of Figure 10-3.

Answer

(a) $$\frac{1224-797}{797} = \underline{0.536\ (53.6\%)}$$

(b) $$\text{Stress} = \frac{797 \times 10^{-3} \times 9.8\ \text{N}}{1.8 \times 16 \times 10^{-6}\ \text{m}^2} = \underline{0.27\ \text{MPa}}$$

$$\text{Stress} = \frac{1224 \times 10^{-3} \times 9.8\ \text{N}}{1.8 \times 16 \times 10^{-6}\ \text{m}^2} = \underline{0.42\ \text{MPa}}$$

(c) $$\frac{0.27}{8} = \underline{0.034\ (\text{or } 3.4\%)},\ \frac{0.42}{8} = \underline{0.052\ (\text{or } 5.2\%)}$$

(d) About the same recovery.

10.2. BODY RESPONSE TO IMPLANTS

The response of the body toward implants varies widely according to the host site and species, the degree of trauma imposed during implantation, and all of the variables associated with a normal wound healing process. On the other hand, the chemical composition and micro- and macrostructure of the implants induce different body response. The response has been studied in two different areas, i.e., local (cellular) and systemic although a single implant should be tested for both aspects. In practice, testing was not done simultaneously except in a few cases such as bone cement.

10.2.1. Cellular Response to Implants

Generally the body's reaction to foreign materials is to get rid of them. The foreign material could be extruded from the body if it can be moved (as in the case case of a wood splinter), or walled-off if it cannot be moved. If the material is particulate or fluid, then it will be ingested by the giant cells (macrophages) and removed.

These responses are related to the healing process of the wound where the implant is added as an additional factor. A typical tissue response is that the polymorphonuclear leukocytes appear near the implant followed by the *macrophages* called *foreign body giant cells.* However, if the implant is chemically and physically inert to the tissue, then the foreign body giant cells may not form. Instead, only a thin layer of collagenous tissue encapsulates the implant. If the implant is either a chemical or a physical irritant to the surrounding tissue, then the inflammation occurs in the implant site. The inflammation (both acute and chronic type) will delay the normal healing process resulting in granular tissues. Some implants may cause necrosis of tissues by chemical, mechanical, and thermal trauma.

It is generally very difficult to assess tissue responses to various implants due to the wide variations in the experimental protocols. This is exemplified by Table 10-2, which gives the tissue reactions for various suture materials in a descriptive way. It is interesting to note that the monofilament nylon suture maintains its tensile strength and elicits minimum tissue reaction. The multifilament suture loses its strength (disintegrates) and provokes high tissue reaction.

The degree of the tissue responses varies according to both physical and chemical nature of the implants. Pure metals (except the noble metals) tend to evoke a severe tissue reaction. This may be related to the high-energy state or large free energy of pure metals, which tends to lower the metal's free energy by oxidation or corrosion. In fact, titanium exhibits the minimum tissue reaction of all the metals commonly used in implants if its oxide layer is intact. This is due to the tenacious oxide layer, which resists further diffusion of metal ions and gas (O_2) at the interface as discussed in Chapter 6. In fact, this oxide layer is a ceramiclike material that is very inert. Corrosion-resistant metal alloys such as cobalt–chromium and 316L stainless steel have a similar effect on the tissue once they are passivated.

Most ceramic materials investigated for their tissue compatibility are oxides such as TiO_2, Al_2O_3, $BaTiO_3$ and multiphase ceramics of $CaO–Al_2O_3$, $CaO–ZrO_2$, and $CaO–TiO_2$. These materials showed minimal tissue reactions with only a thin layer of encapsulation as shown in Figure 10-7. Similar reactions were seen for carbon implants.

Table 10-2. Effect of Implantation on the Properties of Suture Materials[a]

Material	Wound tensile strength	Suture tensile strength	Tissue reaction
Absorbable			
Plain catgut	Impaired	Zero by 3–6 days	Very severe
Chromic catgut	Impaired	Variable	Moderate (less severe than plain cutgut, but more than non-absorbable materials)
Nonabsorbable			
Silk	No effect	Well maintained	Slight
Nylon			
Multifilament	No effect	Very low at 6 months	Moderately severe and prolonged
Monofilament	No effect	Well maintained	Slight
Polyethylene terephthalate	No effect	Well maintained	Very slight
PTFE (Teflon®)	No effect	Well maintained	Almost none

[a] From J. K. Newcombe, "Wound Healing," in *Scientific Basis of Surgery*, 2nd ed., W. T. Irvin (ed.), J. & A. Churchill, London, 1972.

Some glass-ceramics (e.g., 45 wt% SiO_2, 24.5 wt% Ca_2O, 24.5 wt% CaO, and 6.0 wt% P_2O_5) showed a direct bonding between implant and bone. Dissolution of the silica-rich gel film at the interface permits such a bond to form, as shown in Figure 10-8.

The polymers *per se* are quite inert toward tissue if there are no additives such as antioxidants, filters, antidiscoloring agents, plasticizers, etc. On the other

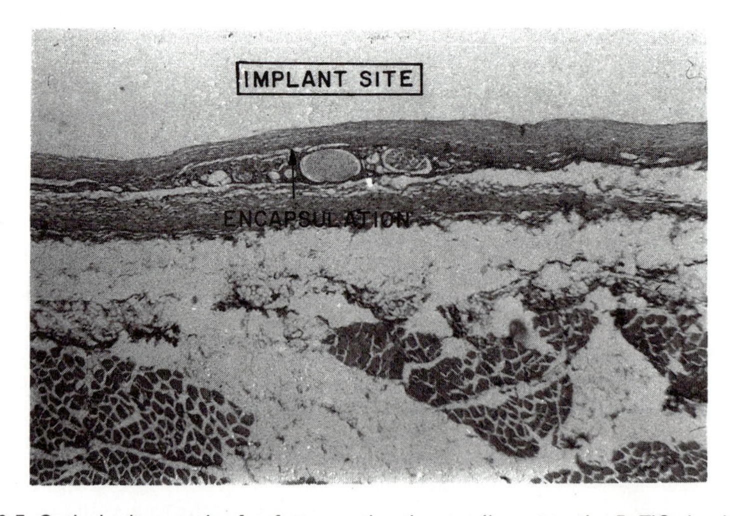

Figure 10-7. Optical micrograph of soft connective tissue adjacent to the $BaTiO_3$ implants after 20 weeks. H & E stain. Arrow indicates encapsulation of dense collagenous tissue. ×400. Courtesy of G. H. Kenner and J. B. Park.

Figure 10-8. Electron micrograph of the interface between 45S5 Bioglass-ceramic (C) and bone (B). The arrows indicate region of gel formation, undecalcified. ×10,000. Courtesy of L. L. Hench.

hand, the monomers can evoke an adverse tissue reaction since they are reactive. Thus, the degree of polymerization is somewhat related to the tissue reaction. Since 100% polymerization is almost impossible to achieve, a range of different size polymer molecules exists and small molecules can be leached out of the polymer. The particulate form of very inert polymeric materials can cause severe tissue reaction. This was amply demonstrated by polytetrafluoroethylene (Teflon®), which is quite inert in bulk form such as rods or woven fabrics but is very reactive in tissue when made into powder form. This is most likely due to the increased surface area and free radical generation by mechanical breakdown into smaller particulates. A schematic summary of tissue responses to implants is given in Figure 10-9.

There has been some concern about the possibility of tumor formation by the wide range of materials used in implantation. Although many implant materials are carcinogenic in rats, there are few well-documented cases of tumors in humans directly related to implants. It may be premature to pass the final judgment since the latency time for tumor formation in humans may be longer than 20 to 30 years. In this case we have to wait longer for the final assessment. However, the number of implants being placed in the body and lack of strong direct evidence for carcinogenicity tend to support the conjecture that the carcinogenesis is species-specific and that no tumors (or an insignificant number) will be formed in man by the implants. Carcinogenicity and its testing are discussed further in Section 10.4.1.

IMPLANT : TISSUE

MINIMAL RESPONSE

Thin Layer of Fibrous Tissue

Silicone rubber, Polyolefins, PTFE (Teflon)
PMMA, most ceramics, Ti- & Co-based alloys

CHEMICALLY INDUCED RESPONSE

Acute, Mild Inflammatory Response

Absorbable sutures, Some thermosetting resins

Chronic, Severe Inflammatory Response

Degradable materials, Thermoplastics with toxic additives, Corrosion metal particles

PHYSICALLY INDUCED RESPONSE

Inflammatory Response to Particulates

PTFE, PMMA, Nylon, Metals

Tissue Growth into Porous Materials

Polymers, Ceramics, Metals, Composites

NECROTIC RESPONSE

Layer of Necrotic Debris

Bone cement, Surgical adhesives

Figure 10-9. A brief summary of tissue response to implants. From D. F. Williams and R. Roaf, *Implants in Surgery*, W. B. Saunders Co., Philadelphia, 1973, p. 233.

Example 10-2

Describe major differences between normal wound healing and the tissue responses to "inert" and "irritant" materials. What factors besides the choice of material can affect the local tissue response to an implant?

Answer

The tissue response to an "inert" material is very much like normal wound healing. No foreign body giant cells appear and a thin fibrous capsule is formed. The tissue in this capsule differs very little from normal scar tissue. In response to irritant materials, foreign body giant cells appear and an inflammatory response is evoked. There is an abundance of leukocytes, macrophages, and granulocytes. Granular tissue will be formed, serving the functions of phagocytosis and organization, and appears only under circumstances of irritation or infection. Healing is slow and a thick capsule forms. If the material is chemically reactive or mechanically irritating, necrosis of surrounding tissue may result. It has been suggested that the size and shape of an implant should be important factors to consider for what type of tissue reaction it could elicit.

10.2.2. Systemic Effects of Implants

The systemic effect of implants is well documented in hip joint replacement surgery. The polymethyl methacrylate bone cement applied in the femoral shaft in dough state is known to lower the blood pressure significantly. There is a concern regarding the systemic effect of the biodegradable implants such as absorbable sutures and surgical adhesives, and the large number of wear and corrosion particles released by the metallic implants. The latter fact is especially

Table 10-3. Concentrations of Metals in Tissue and Organs of the Rabbit after Implantation of Various Metals[a,b]

	Surrounding muscle		Liver		Kidney		Spleen		Lung		Control muscle	
	6 wk	16 wk	6 wk	16 wk	6 wk	16 wk	6 wk	16 wk	6 wk	16 wk	6 wk	16 wk
Vitallium® (61.9% Co, 28–34% Cr, 4.73% Mo, 1.52% Ni, 0.61% Fe) (2 or 3 specimens per experiment)												
Cr	30, 25, 0	0, 0, 15	0, 5, 5	5, 5, 10	0, 5, 5	5, 5, 10	0, 0, 5	5, 5, 10	0, 0, 10	5, 5, 10	0, 5, 5	0, 10, 0
Co	25, 45	0, 30, 0	5, 10	5, 5, 10	0, 0, 105	5, 10, 70	5, 5, 20	5, 5, 300	0, 0, 300	10, 5, 10	0, 0, 50	0, 10, 0
Ni	20, 35	10	5, 10	5, 5, 80	5, 110	5, 5, 30	5, 205	5, 5, 1000	5, 40	5, 0, 45	5, 5	5
Ti	5, 5, 10	0, 10, 0	0, 0, 5	0, 0, 10	0, 0, 10	0, 0, 10	0, 0, 5	0, 0, 15	0, 0, 5	0, 0, 15	0, 0, 10	0, 0, 5
Mo	0, 5, 5	0, 5, 0	0, 15, 80	10, 10, 70	0, 20, 50	20, 20, 75	0, 0, 10	0, 0, 5	0, 0, 10	5, 5, 5	0, 0, 0	0, 0, 0
Iron	0, 40, 90	0, 80, 0	0, 200, 600	100, 160, 80	0, 90, 180	110, 120, 90	0, 250, 270	300, 290, 110	0, 90, 420	110, 100, 50	0, 10, 30	0, 20, 0
316 stainless steel (17.8% Cr, 13.4% Ni, 2.3% Mo, 0.23% Cu, 66.27% Fe)												
Cr	145, 65	115, 295	5, 10	5, 5	5, 5	5, 5	5, 5	5, 20	5, 5	20, 10	0, 20	5, 5
Co	5, 5	0, 10	5, 20	0, 15	5, 10	5, 115	5, 5	15, 600	5, 5	0, 250	0, 0	0, 90
Ni	15, 50	200, 70	20, 20	0, 95	0, 10	10, 10	5, 10	1000, 65	20, 220	20, 45	10, 0	5, 5
Ti	20, 10	25, 10	0, 10	5, 5	0, 5	5, 65	0, 5	10, 50	0, 10	5, 5	5, 10	10, 10
Mo	5, 10	10, 35	10, 75	50, 65	10, 85	75, 80	5, 10	15, 150	0, 10	5, 5	0, 0	0, 0
Fe	80, 70	90, 190	220, 590	550, 520	110, 120	180, 200	180, 420	580, 220	120, 220	150, 300	30, 10	15, 10

Figures are in ppm dry ash, given to nearest 5 ppm.

[a] From A. B. Ferguson, Y. Akahosi, P. G. Laing, and E. S. Hodge, "Characteristics of Trace Ions Released from Embedded Metal Implants in the Rabbit," *J. Bone Joint Surg.*, *44A*, 323–336, 1962.

important in view of the fact that the period of implantation is becoming longer as materials are implanted in younger people.

Table 10-3 indicates that the various organs have different affinities for different metallic elements. These results also indicate that the corrosion-resistant metal alloys are not completely stable chemically and some elements are released into the body. This caused another concern that the elevated ion concentrations in various organs may interfere with normal physiologic activities. The divalent metal ions may also inhibit various enzyme activities.

As mentioned before, the polymeric materials contain additives that cause cellular as well as systemic reaction to a greater degree than the polymer itself. Even the well-accepted polymer dimethylsiloxane (Silastic®, Dow Corning Co.) contains a filler, silica powder, to enhance the mechanical properties. Although the silica powder itself is an irritant when implanted in a concentrated area, there seems to be no problem with this additive. However, it is not certain whether there will be late complications if a large amount of the silica is released into the tissue and retained in various organs.

Example 10-3

A biodegradable suture will have a strength 1 MPa after 6 weeks of implantation. The strength of the implanted suture decreases according to $\sigma = \sigma_0 + b \ln t/a$ as determined by curve fitting to experimental data, where σ_0 is the original strength, $b = -2$ MPa, t = time in weeks, and a is a characteristic time, 1 week. Determine the original strength of the suture.

Answer

$\sigma_0 = \sigma - b \ln t/a = 1 + 2 \ln 6/1 = \underline{4.58 \text{ MPa}}$

10.3. BLOOD COMPATIBILITY

The single most important requirement for the blood-interfacing implants is blood compatibility. Although the blood coagulation is the most important factor for the blood compatibility, the implants should not damage proteins, enzymes, and formed blood elements (red blood cells, white blood cells, and platelets). The latter includes hemolysis (red blood cell rupture) and initiation of the platelet release reaction.

The coagulated blood is called clot. However, sometimes the clot formed inside the blood vessels is referred to as *thrombus* or *embolus* depending on whether the clot is fixed or floating, respectively. The mechanism and route of blood coagulation are not completely understood. A simplified version of blood clotting is proposed as a cascading sequence as shown in Figure 10-10. As discussed earlier, immediately after injury the blood vessels constrict to minimize the flow of blood. Platelets adhere to the vessel walls by contacting the exposed collagen. The aggregation of platelets is achieved through release of adenosine

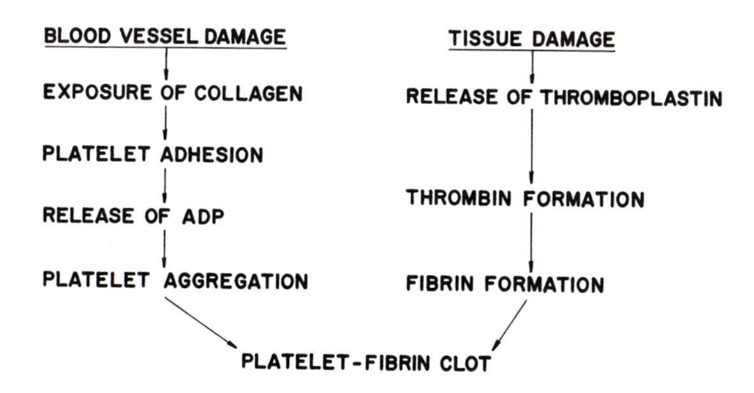

Figure 10-10. Two routes for blood clot formation (note the cascading sequence).

diphosphate (ADP) from damaged red blood cells, vessel walls and from adherent platelets.

10.3.1. Factors Affecting Blood Compatibility

The *surface roughness* is an important factor since the rougher the surface the more area is exposed to blood. Therefore, a rough surface promotes faster blood coagulation than the highly polished surface of glass, PMMA, polyethylene, and stainless steel. Sometimes, thrombogenic (clot-producing) materials with rough surfaces are used to promote clotting in porous interstices to prevent initial leaking of blood and to allow later tissue ingrowth through the pores of vascular implants.

The *surface wettability*, i.e., hydrophilic (wettable) or hydrophobic (nonwettable), was thought of as an important factor. However, the wettability parameter, contact angle with liquids, does not correlate consistently with the blood clotting time.

The surface of the intima of blood vessels is negatively charged (1-5 mV) with respect to the adventitia. This phenomenon is associated partially with the nonthrombogenic or thromboresistant character of the intima since the formed elements of blood are also negatively charged and hence are repelled from the surface of the intima. This was demonstrated experimentally by using a copper tube, which is a thrombogenic material, implanted as an arterial replacement. When the tube was negatively charged, the clot formation was delayed when compared with the control. In connection with this phenomenon, the streaming potential or zeta potential has been investigated since the formed elements of blood are flowing particles *in vivo*. However, it was not possible to establish a one-to-one direct relationship between the clotting time and zeta potential.

The chemical nature of a material surface interfacing with blood is closely related to the electrical nature of the surface since the type of functional groups of a polymer determines the type and magnitude of the surface charge. No intrinsic surface charge exists for metals and ceramics although some ceramics and polymers can be made piezoelectric. The surface of the intima is negatively charged largely due to the presence of polysaccharides, especially chondroitin sulfate and heparin sulfate.

10.3.2. Nonthrombogenic Surfaces

There have been many efforts directed at obtaining nonthrombogenic materials. The empirical approach has been used often. These approaches can be categorized as (1) heparinized or biological surfaces, (2) surfaces with anionic radicals for negative electric charges, (3) inert surfaces, and (4) solution-perfused surfaces. An early approach to nonthrombogenic surfaces is shown in Figure 10-11.

Heparin is a polysaccharide with negative charges due to the sulfate groups as shown:

$$\left[-O-\underset{OH}{\overset{CH_2OH}{(\text{ring: }OH)}}-O-\underset{NH_2}{\overset{CH_2OH}{(\text{ring: }OH)}}-O-\underset{NH_2}{\overset{CH_2OH}{(\text{ring: }OH)}}-O-\underset{NH_2}{\overset{CH_2OH}{(\text{ring: }OH)}}-O- \right]_n + 5 \text{ Sulfate groups} \qquad (10\text{-}1)$$

Initially the heparin was attached to a graphite surface treated with quaternary/salt, benzalkonium chloride (GBH process). Later a simpler heparinization

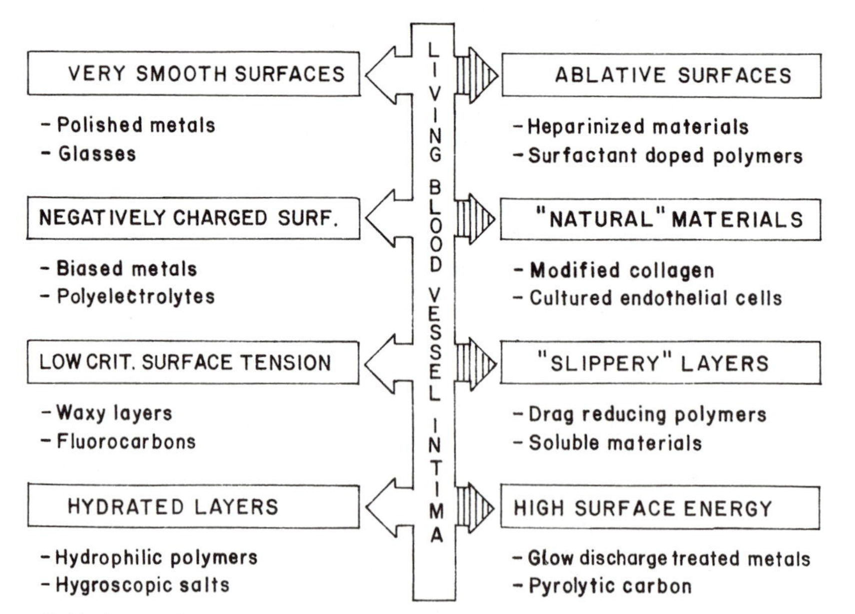

Figure 10-11. Approaches to producing thromboresistant surfaces. From R. E. Baier, "The Role of Surface Energy in the Thrombosis," *Bull. N.Y. Acad. Med., 48,* 257–272, 1972.

was accomplished by exposing the polymer surface to a quaternary salt such as tridodecylmethylammonium chloride (TDMAC). This method was further simplified by making a TDMAC and heparin solution in which the implant can be immersed followed by drying.

The heparinized materials showed a significant increase in thromboresistance compared with untreated control materials. In an interesting application, a polyester fabric graft was heparinized. This reduced the tendency of initial bleeding through the fabric and a thin *neointima* was later formed. Many polymers were tried for heparinization including polyethylene, silicone rubber, etc. Leaching of the heparin into the medium is a drawback although some improvement was seen by cross-linking of the heparin with glutaraldehyde and directly covalent bonding it onto the surface.

Some studies were carried out to coat the cardiovascular implant surface with other biological molecules such as albumin, gelatin (denatured collagen), and heparin. Some reported that the albumin alone can be thromboresistant and decrease the platelet adhesion. Also the vascular grafts were coated with inert materials such as carbon by depositing ULTI (ultra-low-temperature isotropic) pyrolytic carbon. The pyrolytic carbon showed an excellent compatibility and is currently most widely used to make artificial heart valve discs.

Negatively charged surfaces with anionic radicals (acrylic acid derivatives) were made by copolymerization or grafting. Negatively charged *electrets* on the surface of a polymer enhance its thromboresistance.

Hydrogels of both hydroxyethyl methacrylate (poly-HEMA) and acrylamide are classified as inert materials since they contain neither highly negative anionic radical groups nor are negatively charged. These coatings tend to be washed away when exposed to the bloodstream as was also seen with the heparin coatings. Segmental or block polyurethanes also showed some thromboresistance without surface modification.

Another method of making surfaces nonthrombogenic is perfusion of water (saline solution) through the interstices of a porous material that interfaces with blood. This new approach to a nonthrombogenic surface has the advantage of avoiding damage to formed elements of the blood. The disadvantage is the dilution of blood plasma. This is not a serious problem since saline solution is deliberately injected for the kidney and heart/lung machine; the method can be used only in such temporary blood-interfacing applications.

Example 10-4

In the circulation of the blood, the formed elements are being destroyed by the blood pump and tube wall contacts. A bioengineer measured the rate of hemolysis (red blood cell lysis) as 0.1 g/liter pumped. If the normal cardiac output for a dog is 0.1 liter/kg/min, what is the hemolysis rate? If the animal weighs 20 kg and the critical amount of hemolysis is 0.1 g/kg body weight, how long can the bioengineer circulate the blood before reaching a critical condition? Assume a negligible amount of new blood formation.

Answer

$$\text{Hemolysis rate} = 0.1 \text{ g/liter} \times 0.1 \text{ liter/kg/min} \times 20 \text{ kg} = \underline{2 \text{ mg/min}}$$

$$\text{Critical hemolysis} = 0.1 \text{ g/kg} \times 20 \text{ kg} = \underline{2\text{g}}$$

$$\frac{2\text{g}}{2 \text{ mg/min}} = \underline{1000 \text{ min (or 16 hr 40 min)}}$$

10.4. CARCINOGENICITY

A variety of chemical substances are known to induce the onset of cancerous disease in human beings and are known as carcinogens. Carcinogenic agents may act on the body by skin contact, by ingestion, by inhalation, or by direct contact with interior tissue. It is the last possibility that is of primary concern in the context of biomaterials.

Early studies showed that sheets or films of many polymers produced cancer when implanted in animals, especially rats. It was later found that the physical form of the implant was important, and that fibers and fabrics produced fewer tumors than sheets of the same material, and powders produced almost no tumors. Other materials, in contrast, are carcinogenic by virtue of their chemical constitution. In these cases, the risk would be greater in proportion to the surface area exposed, which is greatest for fine powders.

10.4.1. Testing of Carcinogenicity

New materials to which people may be exposed should be tested for possible carcinogenicity. Assessment of materials for possible carcinogenic effect is done as follows.

Chemical structure or function. If a material is similar structurally or pharmacologically to known carcinogenic agents, it may be suspected and evaluated further. Known carcinogens include aromatic amines, polynuclear aromatic hydrocarbons with multiple ring structures, alkylating agents including urethanes (ethylcarbamate, e.g., $H_2NCOOC_2H_5$), aflatoxins, halogenated hydrocarbons including vinyl chloride monomer, chloroform, polychlorinated biphenyls (PCBs), and certain pesticides; metallic nickel, cadmium, and cobalt.

In vitro tests. These tests involve exposing cultured cells to the agent in question. Their usefulness is predicated on the fact that carcinogenicity is correlated with mutagenicity. Following *in vitro* testing, the cells are examined for gene mutations, chromosomal aberrations, and/or deoxyribonucleic acid (DNA) damage and repair. The most well known and widely used of these tests is the *Ames test*, which involves exposing bacteria of the strain *Salmonella typhimurium* to the suspect agent, and looking for reverse mutations. Some mammalian metabolic capacity is provided by incubating the culture with mitochondrial extracts from rat liver. *In vitro* tests have the advantages of being quick and relatively inexpensive, but they are not sensitive to all carcinogenic agents (e.g., asbestos) and they do not reflect the complexities in uptake, organ specificity, distribution, and excretion found in whole animals and in humans.

Long-term animal bioassay. Rats and mice are the animals of choice in view of their relatively low cost and short life span. The short life span permits follow-up in whole life exposure studies; moreover, since the latency of tumor onset is a specified fraction of animal lifetime, the waiting period is not excessive. In a typical bioassay, there are four groups of animals: control (no exposure), maximum tolerated dose, and two intermediate doses. A minimum of 70 rodents per dose group per sex is used. The maximum tolerated dose is that which generates no overt toxicity, which does not reduce survival for reasons other than cancer. Following 2 years, animals that have died are examined, and those surviving are subjected to necropsy and histopathologic study. Control and dose groups are then compared statistically. As for the relevance to human beings, virtually all materials known to be carcinogenic in humans are also carcinogenic in animals.

Animal assays for solids. Cancer may be associated with implanted solids such as shrapnel, rifle bullets, and prosthetic implants—"foreign body" carcinogenesis. This type of situation is the most relevant to the biomaterials area. In animal assays, the suspect material is implanted in the flanks of rodents. In addition to examining for tumors, the investigators look for precancerous changes in the cells around the implant. The rationale is to increase the sensitivity of the test, since it is not feasible to increase the "dose" as done above.

Epidemiology. This approach is clearly the most relevant since it deals directly with humans, but there are difficulties. For example, the latency period between exposure to a carcinogenic agent and development of disease is from 5 to 40 years in adult humans. The risk of a newly introduced agent, therefore, will not be apparent until many years has elapsed. Epidemiologic methods also are relatively insensitive unless a large fraction of the population has been exposed, as in the case of cigarette smoking.

10.4.2. Risk Assessment

Prediction of risk is complicated by the fact that many tests are conducted at high dose levels, but human exposure is ordinarily at very low dose levels. Therefore, in carcinogenicity testing, the *linearity hypothesis* is often referred to in the interpretation of the results. According to this hypothesis, the carcinogenic response is a linear function of the dose.

Example 10-5

One wishes to identify materials that will cause cancer in 1 out of 100,000 human subjects, by experiments on rats. Is it feasible to use the same dose rate the humans would be exposed to? How many rats would one need under those conditions? Suggest an alternative experiment design based on the linearity hypothesis.

Answer

If the carcinogenic potential of the material is the same in humans as in rats, then, out of 100,000 rats, 1 would get cancer from the material, and perhaps 30,000 (30%) would get cancer from other causes by old age. A control group would contain an identical number of rats of which about 30% would also ultimately suffer cancer. There are two problems with this experiment: 200,000 rats is an excessive number representing an excessive expense, and it would be impossible to pick out the one extra cancer case from the 30,000 "naturally" occurring ones. A better approach would be to increase the dosage of the material by a factor of 10,000, so that 10% of the rats would suffer cancer from the exposure. A smaller number of rats could then be used. This approach is warranted if the linearity hypothesis is valid.

Other dose–response curves are conceivable, including a "threshold" or sublinear response in which the low-dose risk is less than that predicted by the linear model, and a supralinear model in which it is greater. Prediction of risk at low dose can vary by orders of magnitude depending on the model chosen. It is standard procedure to use a linear dose–response curve.

As for the potential risk associated with implant materials, pure metallic nickel, cadmium, and cobalt are known carcinogens when injected in solution into rat muscles. Nickel is a known industrial carcinogen, and cobalt is a suspect one. Implants in animals often induce tumors, but the epidemiologic evidence for human implants suggests a rather small risk.

PROBLEMS

10-1. What makes it so difficult to evaluate the tissue and blood compatibility of implants?

10-2. Calculate the corrosion rate in mm/year if a platinum electrode was used in Example 10-1.

10-3. Some materials do not normally induce tissue reaction when implanted in bulk form. However, when implanted in powder form, they become nonbiocompatible. Explain why.

10-4. Sometimes the degree of tissue reaction toward an implant is represented by the thickness of the collagenous capsule (e.g., Figure 10-7).

(a) State what fallacy these experimental results may have with regard to deciding biocompatibility.
(b) What factors affect the thickness of the encapsulation?

10-5. Explain why the metals are generally less biocompatible than ceramics or polymers. What can you do to improve this disadvantage of metals as implant materials?

10-6. The temperature change due to the heat of polymerization of bone cement (PMMA–polystyrene copolymer powder plus MMA monomer liquid) was monitored at the interface between bone and cement. The mixed cement was placed in the canine femur as a 9-mm-diameter plug and the temperature was measured with a thermocouple and results are shown in the accompanying diagram.

(a) What will happen to the adjacent tissues due to the heat generated by the polymerization?
(b) Would the temperature increase or decrease as a result of putting a metal cylinder in the middle, in a way similar to the situation in femoral hip replacement?
(c) What problems will arise if the cement shrinks when it reaches ambient temperature?

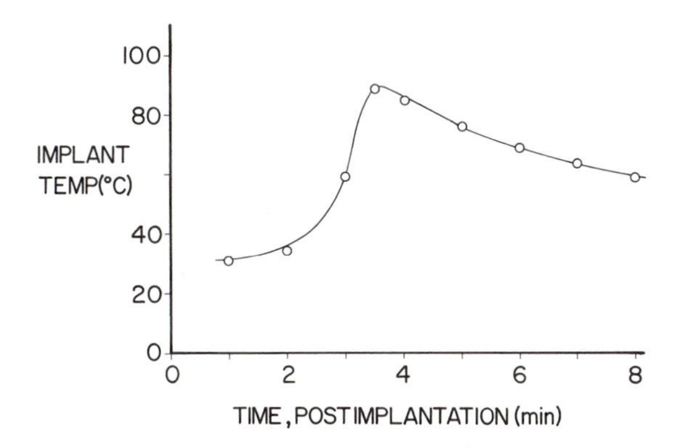

10-7. Explain why the nylon monofilament suture is less prone to lose its strength than multifilament suture material *in vivo*. Also explain why the monofilament suture causes less tissue reaction.

DEFINITIONS

Aflatoxin: A naturally occurring carcinogenic material. It is formed by mold infestation of food crops such as corn or peanuts.

Ames test: A screening test for carcinogenic potential of a material. Genetic mutations are observed in bacteria; carcinogenicity is correlated with mutagenicity.

Biocompatibility: Acceptance of an artificial implant by surrounding tissues and by the body as a whole. The implant should be compatible with tissues in terms of mechanical, chemical, surface, and pharmacological properties.

Callus: Unorganized fibrous collagenous tissue formed during the healing process of bone fracture. It is usually replaced with compact bone.

Carcinogen: Any substance that produces cancer.

Embolus: Any foreign matter, as a blood clot or air bubble, carried in the bloodstream.

Epidemiology: The study of disease incidence in a population of humans. Epidemiologic study constitutes the final test of the potential of a causative agent such as a biomaterial, in inducing disease.

Formed elements of blood: The solid components of blood; red and white blood cells and platelet.

Heparin: A substance found in various body tissues, especially in the liver, that prevents the clotting of blood.

Hydrogel: Highly hydrated (over 30% by weight) polymer gel that is used to make soft contact lenses. Acrylamide and poly-HEMA (hydroxyethylmethacrylate) are two common hydrogels.

Linearity hypothesis: This is the assumption that the incidence of cancer produced by a carcinogen is linearly proportional to the dose.

Macrophage: Any of various phagocytic cells in connective tissue, lymphatic tissue, bone marrow, etc. Sometimes called foreign body giant cell, which is associated with the presence of implants. It becomes multinucleated if the implant is not biocompatible.

Neointima: Sometimes called pseudointima. It is a new lining formed in the inner surface of porous arterial grafts and it has similar nonthrombogenic properties as the intima (natural lining) of the arteries.

PCB: Polychlorinated biphenyl, an industrial carcinogen. It is not used in biomaterials.

Supralinear model: This is the assumption of a nonlinear relation between dose of a carcinogen and the number of cancers produced. More commonly a linear relation is assumed. See linearity hypothesis.

Thrombus: The fibrinous clot attached at the site of thrombosis.

BIBLIOGRAPHY

B. Ames, "Dietary Carcinogens and Anticarcinogens," *Science, 221*, 1256–1264, 1983.

C. O. Bechtol, A. B. Ferguson, and P. G. Laing, *Metals and Engineering in Bone and Joint Surgery*, Balliere, Tindall and Cox, London, 1959.

S. D. Bruck, *Blood Compatible Synthetic Polymers: An Introduction*, Charles C. Thomas, Publ., Springfield, Ill., 1974.

J. Charnley, *Acrylic Cement in Orthopaedic Surgery*, Livingstone, Edinburgh, 1970.

D. R. Clarke and J. B. Park, "Prevention of Erythrocyte Adhesion onto Porous Surfaces by Fluid Perfusion," *Biomaterials, 2*, 9–13, 1981.

L. L. Hench and E. C. Ethridge, *Biomaterials: An Interfacial Approach*, Academic Press, New York, 1982.

A. S. Hoffman, "Applications of Radiation Processing in Biomedical Engineering—A Review of the Preparation and Properties of Novel Biomaterials," *Radiat. Phys. Chem., 9*, 207–219, 1977.

S. F. Hulbert, S. N. Levine, and D. D. Moyle, *Prosthesis and Tissue: The Interfacial Problems*, J. Wiley and Sons, New York, 1974.

S. N. Levine (ed.), "Materials in Biomedical Engineering," *Ann. N.Y. Acad. Sci., 146*, 1968.

H. I. Maiback and D. T. Rovee (eds.), *Epidermal Wound Healing*, Year Book Medical Pubs., Inc., Chicago, Ill., 1972.

H. A. Milman and E. K. Weisburger, *Handbook of Carcinogen Testing*, Noyes Pubns., Park Ridge, N.J., 1985.

A. Reif, "The Causes of Cancer," *Am. Sci., 69*, 437–447, 1981.

E. W. Salzman, "Nonthrombogenic Surfaces: Critical Review," *Blood, 38*, 509–523, 1971.

P. N. Sawyer and S. Srinivasan, "The Role of Electrochemical Surface Properties in Thrombosis at Vascular Interfaces: Cumulative Experience of Studies in Animals and Man," *Bull. N.Y. Acad. Med., 48*, 235–256, 1972.

G. K. Smith and J. Black, "Models for Systemic Effects of Metallic Implants," in *Retrieval and Analysis of Orthopaedic Implants*, A. Weinstein, E. Horowitz, and A. W. Ruff (eds.), National Bureau of Standards, 1977, pp. 23–28.

M. R. Urist (ed.), *Fundamental and Clinical Bone Physiology*, J. B. Lippincott Co., New York, 1980.

L. Vroman, *Blood*, American Museum Science Books, B26, Doubleday & Co., Inc., Garden City, N.Y., 1971.

CHAPTER 11

SOFT TISSUE REPLACEMENT I: SUTURES, SKIN, AND MAXILLOFACIAL IMPLANTS

In soft tissue implants as in other applications that involve engineering, the performance of an implanted device depends on both the materials used and the design of the device or implant. The initial selection of material should be based on sound materials engineering practice. The final judgment on the suitability of the material depends on observation of the *in vivo* clinical performance of the implant. Such observations may require many years. This requirement of *in vivo* observation represents one of the major problems in the selection of appropriate materials for use in the human body. Another problem is that the performance of an implant may also depend on the design rather than the materials *per se.*

The success of soft tissue implants has primarily been due to the development of synthetic polymers. This is mainly because the polymers can be tailor-made to match the properties of soft tissues. In addition, polymers can be made into various physical forms, such as liquid for filling spaces, fibers for suture materials, films for catheter balloons, knitted fabrics for blood vessel prostheses, and solid forms for cosmetic and weight-bearing applications.

It should be recognized that different applications require different materials with specific properties. The following are minimal requirements for all soft tissue implant materials:

1. They should achieve a reasonably close approximation of physical properties, especially flexibility and texture.
2. They should not deteriorate or change properties after implantation with time.
3. They should not cause adverse tissue reaction.
4. They should be noncarcinogenic, nontoxic, nonallergenic, and nonimmunogenic.

5. They should be sterilizable.
6. They should be low cost.

Other important factors are feasibility of mass production, aesthetic quality, etc.

11.1. SUTURES, SURGICAL TAPES, AND ADHESIVES

The most common soft tissue implants are sutures. In recent years, surgical tapes and tissue adhesives have been added to the surgeon's armamentarium. Although their use in actual surgery is limited to some surgical procedures, they are indispensable.

11.1.1. Sutures

There are two types of sutures, classified as to their long-term physical *in vivo* integrity: absorbable and nonabsorbable. They may also be distinguished by their raw material source: natural sutures (catgut, silk, and cotton) and synthetic sutures (nylon, polyethylene, polypropylene, stainless steel, and tantalum). Sutures may also be classified according to their physical form: monofilament and multifilament.

The absorbable suture, catgut, is made of collagen derived from sheep intestinal submucosa. It is usually treated with a chromic salt to increase its strength and is cross-linked to retard resorption. Such treatment extends the life of catgut suture from 3–7 days up to 20–40 days. Table 11-1 gives initial strength data for catgut sutures according to their sizes. The catgut sutures are preserved

Table 11-1. Minimum Breaking Loads for British-Made Catgut[a]

	Diameter (mm)		Minimum breaking load (lbf)	
Size	Minimum	Maximum	Straight pull	Over knot
7/0	0.025	0.064	0.25	0.125
6/0	0.064	0.113	0.5	0.25
5/0	0.113	0.179	1	0.5
4/0	0.179	0.241	2	1
3/0	0.241	0.318	3	1.5
2/0	0.318	0.406	5	2.5
0	0.406	0.495	7	3.5
1	0.495	0.584	10	5
2	0.584	0.673	13	6.5
3	0.673	0.762	16	8
4	0.762	0.864	20	10
5	0.864	0.978	25	12.5
6	0.978	1.105	30	15
7	0.105	1.219	35	17.5

[a] From L. A. G. Rutter, "Natural Materials," in *Modern Trend in Surgical Materials*, L. Gills (ed.), Butterworths, London, 1958, p. 208.

with needles in a physiological solution in order to prevent drying, which would make the sutures very brittle and thus not easily usable.

It is interesting to note that the stress concentration at a surgical knot decreases the suture strength of catgut by half, no matter what kind of knotting technique is used. It is suggested that the most effective knotting technique is the square knot with three ties to prevent loosening. According to one study there is no measurable difference in the rate of wound healing whether the suture is tied loosely or tightly. Therefore, loose suturing is recommended because it lessens pain and reduces cutting soft tissues.

Catgut and other absorbable sutures (e.g., polyglycolic acid, PGA: polylactic acid, PLA) invoke tissue reactions although the effect diminishes as they are being absorbed. This is true of other natural, nonabsorbable sutures like silk and cotton, which *showed more* reaction than synthetic sutures like polyester, nylon, polyacrylonitrile, etc. as shown in Figure 11-1. As in the case of the wound healing process (discussed in Chapter 10), the cellular response is most intensive one day after suturing and subsides in about a week.

As for the risk of infection, if the suture is contaminated even slightly the incidence of infection increases manyfold. The most significant factor in infection is the chemical structure, not the geometric configuration of the suture. Polypropylene, nylon, and PGA sutures developed lesser degrees of infection than sutures made of stainless steel, plain and chromic catgut, and polyester. The ultimate

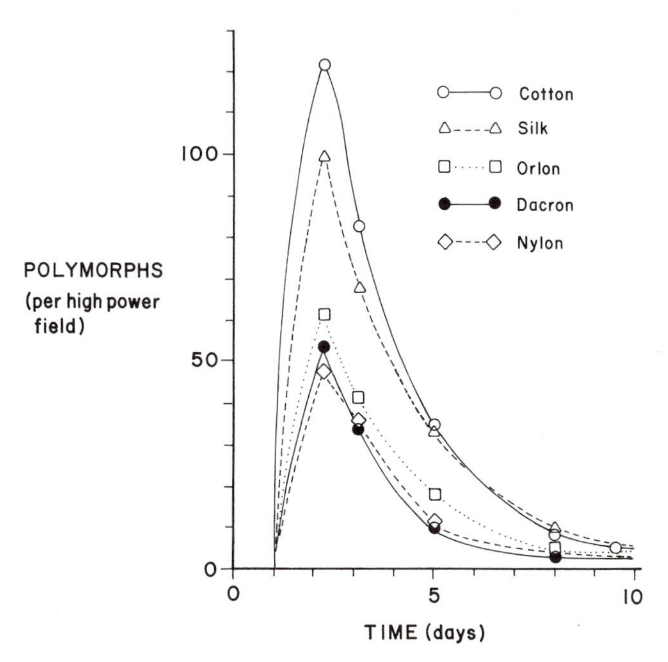

Figure 11-1. Cellular response to sutured materials. From R. W. Postlethwait, J. F. Schaube, M. L. Dillon, and J. Morgan, "Wound Healing, II. An Evaluation of Surgical Suture Material," *Surg. Gynecol. Obstet.*, *108*, 555–566, 1959.

cause of infection is a pathogenic microorganism, not the biomaterial. The role of the suture in infection is to provide a conduit for ingress of bacteria, to chemically or physically modify the body's immune response, or to provide an environment favorable to bacterial growth.

11.1.2. Surgical Tapes and Staples

Surgical tapes are intended to offer a means of avoiding pressure necrosis, scar tissue formation, problems of stitch abscesses, and weakened tissues. The problems with surgical tapes are similar to those experienced with Band-Aids: (1) misaligned wound edges, (2) poor adhesion caused by moisture or dirty wounds, (3) late separation of tapes when hematoma, wound drainage, etc. occur.

The wound strength and scar formation in the skin may depend on the type of incision made. If the subcutaneous muscles in the fatty tissue are cut and the overlying skin is closed with tape, then the muscles retract. This in turn increases the scar area, resulting in poor cosmetic appearance compared to a suture closure. However, with the higher strength of scar tissue, the taped wound has a higher wound strength than the sutured wounds only if the muscles were not cut. Because of this, tapes have not enjoyed the success that was anticipated when they were introduced.

Tapes have been used successfully for assembling scraps of donor skin for skin graft, correcting nerve tissues for neural regrowth, etc.

Staples made of metals (Ta, stainless steel, and Ti–Ni alloy) can be used to facilitate closure of large surgical incisions produced in procedures such as cesarean sections, intestinal surgery, and surgery for bone fractures. The tissue response to the staples is the same as that of synthetic sutures but they are not used in places where esthetic outlook is important.

11.1.3. Tissue Adhesives

The special environment of tissues and their regenerative capacity make the development of an ideal tissue adhesive difficult. Past experience indicates that the ideal tissue adhesive should be able to be wet and bond to tissues, be capable of rapid polymerization without producing excessive heat or toxic by-products, be resorbable as the wounds heal, not to interfere with the normal healing process, have ease of application during surgery, be sterilizable, have adequate shelf life, and ease of large-scale production.

The main strength of tissue adhesion comes from the covalent bonding between amine, carboxylic acid, and hydroxyl groups of tissues, and the functional groups of adhesives such as

$$\mathrm{R{-}\overset{|}{C}{-}\overset{|}{C}{-}}\ (\text{epoxide, bridged by O}) \qquad \mathrm{{-}\overset{|}{C}{-}\overset{|}{C}{-}}\ (\text{aziridine, bridged by NH}) \qquad \mathrm{R{-}\overset{|}{\underset{|}{C}}{-}}\ \mathrm{N{=}O} \tag{11.1}$$

There are several adhesives available of which alkyl-α-cyanoacrylate is best known. Among the homologues of alkyl-cyanoacrylate, the methyl- and ethyl-2-

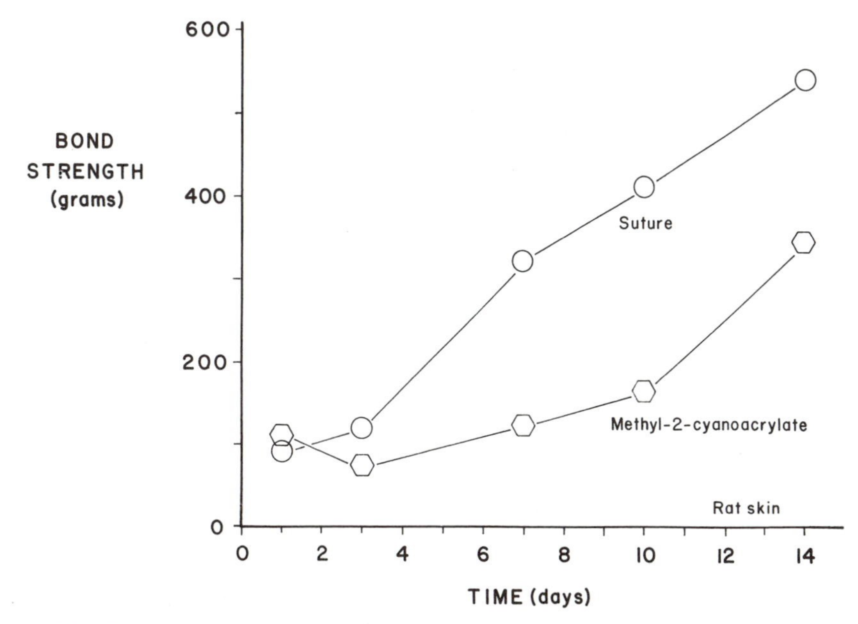

Figure 11-2. Bond strength of wounds with different closure material. From S. Houston, J. W. Hodge, Jr., D. K. Ousterhour, and F. Leonard, "The Effect of α-Cyanoacrylate on Wound Healing," *J. Biomed. Mater. Res.*, *3*, 281–289, 1969.

cyanoacrylate are most promising. With the addition of some plasticizers and fillers they are commercially known as Eastman 910®, Crazy Glue®, etc. An interesting comparison is illustrated in Figure 11-2, which shows that the bond strength of adhesive-treated wounds is about half that of sutured wounds after 10 days. Because of the lower strength and lesser predictability of *in vivo* performance of adhesives, the application is limited to use after trauma on fragile tissues such as spleen, liver, and kidney or after an extensive surgery on soft tissues such as lung. The topical use of adhesives in plastic surgery and fractured teeth has been moderately successful. As with many other adhesives, the end results of the bond depend on many variables such as thickness, open porosity, and flexibility of the adhesive film, as well as the rate of degradation.

Some workers have tried to use adhesives derived from fibrinogen, which is one of the clotting elements of blood. This material has sufficient strength (0.1 MPa) and elastic modulus (0.15 MPa) to sustain the adhesiveness for the anastomoses of nerve, microvascular surgery, dural closing, bone graft fixation, skin graft fixation, and other soft tissue fixation. This material is available commercially in Europe and will be in the United States pending FDA approval.

Example 11-1

A nylon suture was implanted in the abdominal cavity of a dog. The suture was removed after 10 and 20 days and its average tensile strength was measured. The strength decreased by 40 and 50%, respectively. How long will it take for the strength to decay 60% of its original strength? Assume an exponential decay of strength.

Answer

Since the strength decreases exponentially we can assume

$$\frac{\sigma_t}{\sigma_0} = A\exp(-Bt)$$

where A, B are constants, t is time (days), and σ_t is the strength at time t and σ_o is the original strength. Therefore,

$$0.6 = A\exp(-10B)$$

$$0.5 = A\exp(-20B)$$

By solving simultaneously

$$\frac{\sigma}{\sigma_0} = 0.72\exp(-0.018t)$$

$$0.4 = 0.72\exp(-0.018t)$$

$$t = \underline{33\text{ days}}$$

11.2. PERCUTANEOUS AND SKIN IMPLANTS

The need for percutaneous (trans or through the skin) implants has been accelerated by the advent of artificial kidneys and hearts, and by the need for prolonged injection of drugs and nutrients. Artificial skin (or dressing) is urgently needed to maintain the body temperature of severely burned patients. Actual permanent replacement of skin by biomaterials is beyond the capability of today's technology.

11.2.1. Percutaneous Devices

The problem of obtaining a functional and a viable interface between the tissue (skin) and an implant (percutaneous) device is primarily due to the following factors. First, although initial attachment of the tissue into the interstices of the implant surface occurs, it cannot be maintained for a long period of time, since the dermal tissue cells turn over continuously and dynamically. Furthermore, downgrowth of epithelium around the implant (extrusion) or overgrowth of implant (invagination) occurs. Second, any openings large enough for bacteria to penetrate may result in *infection* even though initially there is complete sealing between skin and implant.

Many variables and factors are involved in the development of percutaneous devices. These are:

1. End-use factors: Transmission of information (biopotentials, temperature, pressure, blood flow rate), energy (electrical stimulation, power for heart assist devices), matter (cannula for blood), and load (attachment of prosthesis)
2. Engineering factors
   a. Materials selection: polymers, ceramics, metals, and composites
   b. Design variation: button, tube with and without skirt, porous or smooth surface, etc.
   c. Mechanical stresses (soft or hard tissue interface, porous or smooth interface)
3. Biological factors
   a. Implant host: man, dog, hog, rabbit, sheep, etc.
   b. Implant location: abdominal, dorsal, forearm, etc.
4. Human factors
   a. Postsurgical care
   b. Implantation technique
   c. Esthetic outlook

Figure 11-3 shows a simplified cross-sectional view of a generalized percutaneous device (PD), which can be broken down into five regions:

A. Interface between epidermis and PD should be completely sealed against invasion by foreign organisms.
B. Interface between dermis and PD should reinforce the sealing of (A), as well as resist mechanical stresses. Due to the relatively large thickness of the dermis, the mechanical aspect is more important at this interface.
C. Interface between hypodermis and PD should reinforce the function of (B). The immobilization of the PD against piston action is a primary function of (C).
D. Implant material *per se* should meet all of the requirements of an implant for soft tissue replacement.
E. The line where epidermis, air, and PD meet is called a three-phase line which is similar to (A).

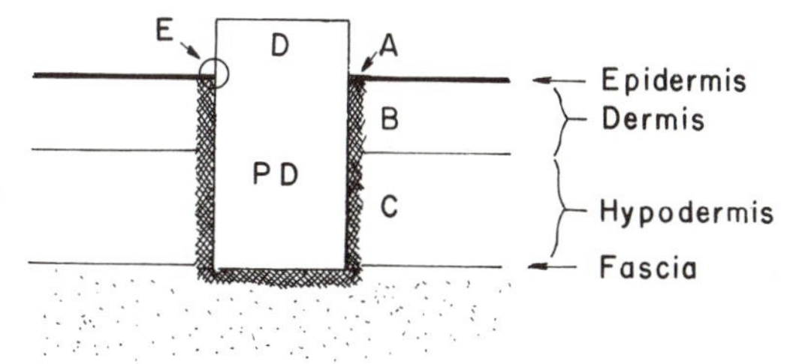

Figure 11-3. Simplified cross-sectional view of PD–Skin interfaces. From A. F. von Recum and J. B. Park, "Percutaneous Devices," *CRC Crit. Rev. Bioeng.*, *5*, 37–77, 1979.

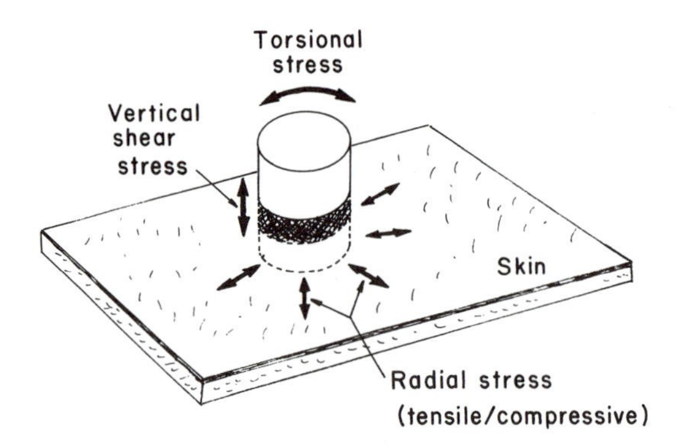

Figure 11-4. Various mechanical stresses acting at the PD–skin interface. From A. F. von Recum and J. B. Park, "Percutaneous Devices," *CRC Crit. Rev. Bioeng.*, *5*, 37–77, 1979.

The stresses generated between a cylindrical PD and skin tissue can be simplified as shown in Figure 11-4. The relative motion of the skin and implant results in shear stresses, which can be avoided if the implant floats (or moves) freely. For this reason, PDs without connected leads or catheters function longer. There have been many different PD designs to minimize shear stresses. All designs have centered around creating a good skin tissue/implant attachment in order to stabilize the implant. This is done by providing felts, velours, and other porous materials at the interface. Figure 11-5 shows a design to minimize the transfer of stresses and strains to the skin. The device includes an air chamber made of a rubber balloon (a) interposed between skin and PD, and firmer fixation of the cannula by providing a large surface for tissue ingrowth (b and c). Some designs have tried to minimize the trauma imposed by the external tubes and wires by

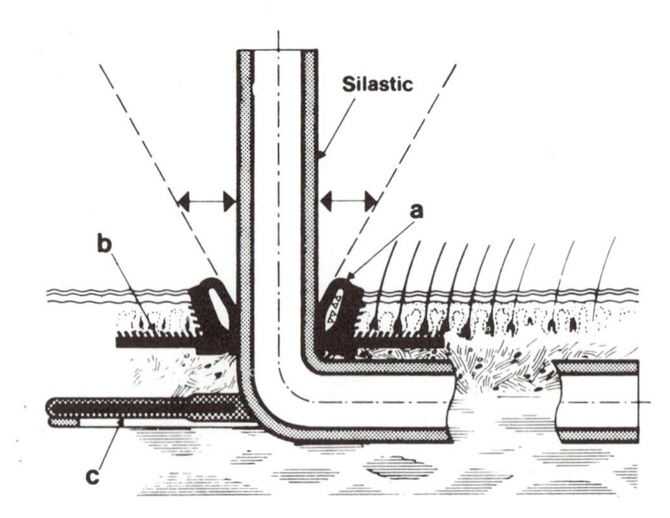

Figure 11-5. Schematic drawing of a Grosse-Siestrup PD. From C. Grosse-Siestrup, Entwicklung und Klinische Erprobung von Hautdurchleitungen Veterinaermedizin, Dissertation, Free University, Berlin, 1978.

use of a pin connector with good provision for firm tissue attachment subcutaneously.

No PDs have been completely satisfactory. Nevertheless, some researchers believe that hydroxyapatite may be a solution to the problem. In one experimental trial, hydroxyapatite-based PDs showed very little epidermal downgrowth (1 mm after 17 months versus 4.6 mm after 3 months for the silicone rubber control specimens in dorsal skin of canines, see Figure 11-6) and a high level of success rate (over 80% versus less than 50% for the control). The amino acid contents of the tissue capsules formed over the subcutaneous implants of the same materials showed that the hydroxyapatite site had the same composition as the periosteum of the femur while the control site showed a similar composition to that found in pathological tissues. Some researchers have tried to switch to subcutaneous implants that can be accessed by a needle for peritoneal dialysis as shown in Figure 11-7.

11.2.2. Artificial Skins

Artificial skin is another example of a percutaneous implant, and hence the problems are similar to those described in the previous section. Most needed for this application is a material that can adhere to a large (burned) surface and thus prevent the loss of fluids, electrolytes, and other biomolecules until the

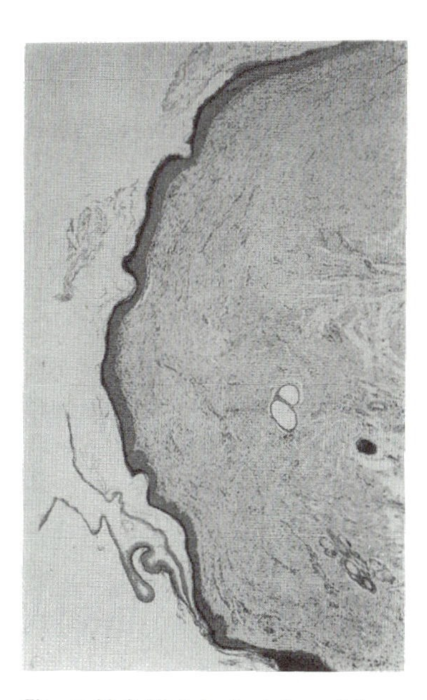

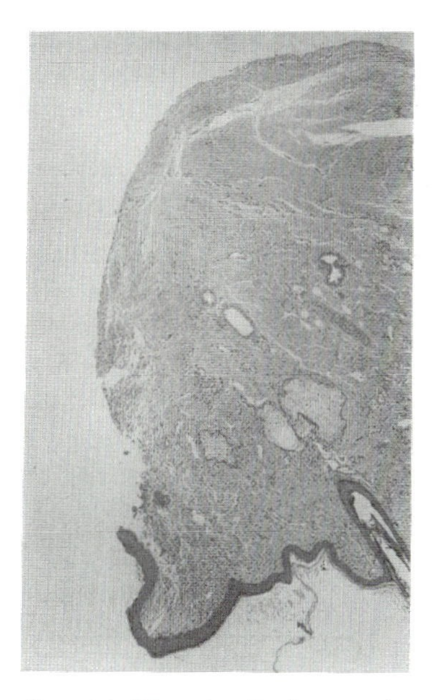

Figure 11-6. Histological view of the canine dermal tissues adjacent to PD made of hydroxyapatite (left) and silicone rubber (right) 3 months postimplantation (ca. 100× magnification). From H. Aoki, M. Akao, Y. Shin, T. Tsuki, and T. Togawa, "Sintered Hydroxyapatite for a Percutaneous Device and Its Clinical Applications," *Med. Prog. Technol. 12,* 213–220, 1987.

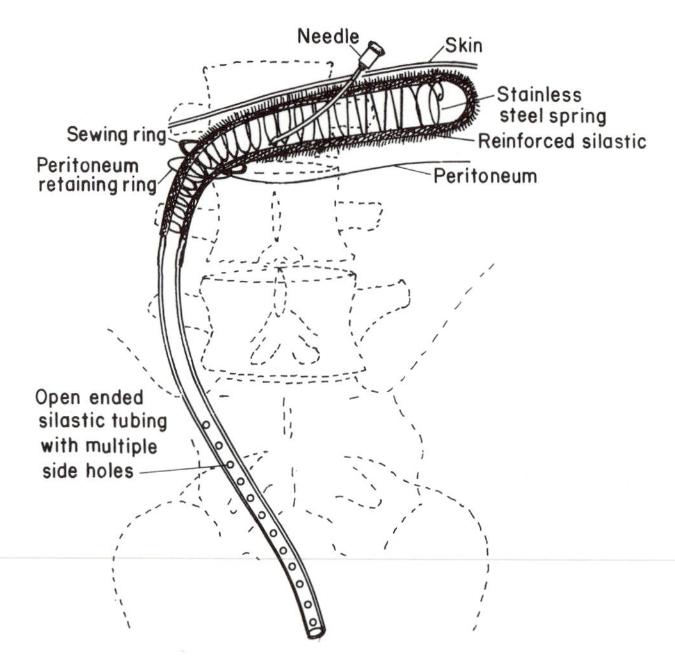

Figure 11-7. Subcutaneous peritoneal dialysis assist device. From C. Kablitz, T. Kessler, P. A. Dew, R. L. Stephen, and W. J. Kolff, "Subcutaneous Peritoneal Catheter: $2\frac{1}{2}$ Years Experience," *Artif. Organs, 3,* 210–217, 1979.

wound has healed. Although a permanent skin implant is needed, it is a long way from being realized for the same reasons given in the case of percutaneous implants proper. Presently, autografting and homografting (skin transplants) are the only methods available as a permanent solution.

In one study, wound closure was achieved by controlling the physicochemical properties of the wound-covering material (membrane). Six ways were suggested to improve certain physicochemical and mechanical requirements necessary in the design of artificial skin. These are shown schematically in Figure 11-8. Biomechanical and chemical analysis conducted in this study led to the design of a cross-linked collagen–polysaccharide (chondroitin 6-sulfate) composite membrane chosen for the ease in controlling porosity (5- to 150-μm diameter), flexibility (by varying cross-link density), and moisture flux rate.

Several polymeric materials including reconstituted collagen have also been tried as burn dressings. Among them are the copolymers of vinyl chloride and acetate and methyl-2-cyanoacrylate. The methyl-2-cyanoacrylate was found to be too brittle and histotoxic for use as a burn dressing. The ingrowth of tissue into the pores of sponge (Ivalon®, polyvinyl alcohol), and woven fabric (nylon and silicone rubber velour) was also attempted without much success. Plastic tapes have sometimes been used to hold skin grafts during microtoming (ultrathin sectioning) and grafting procedures. For severe burns the immersion of the patient

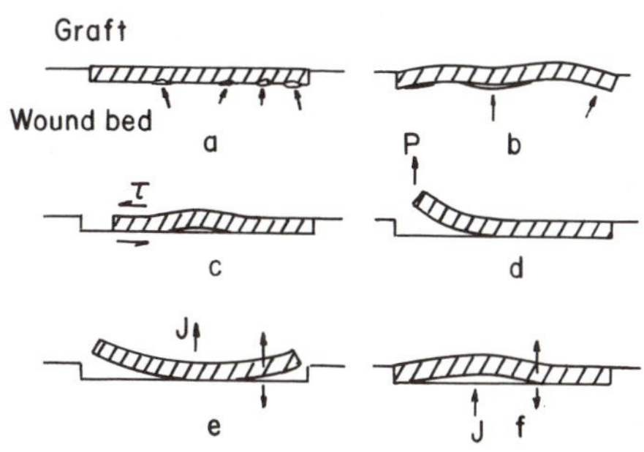

Figure 11-8. Schematic representation (not drawn to scale) of certain physicochemical and mechanical requirements in the design of an effective wound closure. (a) Skin graft (hatching) does not displace air pockets (arrows) efficiently from the graft–wound bed interface. (b) Flexural rigidity of graft is excessive; graft does not deform sufficiently under its own weight to make contact with depressions in wound bed surface, and air pockets (arrows) result. (c) Shear stresses (arrows) cause buckling of graft, ruptures of the graft–wound bed bond, and formation of air pocket. (d) Peeling force P lifts graft away from wound bed. (e) Excessively high moisture flux rate through graft causes dehydration and development of shrinkage stresses at edges (arrows), which cause lift-off away from wound bed. (f) Very low moisture flux J causes accumulation (edema) at the graft–wound bed interface and peeling off (arrows). From I. V. Yannas and J. F. Burke, Design of an artificial skin—I. Basic design principles, *J. Biomed. Mater. Res.*, *14*, 65–81, 1980.

into silicone fluid was found to be beneficial for prevention of early fluid loss, decubitus ulcers, and reduction of pain.

Rapid epithelial layer growth by culturing cells *in vitro* from the skin of the burn patient for covering the wound area may offer a better solution.

11.3. MAXILLOFACIAL AND OTHER SOFT-TISSUE AUGMENTATION

In the previous section we have dealt with problems associated with wound closing and wound/tissue interfacial implants. In this section we will study (cosmetic) reconstructive implants. Although soft-tissue implants can be divided into (1) space filler, (2) mechanical support, and (3) fluid carrier or storer, most have two or more combined functions. For example, breast implants fill space and provide mechanical support.

11.3.1. Maxillofacial Implants

There are two types of maxillofacial implant (often called prosthetics, which implies extracorporeal attachment) materials: extraoral and intraoral. The latter is defined as "the art and science of anatomic, functional or cosmetic reconstruction by means of artificial substitutes of those regions in the maxilla, mandible, and face that are missing or defective because of surgical intervention, trauma, etc."

There are many polymeric materials available for the extraoral implant, which requires: (1) color and texture should be matched with those of the patient, (2) it should be mechanically and chemically stable, i.e., it should not creep or change colors or irritate skin, and (3) it should be easily fabricated. Polyvinyl chloride and acetate (5–20%) copolymers, polymethyl methacrylate, silicone, and polyurethane rubbers are currently used.

The requirements for the intraoral implants are the same as for other implant materials, since they are in fact implanted. For maxillary, mandibular, and facial bone defects, metallic materials such as tantalum, titanium, and Co–Cr alloys, etc. are used. For soft tissues like gum and chin, polymers such as silicone rubber, PMMA, etc. are used for the augmentation.

The use of injectable silicones that polymerize *in situ* has proven partially successful for correcting facial deformities. Although this is obviously a better approach in terms of the minimal initial surgical damage, this procedure was not accepted due to the tissue reaction and the eventual displacement or migration of the implant.

11.3.2. Ear and Eye Implants

The use of implants can restore the conductive hearing loss from otosclerosis (a heredity defect that involves a change in the bony tissue of the ear) and chronic otitis media (the inflammation of the middle ear, which may cause partial or complete impairment of the ossicular chain: malleus, incus, and stapes). Many different prostheses are available to correct the defects, some of which are shown in Figure 11-9. The porous polyethylene total ossicular replacement implant is used to obtain a firm fixation of the implant by tissue ingrowth. The tilt-top implant is designed to retard tissue ingrowth into the section of the shaft which may diminish sound conduction.

Many different materials have been tried in fabricating implants: polytetrafluoroethylene, polyethylene, silicone rubber, stainless steel, and tantalum. More recently, polytetrafluoroethylene–carbon composite (Proplast®), porous polyethylene (Plastipore®), and pyrolytic carbon (Pyrolite®) have been shown to be suitable materials for cochlear (inner ear) implants.

Artificial ear implants capable of processing speech have been developed and are undergoing clinical evaluation. These types of cochlear implants have electrodes that stimulate the cochlear nerve cells. The implant also has a speech processor that transforms sound waves into electrical impulses that can be conducted through coupled external and internal coils as shown in Figure 11-10. The electrical impulses can be transmitted directly by means of a PD.

Eye implants are used to restore the functionality of the cornea and lens when they are damaged or diseased. Usually the cornea is transplanted from a suitable donor rather than implanted since the longevity of the cornea implant is uncertain because of fixation problems and infection. Figure 11-11 shows some of the eye implants tried clinically. They are made from "transparent" acrylics, especially PMMA, which has a comparatively high refractive index (1.5). In cataract, the lens of the eye becomes cloudy; the lens can then be removed surgically. The lost optical power can be restored with thick-lens spectacles, but these cause distortion and restriction of the field of view, and some people object to their appearance. Intraocular lenses are implanted surgically to replace the original eye lens, and they restore function without the problems associated with thick spectacles. The problems of infection and fixation of the lens to the tissues

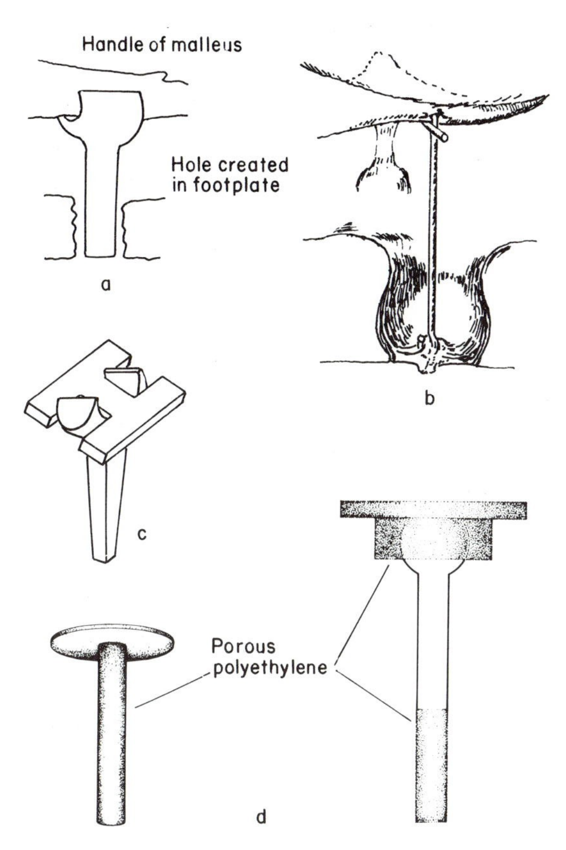

Figure 11-9. Prostheses for the reconstruction of the ossicles. (a) PTFE "piston" stapes prosthesis of J. J. Shea, F. Sanabria, and G. D. L. Smyth (*Arch. Otolaryngol.*, *76*, 516–521, 1962). (b) Incus replacement prosthesis of Sheehy [in *Hearing Loss—Problems in Diagnosis and Treatment*, L. R. Boies (ed.), W. B. Saunders Co., Philadelphia, 1969, p. 141]. (c) J. R. Tabor prosthesis for replacement of whole ossicular chain (*Arch. Otolaryngol.*, *92*, 141–146, 1970). (d) Porous polyethylene total ossicular replacement prosthesis (Smith & Nephew, Richard Medical Co., Technical Publ. No. 4240, Memphis, Tenn., 1980).

are again the major drawbacks of intraocular lens implants, as for the corneal implants. The intraocular lens can damage the soft structures to which it is attached, and it can become dislodged. Nevertheless, this type of cataract surgery has become commonplace, and many such implantation procedures are successfully conducted.

Recently, some researchers have tried to develop an artificial eye for people who have lost all of the conductive functions of the optic nerve. The device provides stimulation to the brain cells as shown in Figure 11-12. One of the major problems with this type of total organ replacement is the development of suitable electrode materials that will last a long time *in vivo* without changing their characteristics electrochemically. Another difficulty with the artificial eye is that significant image processing goes on in the retina. Consequently, simple electrical stimulation of the visual cortex of the brain yields a very poor image.

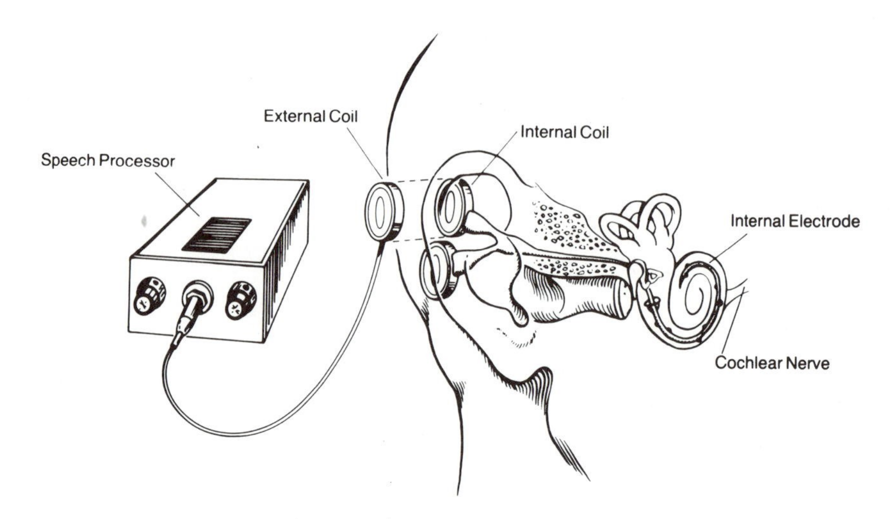

Figure 11-10. Basic components of artificial ear implants. From B. J. Gantz, "Cochlear Implants: An Overview," *Acta Otolaryngol. Head Neck Surg.*, *1*, 171–200, 1987.

11.3.3. Fluid Transfer Implants

Fluid transfer implants are required for cases such as hydrocephalus, urinary incontinence, and chronic ear infection. Hydrocephalus, caused by abnormally high pressure of the cerebrospinal fluid in the brain, can be treated by draining the fluid (essentially an ultrafiltrate of blood) through a cannula as shown in Figure 11-13. The earlier shunt had two one-way valves at the ends while the

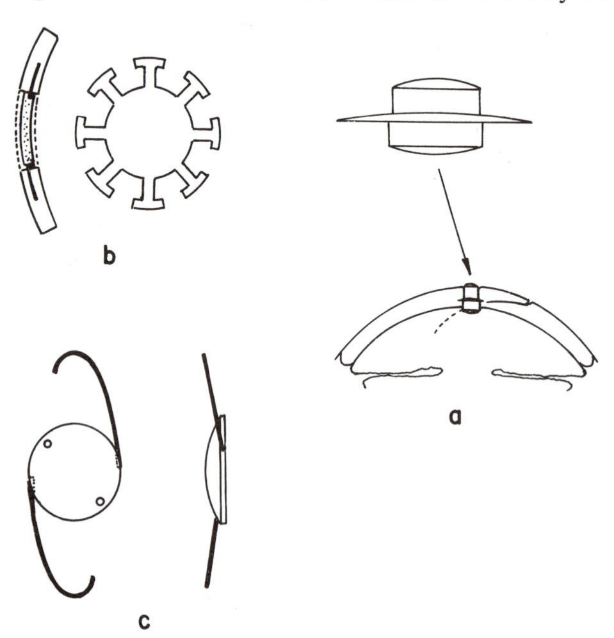

Figure 11-11. (a) Corneal implant of D. G. McPherson and J. M. Anderson (*Br. Med. J.*, *1*, 330, 1953). (b) Corneal implant of H. Cardona (*Am. J. Ophthalmol.*, *54*, 284, 1962). (c) Intraocular lens. (Courtesy of Intra-Intermedics, Inc., Pasadena, Calif.)

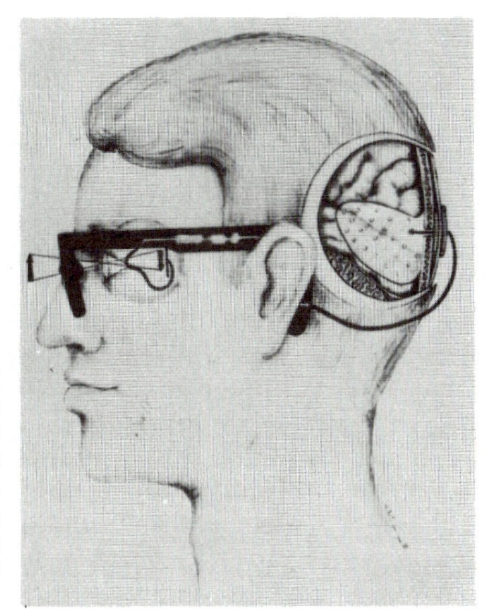

Figure 11-12. Diagram of concept of artificial eye. Television cameras in the glasses relay the message via microcomputers with radio waves to the array of electrodes on the visual cortex of the brain. From W. H. Dobelle, M. G. Mladejovsky, and J. P. Girvin, "Artificial Vision for the Blind: Electrical Stimulation of Visual Cortex Offers Hope for a Functional Prosthesis," *Science, 183,* 440–444, 1974.

Ames shunt has simple slits at the discharging end, which opens when enough fluid pressure is exerted. The Ames shunt empties the fluid in the peritoneum while others drain into the bloodstream through the right internal jugular vein or right atrium of the heart. The simpler peritoneal shunt showed less incidence of infection.

The use of implants for correcting the urinary system has not been successful because of the difficulty of joining a prosthesis to the living system to achieve fluid tightness. In addition, blockage of the passage by deposits from urine and constant danger of infection have been problematical. Many materials have been tried including glass, rubber, silver, tantalum, Vitallium®, polyethylene, Dacron®, Teflon®, polyvinyl alcohol, etc. without much long-term success.

The drainage tubes for chronic ear infection can be made from polytetrafluoroethylene (Teflon®) or other inert materials. These are not permanent implants.

11.3.4. Space-Filling Implants

Breast implants are quite common space-filling implants. At one time, the enlargement of breasts was done with various materials such as paraffin wax, beeswax, silicone fluids, etc. by direct injection or by enclosure in a rubber balloon. There have been several problems associated with directly injected implants, including progressive instability and ultimate loss of original shape and texture, as well as infection, pain, etc. In the 1960s the FDA banned such practices by classifying injectable implants such as silicone gel, as drugs.

One of the early efforts in breast augmentation was to implant a sponge made of polyvinyl alcohol. However, soft tissues grew into the pores and then

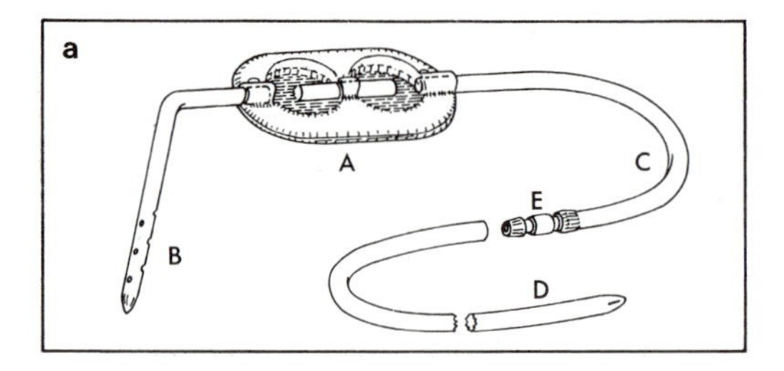

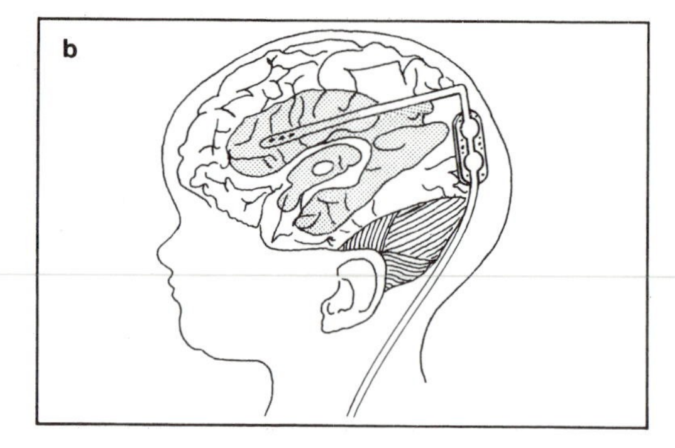

Figure 11-13. (a) Ames design hydrocephalus shunt; (b) Ames shunt *in situ.* (c) Valve for another shunt. The Ames shunt is made of silicone rubber (Silastic®) and consists of: (A) translucent double-chamber flushing device, (B) radiopaque ventricular catheter, (C) radiopaque connector tubing, (D) radiopaque peritoneal catheter, (E) stainless steel connector. (a, b) from Dow Corning Co., Silastic, Hospital-Surgical Products, Bulletin No. 51–051A, Midland, Mich., Dec. 1972; (c) from F. E. Nulsen and E. B. Spitz, "Treatment of Hydrocephalus by Direct Shunt from Ventricle to Jugular Vein," *Surg. Forum, 2,* 399–403, 1951.

calcified with time and the so-called marble breast resulted. Although the enlargement or replacement of breasts for cosmetic reasons alone is not recommended, prostheses have been developed for the patient who has undergone radical mastectomy or who has nonsymmetrical deformities. They are probably beneficial for psychological reasons. In this case a silicone rubber bag filled with silicone gel and backed with polyester mesh to permit tissue ingrowth for fixation, is a widely accepted prosthesis as shown in Figure 11-14. The artificial penis, testicles, and vagina fall into the same category as breast implants in that they make use of silicones and are implanted for psychological reasons rather than to improve physical health.

Example 11-2

Experience has shown that the silicone membrane used in the breast implants leaked the silicone fluid contained within to the surrounding tissue. Calculate the amount of leakage

Figure 11-14. An example of an artificial breast.

in 1 year. Assume that the leakage is entirely by diffusion rather than by macroscopic pores. Assume the silicone oil has a molecular weight of 740 amu. Assume the membrane is 1 mm thick and the surface area is 400 cm^2. The membrane has a diffusion constant of $D = 5 \times 10^{-17}$ cm^2/sec. Assume, moreover, that the implant has a volume of 1000 cm^3 and a density of $\rho = 1.5$ g/cm^3. Discuss other ways silicone fluid or gel could escape. Discuss implications.

Answer

From Fick's first law for diffusion, the flux F is $F = -D(dc/dx)$ in which D is the diffusion coefficient and c is the concentration. The flux is mass per unit area per time, so that if the concentration is initially zero in the tissue,

$$\text{mass/time} = \text{flux} \times \text{area} = FA = D\frac{dc}{dx}\,400\ \text{cm}^2$$

$$= 5 \times 10^{-17}\ \text{cm}^2/\text{sec}\,\frac{1.5\ \text{g/cm}^3 - 0}{0.1\ \text{cm}}\,400\ \text{cm}^2$$

$$= \underline{3 \times 10^{-13}\ \text{g/sec}}, \text{ or } \underline{9.6 \times 10^{-6}\ \text{g/yr}}$$

The body normally tolerates silicones well; problems do not usually arise unless gross amounts are lost and migrate through the tissues. Silicone fluid or gel could also escape into surrounding tissue through pores allowed by inadequate quality control, or from cracks due to fatigue from repeated flexure of the membrane.

As of 1992, there have been some reports of allergic reactions to gross amounts of escaped silicone gel. Concern over such cases has led to revised recommendations that silicone breast implants not be used solely for cosmetic augmentation.

We remark that the given volume corresponds to a mass $m = \rho V = 1.5$ kg corresponding to a weight of about 3.3 lb for each breast. If the shape is hemispherical, the volume is $2\pi r^3/3 = 1000\ \text{cm}^3$ so that the diameter (twice the radius) is 15.6 cm. The area of the curved surface is $2\pi r^2 = 384\ \text{cm}^2$. Commercially available implants are not quite hemispherical. The largest one made by one manufacturer has a diameter of 18 cm with a volume of 600 cm^3.

PROBLEMS

11-1. A bioengineer is trying to understand the biomechanics of a hole created in the skin for a transcutaneous implant. He made a hole using a circular biopsy drill in the dorsal skin of a dog. The diameter of the drill is 5 mm. If the hole became an ellipse with a minor and major axis of 3 and 7 mm, answer the following questions:

(a) In which direction is the internal stress in the skin greater?
(b) In which direction are the collagen fibers more oriented?
(c) How can the bioengineer obtain a circular rather than elliptical hole for his implant?
(d) Assuming the implant is nondeformable compared to the skin, what problems will arise between skin and implant when a load or force is applied to the skin or implant by handling accidentally?

11-2. Design a blood access device for kidney dialysis or other long-term use and give specific materials selected for each part and explain why you chose each particular material.

11-3. Draw the anatomy of the eye and label salient features.

11-4. Draw the anatomy of the ear and label salient features.

11-5. A breast implant is made of silicone rubber membrane filled with silicone rubber foam. Discuss the advantages and disadvantages of this design in comparison with an oil-filled implant.

11-6. Proplast® is a composite of PTFE and carbon (graphite). If it is made of 50% by volume each and has 20% porosity, what is its density? Estimate its Young's modulus.

11-7. Compare the breaking strength of catgut sutures (Table 11-1) of sizes between 7/0 and 7. What conclusions can you draw?

11-8. Design a penile implant that can carry out the erectile function for a person who has lost that capability due to disease or injury. What kind of materials would you need for its construction?

11-9. The retention of tensile breaking strength of absorbable sutures after implantation is shown for chromic catgut and PGA suture in the accompanying figure.

(a) Express the rate of strength decrease mathematically for both sutures.
(b) From the mathematical expression, calculate the zero strength times.

11-10. An ideal suture is defined as "handles comfortably and naturally, minimum tissue reaction, adequate tensile strength and knot security, be not favorable for bacterial growth and easily sterilizable, nonelectrolytic, noncapillary, nonallergenic and noncarcinogenic" [C. C. Chu, in *Biocompatible Polymers, Metals and Composites*, M. Szycher (ed.), Chapter 22, Technomic Pub., Westport, Conn., 1983]. Can you add more to the list?

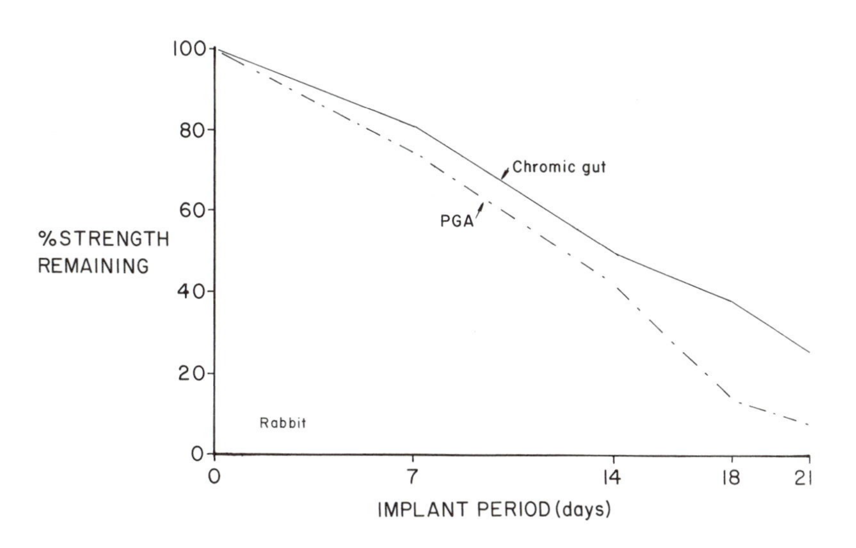

DEFINITIONS

Catgut:	Absorbable suture material prepared from collagen from healthy mammals.
Chromic salt:	Chemical compound that is used to treat collagen to achieve cross-linking between molecular chains of collagen. Such treatment increases its strength, but decreases its flexibility.
Cyanoacrylate:	A polymer used as a tissue adhesive since it can polymerize fast in the presence of water.
Dacron®:	Polyethylene terephthalate polyester that is made into fibers. If the same polymer is made into a film, it is called Mylar®.
FDA:	Food and Drug Administration, which regulates the use of medical devices in the United States.
Fibrinogen:	A plasma protein of high molecular weight that is converted to fibrin through the action of thrombin. This material is used to make (absorbable) tissue adhesives.
Percutaneous device (PD):	An implant designed to transfer matter, information, etc. from the body to the outside of the body transcutaneously.
Plastipore®:	Porous polyethylene.
Polyglycolic acid (PGA):	Polymer made from glycolic acid and used to make absorbable sutures or other products.
Polylactic acid (PLA):	Polymer made from lactic acid and used to make absorbable sutures or other products.
Proplast®:	A composite material made of fibrous polytetrafluoroethylene and carbon. It is usually porous and has low modulus and low strength.
Pyrolite®:	Pyrolytic carbon.
Suture:	Material used in closing a wound with stitches.
Teflon®:	Polytetrafluoroethylene.
Vitallium®:	Co–Cr alloy.

BIBLIOGRAPHY

J. Black, *Biological Performance of Materials*, Marcel Dekker, New York, 1981.

A. H. Bulbulian, *Facial Prosthetics*, Charles C. Thomas Pub., Springfield, Ill., 1973.

M. Chvapil, "Considerations on Manufacturing Principles of a Synthetic Burn Dressing: A Review," *J. Biomed. Mater. Res.*, Vol. 16, 245-263, 1982.

W. S. Edwards, *Plastic Arterial Grafts*, Charles C. Thomas Pub., Springfield, Ill., 1965.

H. Lee and K. Neville, *Handbook of Biomedical Plastics*, Chapters 4 and 13, Pasadena Technology Press, Pasadena, 1971.

W. Lynch, *Implants: Reconstructing the Human Body,* Van Nostrand Reinhold Co., Princeton, N.J., 1982.

G. B. Park, "Burn Wound Coverings: A Review," *Biomater. Med. Devices Artif. Organs*, *6*, 1-35, 1978.

A. F. von Recum and J. B. Park, "Percutaneous Devices," *CRC Crit. Rev. Bioeng.*, *5*, 37-77, 1979.

D. F. Williams (ed.), *Fundamental Aspects of Biocompatibility*, Vols. I and II, CRC Press, Boca Raton, Fla., 1981.

D. F. Williams (ed.), *Biocompatibility in Clinical Practice*, Vols. I and II, CRC Press, Boca Raton, Fla., 1982.

CHAPTER 12

SOFT TISSUE REPLACEMENT II: BLOOD-INTERFACING IMPLANTS

Blood-interfacing materials can be divided into two categories: short-term extracorporeal devices such as membranes for artificial organs (kidney and heart/lung machine), tubes and catheters for the transport of blood, and long-term *in situ* implants such as vascular implants and implantable artificial organs. Although pacemakers for the heart are not interfaced with blood directly, they are considered here since they are devices that help to circulate blood throughout the body.

The single most important requirement for blood-interfacing implants is blood compatibility (review Section 10.3). Blood coagulation is the most important aspect of blood compatibility: the implant should not cause the blood to clot. In addition, the implant should not damage proteins, enzymes, and formed elements of blood (red blood cells, white blood cells, and platelets). The implant should not cause hemolysis (red blood cell rupture) or initiation of the platelet release reaction.

Blood is circulated throughout the body according to the sequence shown in Figure 12-1. Implants are usually used to replace or patch large arteries and veins as well as the heart and its valves. Surgical treatment without using implants is usually preferred. However, there are many unavoidable situations when it is necessary to anastomose or replace a large segment with implants.

12.1. VASCULAR IMPLANTS

Implants have been used in various circumstances to treat vascular maladies. Examples include simple sutures for anastomosis after removal of vessel segments, vessel patches for aneurysms, as well as total replacements for large arteries. Vein

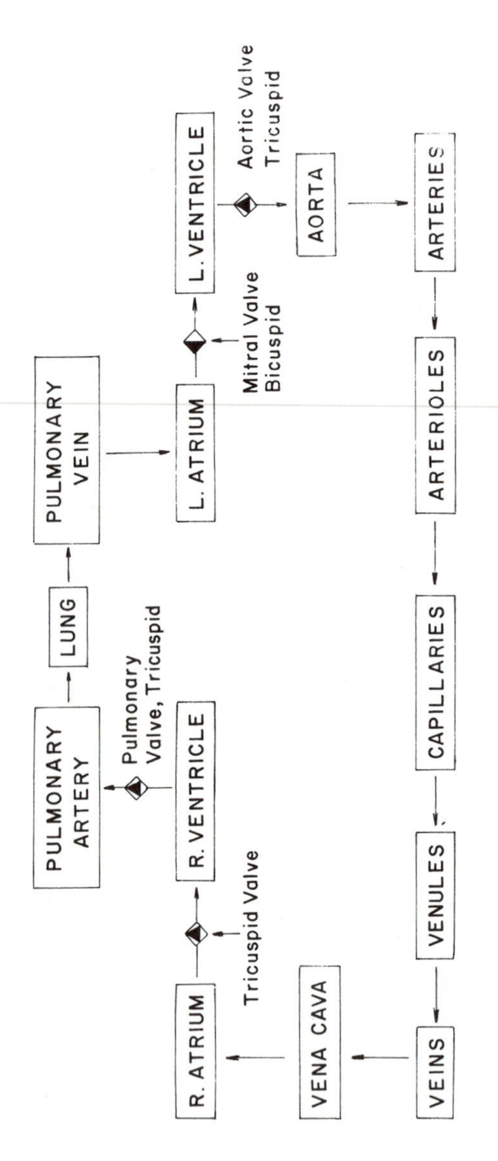

Figure 12-1. Schematic diagram of blood circulation in the body.

implants have encountered some difficulties because of the collapse of an adjacent vein or clot formation, which in turn is due to low blood pressure and stagnant blood flow in veins as compared to arteries. Vein replacements have not been a major concern since autografting can be performed for the majority of cases. Nonetheless, many materials including nylon, PTFE, polyester, etc. were fabricated for clinical applications.

Early designs for arterial replacements were solid wall tubes made of glass, aluminum, gold, silver and PMMA. All of the implants developed clots and became useless. In the early 1950s, porous implants made of fabrics were introduced, which allowed tissue growth into the interstices as shown in Figure 12-2. The new tissues interface well with blood, and thus minimize clotting. Ironically, for this type of application thrombogenic materials were found to be more satisfactory. Another advantage of tissue ingrowth is the fixation of the implant by the ingrown tissues, which make a viable anchor. The initial leakage through pores is disadvantageous but this can be prevented by preclotting the outside surface of the implant prior to placement. Crimping of the prosthesis, as shown in Figure 12-3, is done to prevent kinking when the implant is flexed. Also, the crimping allows expansion of the graft in the longitudinal direction, which reduces strain on the prosthesis wall. Arteries can expand circumferentially and longitudinally to accommodate the pulsatile flow of the blood.

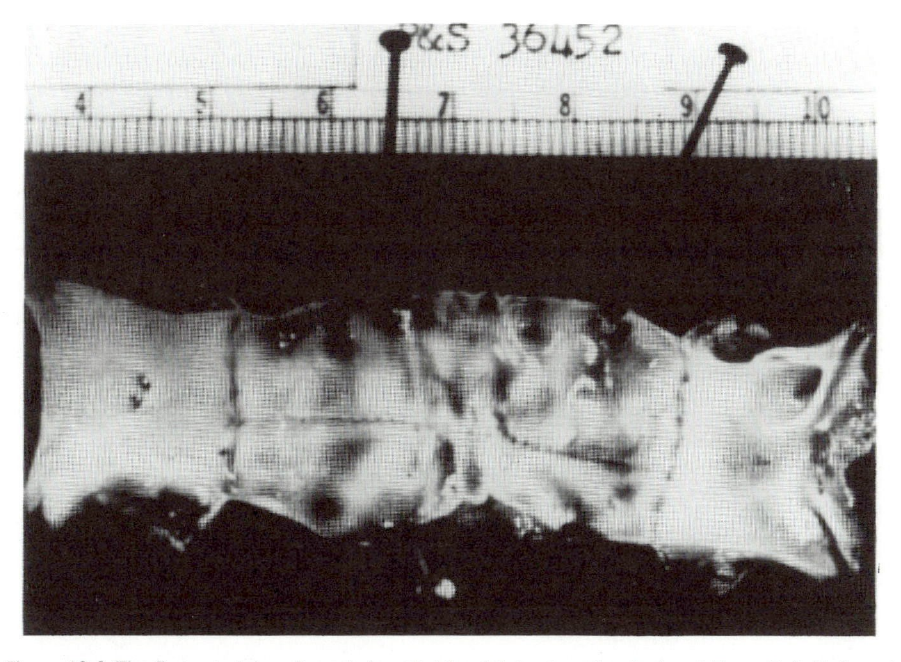

Figure 12-2. The first arterial graft made by stitching fabrics together by hand. From C. A. Hufnagel, "History of Vascular Grafting," in *Vascular Grafting: Clinical Applications and Techniques,"* C. W. Wright *et al.* (eds.), J. Wright, Boston, 1983, pp. 1–12.

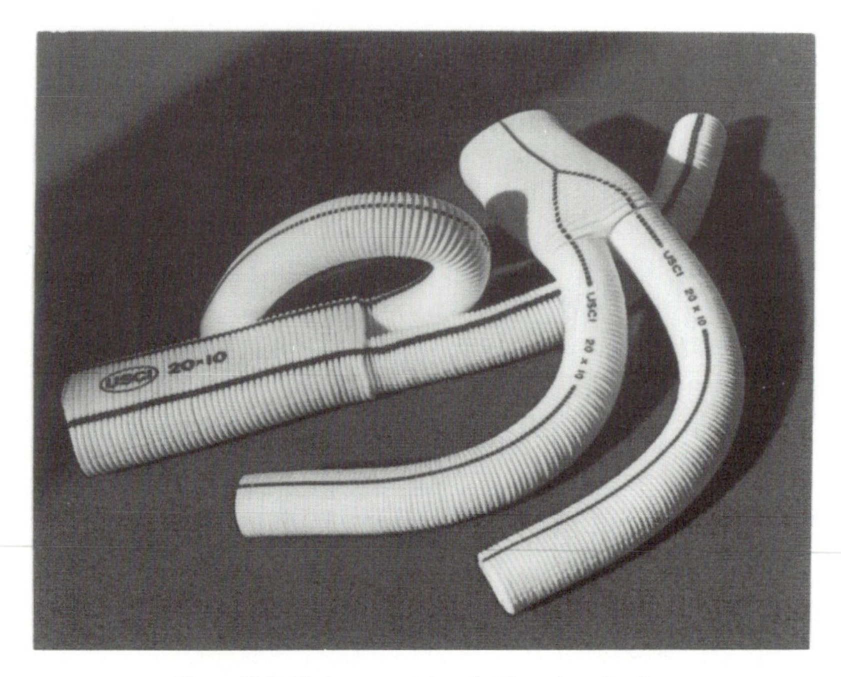

Figure 12-3. Modern arterial graft. Note the crimping.

Although the exact sequence of tissue formation in implants in humans is not fully documented, quite a bit is known about reactions in animals. Generally, soon after implantation the inner and outer surface of the implant are covered with fibrin and fibrous tissue, respectively. A layer of fibroblasts replaces the fibrin, becoming *neointima,* which is sometimes called *pseudointima* or *pseudoneointima.* The long-term fate of the neointima varies with the species of animals; in dogs it stabilizes into a constant thickness while for pigs it will grow until it occludes the vessel. In man the initial phase of the healing is the same as for animals but in later stages the inner surface is covered by both fibrin and a cellular layer of fibroblasts. The sequence for healing of arterial implants in humans, dogs, and pigs is given in Figure 12-4.

The types of materials and the geometry of the implant influence the rate and nature of tissue ingrowth. A number of polymer materials are currently used to fabricate implants, including nylon, polyester, PTFE, polypropylene, polyacrylonitrile, and silicone rubber. However, PTFE, polyester, polypropylene and silicone rubber are the most favorable materials due to the minimal deterioration of their physical properties *in vivo* as discussed in Chapter 8. Polyester (particularly polyethylene terephthalate, Dacron®) is usually preferred because of its superior handling properties.

Recently a pyrolytic carbon-coated arterial graft has been developed by the technique of ultra-low-temperature isotropic (ULTI) deposition. The nonthrombogenic properties of the pyrolytic carbon may enhance the patency of the graft

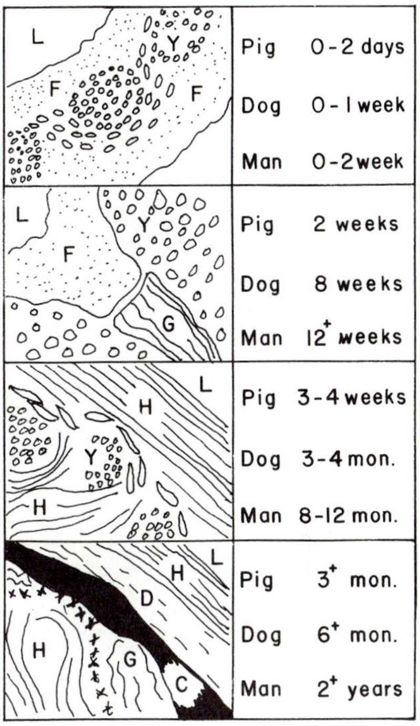

Pig	0-2 days
Dog	0-1 week
Man	0-2week
Pig	2 weeks
Dog	8 weeks
Man	12^+ weeks
Pig	3-4 weeks
Dog	3-4 mon.
Man	8-12 mon.
Pig	3^+ mon.
Dog	6^+ mon.
Man	2^+ years

Figure 12-4. Basic healing pattern of arterial prosthesis. L, lumen of prosthesis; F, fibrin, Y, yarn bundle; G, organizing granulation tissue; H, healed fibrous capsular tissue; D, degenerative fibrous capsular tissue; C, calcified capsular tissue. From S. A. Wesolowski, C. C. Fries, A. Martinez, and J. D. McMahan, "Arterial Prosthetic Materials," *Ann. N.Y. Acad. Sci., 146*, 325–344, 1968.

made from this material and decrease the need for postsurgical anticoagulant drugs.

Another interesting arterial graft is made by pressure-injecting Silastic rubber into premachined molds made of tentacles of sea urchins. The objective is to achieve a microporous structure for the tissues to grow into. After the Silastic rubber is cured, the mold is dissolved away by acid treatment leaving replamineforms of the ultrastructure as shown in Figure 12-5. Animal experiments showed promising results.

The geometry of fabrics and porosity have a great influence on healing characteristics. The preferred porosity is such that 5000 to 10,000 ml of water is passed per cm^2 of fabric per min at a pressure of 120 mm Hg. The fluid permeability depends not only on the porosity (volume fraction of pores) but also on the size, shape, and connectivity of pores. The lower limit will present excessive leakage of blood and the higher limit is better for tissue ingrowth and healing characteristics. Thickness of the implant is directly related to the amount of thrombus formation: the thinner the wall, the smaller or the thinner the thrombus deposited. Less thrombus results in faster organization of the neointima. Also, smaller-caliber (<5-mm diameter) prostheses can be made more easily with thinner walls.

The long-term testing of vascular prostheses is as important as it is with any other implants. A simple *in vivo* testing machine is shown in Figure 12-6 in which the pseudoextracellular fluid is drawn through valve 'A' and pushed out through valve 'B' of the graft at 96 cycles per minute with a peak pressure at 150 mm Hg

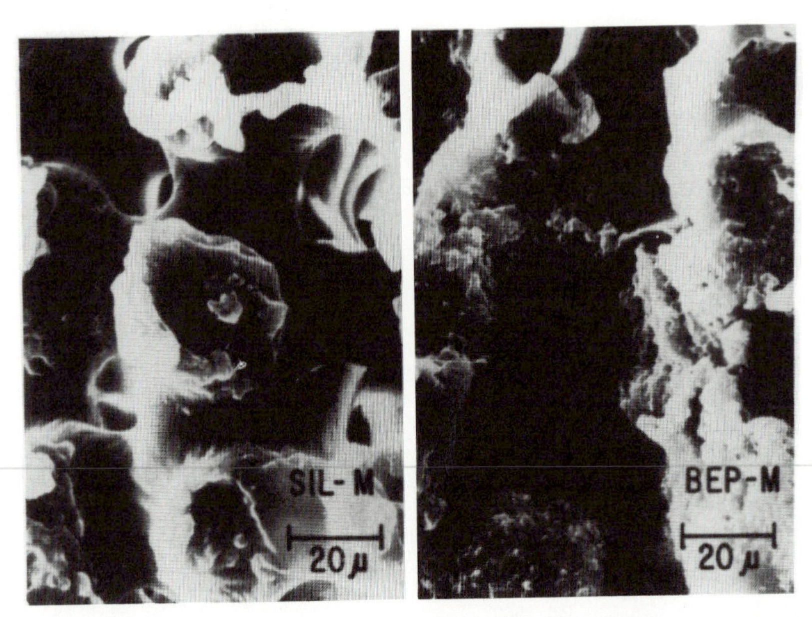

Figure 12-5. Scanning electron microscopic view of the replamineform Silastic arterial graft. From L. F. Hiratzka, J. A. Goeken, R. A. White, and C. B. Wright, "*In Vivo* Comparison of Replamineform, Silastic and Bioelastic Polyurethane Arterial Grafts," *Arch. Surg.*, *114*, 698–702, 1979.

at 37°C. Various grafts were tested and compared with *in vivo* implantation results as shown in Figure 12-7. It can be seen that the Teflon® knit graft did not lose its tenacity (a measure of normalized strength). The initial values of tenacity for Teflon® and Dacron® knit prostheses are about 1.3 and 3.0 g/denier, respectively. The Dacron® grafts showed initial decreases and stabilized after 6 months under both *in vivo* and *in vitro* conditions.

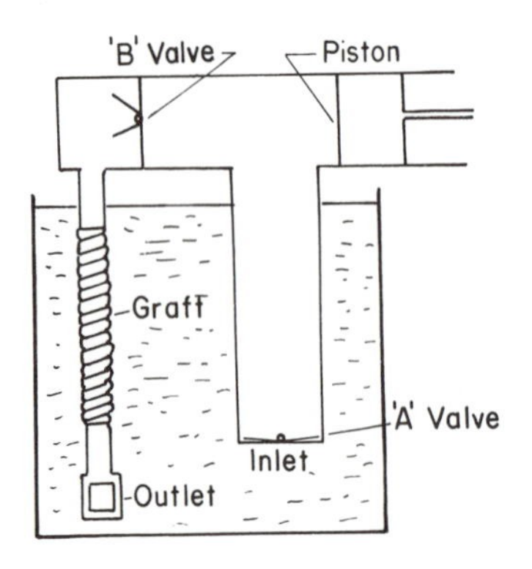

Figure 12-6. Schematic diagram of arterial graft life tester. From K. Botzko, R. Snyder, J. Larkin, and W. S. Edwards, "*In Vivo/in Vitro* Life Testing of Vascular Prostheses," in *Corrosion and Degradation of Implant Materials*, ASTM STP 684, B. C. Syrett and A. Acharya (eds.), ASTM, Philadelphia, 1979, pp. 76–88.

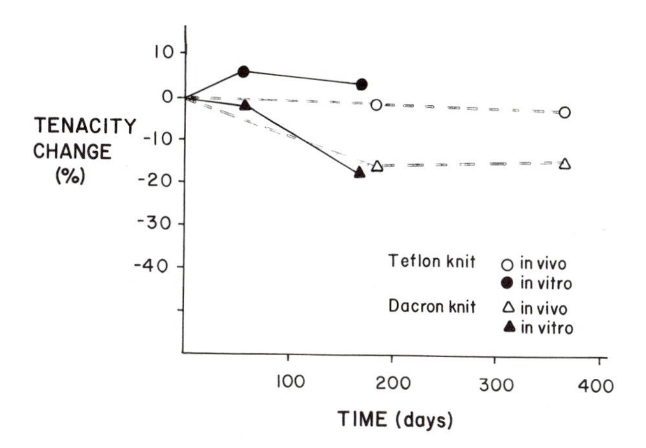

Figure 12-7. Percent change in tenacity for two types of prostheses after life testing of canine implant. From K. Botzko, R. Snyder, J. Larkin, and W. S. Edwards, "*In Vivo/in Vitro* Life Testing of Vascular Prostheses," in *Corrosion and Degradation of Implant Materials*, ASTM STP 684, B. C. Syrett and A. Acharya (eds.), ASTM, Philadelphia, 1979, pp. 76–88.

Example 12-1

The material properties of arterial prostheses change following ingrowth of tissues *in vivo*. A porous silicone rubber arterial prosthesis was implanted in dogs and it was found that one-half of the pores became filled with tissue after 3 months of implantation. The prosthesis has a 5-mm inside diameter, a wall thickness of 1 mm, and a porosity of 30%. The solid silicone from which the prosthesis is derived has a Young's modulus of 10 MPa. Answer the following:

(a) Determine the Young's modulus of the porous silicone.
(b) Find the elastic modulus of the prosthesis following tissue ingrowth. Assume that the ingrown tissue is similar to the natural arterial wall ($E = 0.1$ MPa).
(c) Determine the wall tension assuming a (high) blood pressure of 200 mm Hg.

Answer

(a) There are several relationships that can be used for the determination of properties of porous materials. For example, we may consider the empirical relationship

$$E = E_{\text{solid}}(1 - 1.9V + 0.9V^2)$$

in which V is the porosity (volume fraction of pores) and E_{solid} is the elastic modulus of the solid without porosity. Thus,

$$E = 10(1 - 1.9 \times 0.3 + 0.9 \times 0.3^2) = \underline{5.11 \text{ MPa}}$$

Alternatively, we may consider the model of Gibson and Ashby, which has both empirical and theoretical justification:

$$E = E_{\text{solid}}(\rho/\rho_{\text{solid}})^2$$

The density ratio is $\rho/\rho_{\text{solid}} = 1 - V$ so that

$$E = 10(0.49) = \underline{4.9 \text{ MPa}}$$

The actual elastic modulus will depend on the shape of the pores and their orientation.

(b) This is a rather complicated system. Obviously, the maximum elastic modulus for this problem would be 10 MPa if the pores were filled with an identical silicone rubber. We may approximate the modulus (using the second value above) as $E = 4.90 + 0.1 \times 0.3 = \underline{4.93 \text{ MPa}}$

(c)
$$T = P \times r = 200 \text{ mm Hg} \times \frac{1333 \text{ dynes/cm}^2}{\text{mm Hg}} \times \frac{6 \text{ mm} + 1 \text{ mm}}{2}$$

$$= \underline{80{,}000 \text{ dynes/cm}}$$

12.2. HEART VALVE IMPLANTS

There are four valves in the ventricles of the human heart as shown in Figure 12-8 (cf. Fig. 12-1). In the majority of cases, the left ventricular valves (mitral and aortic) become incompetent more frequently than the right ventricular valves as the result of higher left ventricular pressure. Most important and frequently critical is the aortic valve since it is the last gate the blood has to go through before being circulated in the body.

There have been many different types of valve implants. The early ones in the 1960s were made of flexible leaflets that mimicked the natural valves. Invariably the leaflets could not withstand fatigue for more than 3 years. In addition

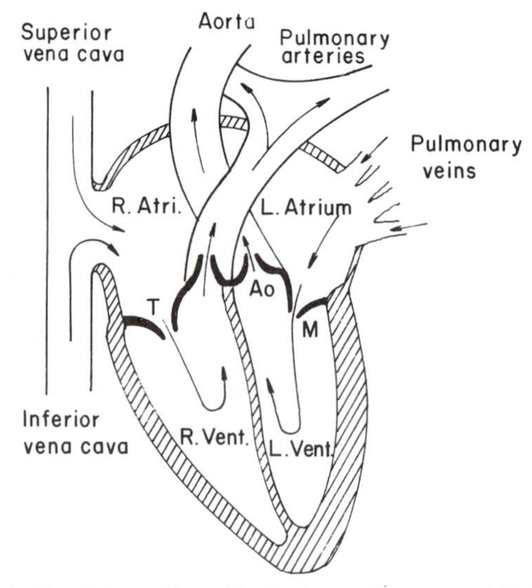

Figure 12-8. Circulation of blood in the heart. Compare with Figure 12-1.

to hemolysis, regurgitation and incompetence were major problems. Later, butterfly leaflets and ball- or disk-in-the-cage were introduced. Some of them are shown in Figure 12-9. The material requirements for valve implants are the same as for vascular implants. Some additional requirements are related to the blood flow and pressure, i.e., the formed elements of blood should not be damaged and the blood pressure should not drop below a clinically significant value. Also, valve noise should be minimal, for psychological reasons.

Figure 12-10 shows a tissue valve made from collagen-rich material such as pericardial tissues. Basically the pericardium is made up of three layers of collagen fibers oriented 60° from each layer and about 0.5 mm thick in the case of bovine pericardium. It can be cross-linked by formaldehyde. During this treatment, the cell viability is destroyed and the proteins are denatured. Therefore, the implant does not provoke immunological reactions. Also, porcine xenograft valves have been used. They are treated with a chemical process that denatures the proteins and kills any living cells.

All valves have a sewing ring that is covered with various polymeric fabrics. This helps during initial fixation of the implant. Later, the ingrown tissue will render the fixation viable in a manner similar to the porous vascular implants. The cage itself is usually made of metals and covered with fabrics, to reduce noise, or with pyrolytic carbons for a nonthrombogenic surface (the disk or ball is also coated with pyrolytic carbon at the same time). The practice of covering

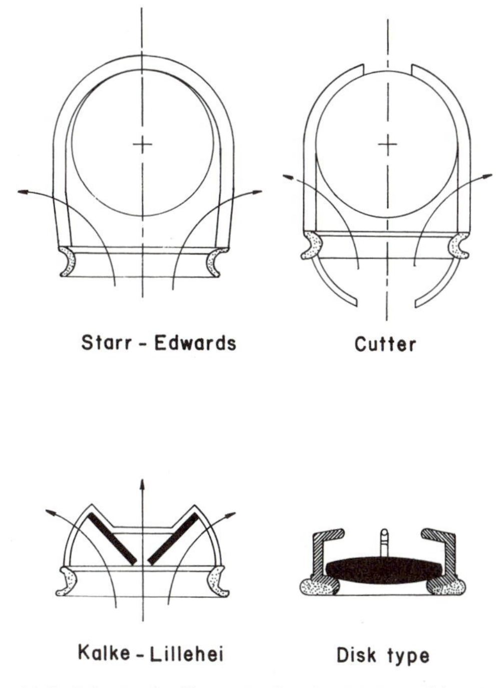

Figure 12-9. Schematic diagram of various types of heart valves.

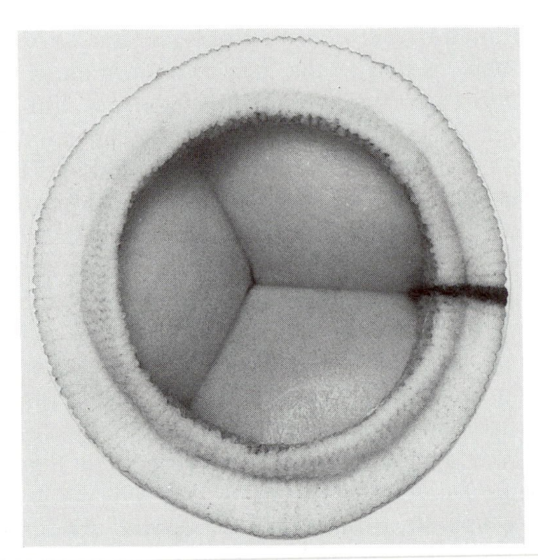

Figure 12-10. Ionescu-Shiley pericardial xenograft heart valve. (Courtesy of Shiley, Inc., Irvine, Calif.)

the struts with fabrics has been abandoned since the fabric fatigued and broke into pieces.

The ball (or disk) is usually made as a hollow structure composed of solid (nonporous) polymers (polypropylene, polyoxymethylene, polychlorotrifluoroethylene, etc.), metals (titanium, Co–Cr alloy), or pyrolytic carbon deposited on a graphite substrate. The early use of a silicone rubber poppet was found undesirable due to lipid absorption and subsequent swelling and dimensional changes. This problem has been corrected in modern valves. Although this was an unfortunate episode (some were fatal), it helped to reinforce the realization that the *in vitro* experiment alone is not sufficient to predict all circumstances that arise during *in vivo* use, no matter how carefully one tries to predict. This is true of any implant research even with very simple devices.

Example 12-2

The bovine pericardium has been tested for its mechanical properties. The stress–strain curve of bovine pericardium is shown in the diagram opposite. Answer the following:

(a) Calculate the initial modulus.
(b) Calculate the secondary modulus.
(c) What is the toughness?

Answer

(a) The initial modulus is, from the slope of the graph,

$$E_i = \frac{15 - 0}{0.75 - 0} = \underline{2.0 \text{ MPa}}$$

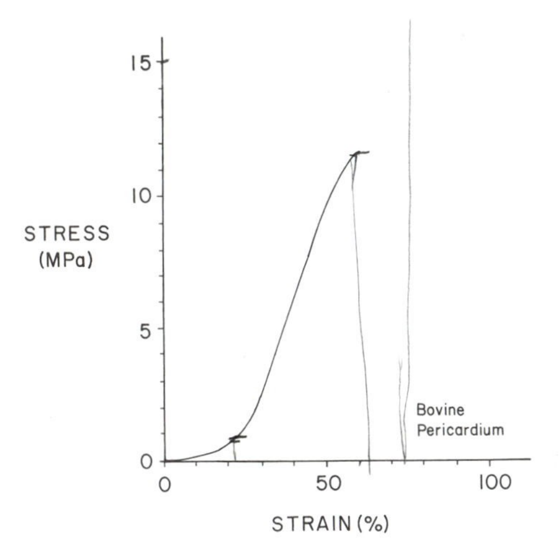

(b) The secondary modulus is, from the slope of the graph,

$$E_s = \frac{15 - 0}{0.62 - 0.21} = \underline{37\text{ MPa}}$$

(c) The toughness is the area under the curve up to the failure strain. It can be approximated by a triangle.

$$\text{Toughness} = 15\text{ MPa} \times \frac{0.62 - 0.21}{2} = 3.1\text{ MPa}\frac{\text{m}}{\text{m}}$$

$$= 3.1\frac{\text{MNm}}{\text{m}^3} = \underline{3.1 \times 10^6}\frac{\text{Nm}}{\text{m}^3}$$

12.3. HEART AND LUNG ASSIST DEVICES

The function of the heart in pumping blood can be temporarily taken over by a mechanical pump. This procedure is most useful in cardiac surgery in which a surgical field free of blood is required. The pump must propel the blood at the correct pressure and flow rate, and its internal parts must be compatible with blood. Moreover, damage to the red blood corpuscles should be minimized. Use of artificial devices to take over the function of the lungs is also common on thoracic surgery. The human heart actually contains two pumps, one for the systemic circulation and one for the pulmonary (lung) circulation. Therefore, even if a cardiac patient has normal lung function, it is usual to assume lung function by a machine, to simplify connections between the pump and the patient's circulatory system. The combination of a blood pump and an oxygenator is known as the *heart and lung* machine.

There are basically three types of oxygenators as shown in Figure 12-11. In all cases oxygen gas is allowed to diffuse into the blood and simultaneously waste gas (CO_2) is removed. In order to increase the rate of gas exchanges at the blood/gas surface of the bubble oxygenator, the gas is broken into small bubbles (about 1-mm diameter; if smaller, it is hard to remove them from blood) to increase the surface contact area. Sometimes the blood is spread thinly as a film and exposed to the oxygen. This is called a film oxygenator. A membrane oxygenator is similar to the membrane-type artificial kidney to be discussed later. The main difference is that the oxygenator membrane is permeable to gases only, while the kidney membrane is also permeable to liquids.

Membrane and bubble oxygenators each have advantages. The membrane oxygenator is considered physiologically superior in view of the fact that there is no blood–gas interface and as a result there is less hemolysis induced by turbulence and better platelet function than with the bubble oxygenator. Moreover, the membrane oxygenator does not introduce any microbubbles or microemboli in the blood. Control of exchange of oxygen and carbon dioxide according to the needs of the patient is readily achieved, and there is no need for antifoam agents such as with the bubble oxygenator. Membrane oxygenators

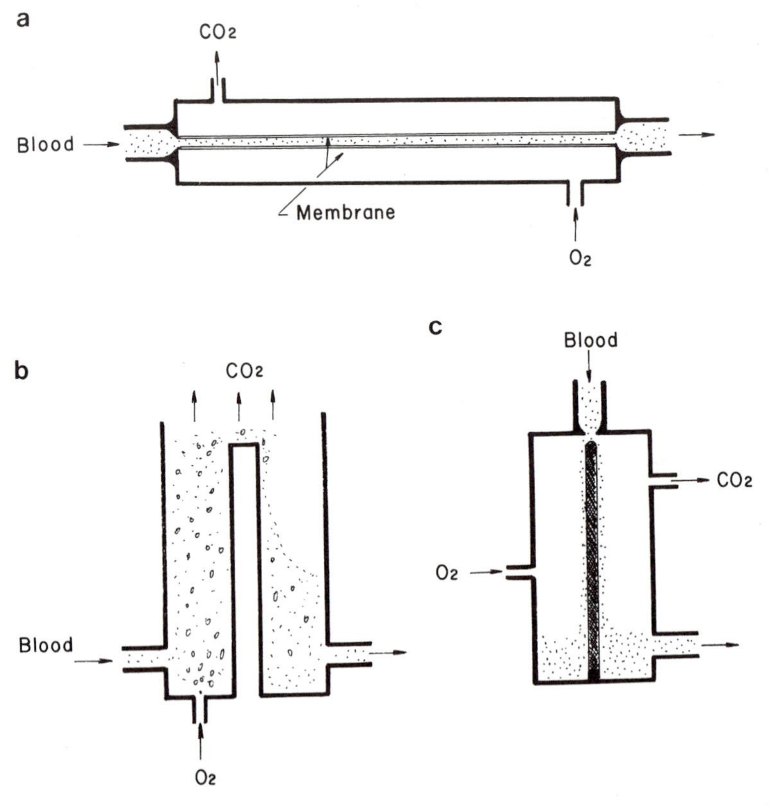

Figure 12-11. Schematic diagrams of different oxygenators: (a) membrane, (b) bubble, and (c) film.

Table 12-1. Physical Characteristics of Natural versus Artificial Lung[a]

Characteristics	Natural lung	Artificial lung
Pulmonary flow (liters/min)	5	5
Head pressure (mm Hg)	12	0–200
Pulmonary blood volume (liters)	1	1–4
Blood transit time (sec)	0.1–0.3	3–30
Blood film thickness (mm)	0.005–0.010	0.1–3.0
Length of capillary (mm)	0.1	20–200
Pulmonary ventilation (liters/min)	7	2–10
Exchange surface (m^2)	50–100	2–10
Veno-alveolar O_2 gradient (mm Hg)	40–50	650
Veno-alveolar CO_2 gradient (mm Hg)	3–5	30–50

[a] From D. O. Cooney, *Biomedical Engineering Principles*, Marcel Dekker, New York, 1976.

are used for prolonged procedures, however; for short surgeries, bubble oxygenators are preferred since they are simpler to operate and consequently less expensive.

Some of the mechanical and chemical characteristics of the natural and artificial lung (oxygenator) are compared in Table 12-1. The surface area of the artificial membrane is about ten times larger than the natural lung since the amount of oxygen transfer through a membrane is proportional to the surface area, pressure, and transit time but inversely proportional to the (blood) film thickness. The blood film thickness for the artificial membrane is about 30 times larger than in the natural lung. This has to be compensated for by increased transit time (16.5 sec) and higher pressure (650 mm Hg) to achieve the same amount of oxygen transfer through the artificial lung.

Table 12-2. Gas Permeability of Teflon and Silicone Rubber Membranes[a,b]

	Thickness (mil)	Oxygen	Carbon dioxide	Nitrogen	Helium
Teflon	1/8	239	645	106	1425
	1/4	117	302	56	730
	3/8	77	181	35	430
	1/2	61	126	30	345
	3/4	41	86	23	240
	1	29			
Silicone rubber	3	391	2072	184	224
	4	306	1605	159	187
	5	206	1112	105	133
	7	159	802	81	94
	12	93	425	48	51
	20	59	279	31	43

[a] From D. O. Cooney, *Biomedical Engineering Principles*, Marcel Dekker, New York, 1976.
[b] Permeation rates of oxygen, carbon dioxide, nitrogen, and helium across Teflon and silicone rubber membranes of a given thickness, in ml/min-m^2-atm (STP).

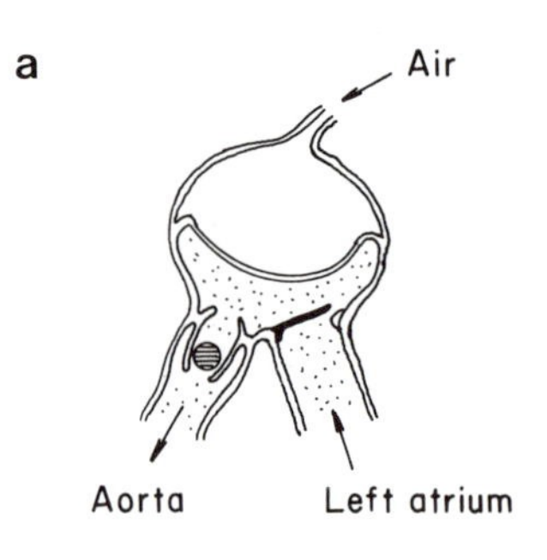

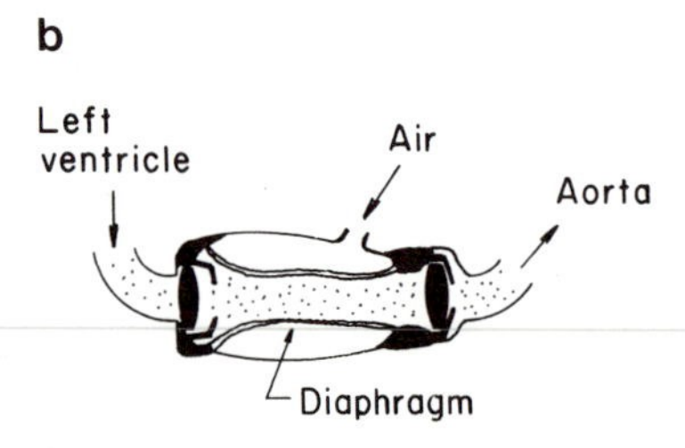

Figure 12-12. Schematic diagram of heart assist devices. (a) DeBakey left ventricular bypass. (b) Bernard–Teco assist pump. From H. Lee and K. Neville, *Handbook of Biomedical Plastics*, Pasadena Technology Press, Pasadena, Calif., 1971.

The membranes are usually made of silicone rubber or PTFE. The gas permeability of these materials is given in Table 12-2. Silicone rubber is 40 and 80 times more permeable to O_2 and CO_2, respectively, than PTFE but the latter can be made about 20 times thinner due to its higher strength. Therefore, silicone rubber is only 2 and 4 times better than PTFE for O_2 and CO_2 transfer, respectively.

Polyurethane, natural and silicone rubbers have been used for constructing balloon-type assist devices such as the left ventricular assist device (LVAD) as shown in Figure 12-12 as well as for coating the inner surfaces of the total artificial heart. This is because these materials are thromboresistant. Sometimes the surfaces are coated with heparin and other nonthrombogenic molecules. The feltlike velour surface was tried but was not successful due to the uneven or minimal tissue attachment.

12.4. ARTIFICIAL ORGANS

The ultimate triumph of biomaterials science and technology would be to make implants behave or function the same way as the organs or tissues they replace without affecting other tissues or organs, and without any negative effect on the patient's mental condition. True regeneration of the natural organ would of course be superior to any artificial one, but that is beyond the scope of biomaterials. As mentioned in the introduction, most implants are designed to substitute mechanical functions, or passive diffusive functions. The electrical functions can be taken over by some implants (pacemakers) and some primitive yet vital chemical functions (passive diffusion) can also be delegated to implants (kidney dialysis machine and oxygenator). Most of the *artificial heart* and *heart assist devices* use a simple balloon and valve system. In all cases a balloon or membrane is used to displace blood. A simpler heart device is the intra-aortic balloon, which is placed in the descending aorta. During the diastolic phase of the heart, the balloon is inflated to prevent back flow.

12.4.1. Artificial Hearts

Several artificial hearts are shown in Figure 12-13. Although the design principle and material requirements are the same as those for assist devices, the power consumption (about 6 watts) is too high for the device to be completely implanted at this time. In current devices the power is introduced through a percutaneous device (Section 11.2.1) in the form of compressed air or electricity. Such an external power unit was used with the first heart replacement done for Dr. B. Clark at the University of Utah in 1982. The artificial heart kept the patient alive for 112 days, but the external power unit restricted his movements. Current artificial hearts are not considered practical as permanent implants since they damage the blood and release emboli into the circulation. They may be useful for short-term use in keeping patients with end-stage heart disease alive until a transplant heart becomes available.

12.4.2. Cardiac Pacemaker

A *cardiac pacemaker* is used to assist the regular contraction rhythm of heart muscles. The sinoatrial (SA) node of the heart originates the electrical impulses that pass through the bundle of His to the atrioventricular (AV) node. In the majority of cases, the pacemakers are used to correct the conduction problem in the bundle of His. Basically the pacemakers should deliver an exact amount of electrical stimulation to the heart at varying heart rates. The pacemaker consists of conducting electrodes attached to a stimulator as shown in Figure 12-14.

The electrodes are well insulated with rubber (usually silicone or polyurethane) except for the tips, which are sutured or directly embedded into the cardiac wall as shown in Figure 12-15. The tip is usually made of a noncorrosive noble metal with reasonable mechanical strength such as Pt-10%Ir alloy. The most significant problems are the fatigue of the electrodes (they are coiled like springs to prevent this) and the formation of collagenous scar tissue at the tip, which increases threshold electrical resistance at the point of tissue contact. The battery and electronic components are sealed hermetically by a titanium case while the electrode outlets are sealed by a polypropylene cuff.

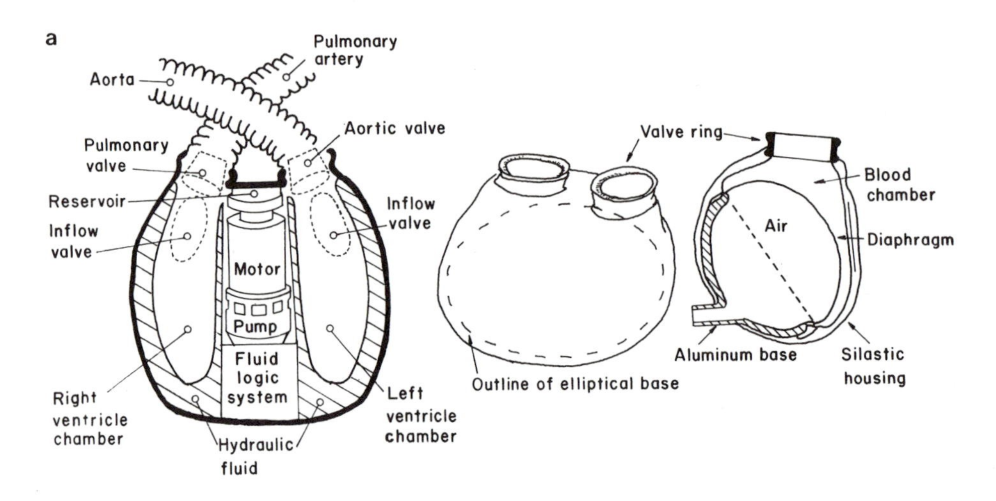

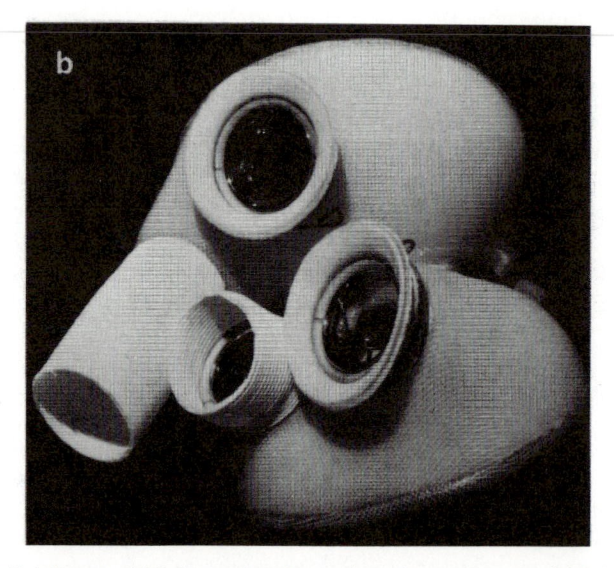

Figure 12-13. Artificial hearts. (a) Schumacker–Burns electrohydraulic heart. (b) Schematic diagram and photograph of the Jarvik heart. (a) From H. Lee and K. Neville, *Handbook of Biomedical Plastics*, Pasadena Technology Press, Pasadena, Calif., 1971; (b) from W. J. Kolff, *Artificial Organs*, J. Wiley and Sons, New York, 1976, and W. J. Kolff, "Artificial Organs and Their Impact," in *Polymers in Medicine and Surgery*, R. L. Kronenthal, Z. Oser, and E. Martin (eds.), Plenum Press, New York, 1975, pp. 1–28.

Pacemakers are usually changed after 2–5 years due to the limitation of the power source. A nuclear energy-powered pacemaker is commercially available. Although this and other new power packs (such as lithium battery) may lengthen the life of the power source, the fatigue of the wires and diminishing conductivity due to tissue thickening limit the maximum life of the pacemaker to less than 10

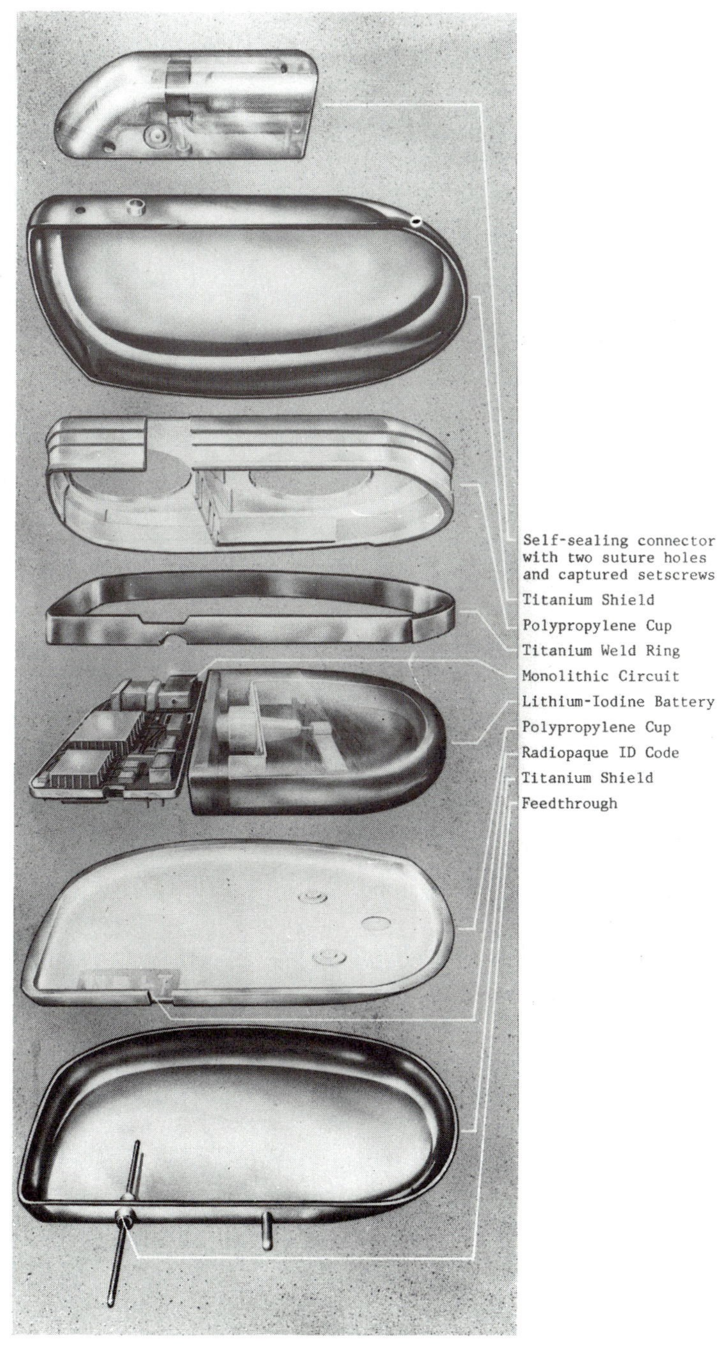

Figure 12-14. A typical pacemaker consists of a power source and electronic circuitry encased in solid plastic. The electrical wires are coated with a flexible polymer, usually a silicone rubber. (Courtesy of Medtronic, Inc.)

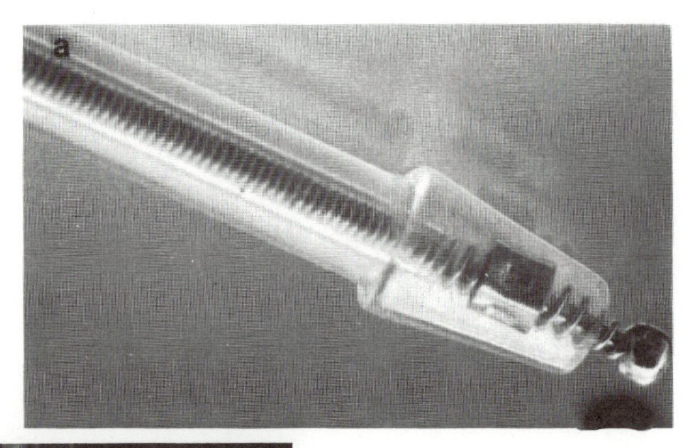

Figure 12-15. Different types of pacemaker electrodes. (a) Ballpoint electrode (Cordis). The ball has a 1-mm diameter and the surface area is 8 mm^2. (b) Screw-in electrode (Medtronic). (c) Details of an arterial electrode (Medtronic 6991). From W. Greatbatch, "Metal Electrodes in Bioengineering," *CRC Crit. Rev. Bioeng., 5,* 1–36, 1981.

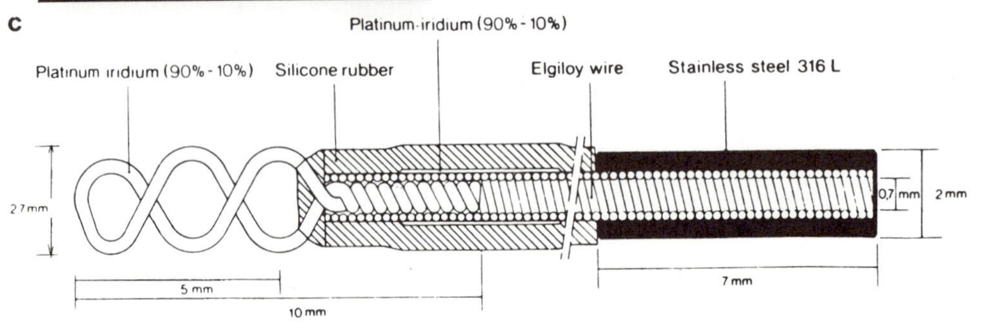

Figure 12-16. Amundsen/CPI porous electrode. From D. Amundsen, W. McArthur, and M. Mosharrafa, "A New Porous Electrode for Endocardial Stimulation," *Pace, 2,* 40, 1979.

years. A porous electrode at the tip of the wires may be fixed to the cardiac muscles by tissue ingrowth as in the case of a vascular prosthesis. This may diminish the interfacial problems as shown in Figure 12-16.

12.4.3. Artificial Kidney Dialysis Membrane

The primary function of the kidney is to remove metabolic waste products. This is accomplished by passing blood through the glomerulus (Figure 12-17) under a pressure of about 75 mm Hg. The glomerulus contains up to 10 primary branches and 50 secondary loops to filter the load. The glomeruli are contained

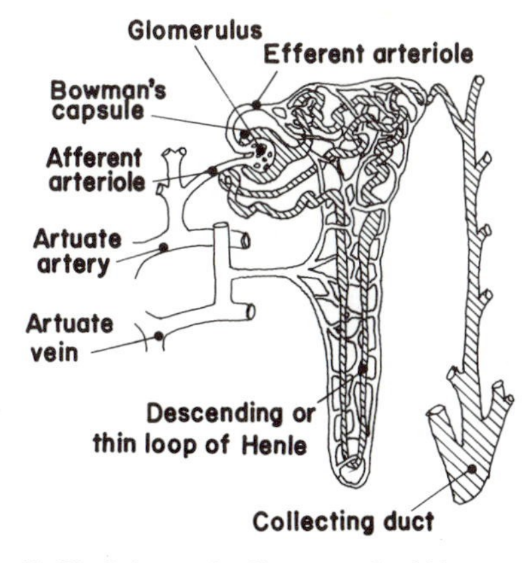

Figure 12-17. Schematic diagram of a kidney nephron.

in Bowman's capsule, which in turn is a part of the nephron, the functional unit of the kidney (see Figure 12-17). The main filtrate is urea (70 times the urea content of normal blood), sodium, chloride, bicarbonate, potassium, glucose, creatinine, and uronic acid.

The artificial kidney uses a synthetic semipermeable membrane to perform the filtering action in a way similar to that of a natural kidney. The membrane is the key component of the artificial kidney machine. In fact, the first attempt to filter or dialyze blood with a machine failed due to an inadequate membrane. In addition to a membrane filter, the kidney dialyzer consists of a bath of saline fluid into which the waste products diffuse out from the blood, and a pump to circulate blood from an artery and return the filtered blood to a vein as shown in Figure 12-18.

There are basically three types of membranes for the kidney dialyzer, shown in Figure 12-19. The flat plate-type membrane was the first to be developed, and can have two or four layers. The blood passes through the spaces between the membrane layers while the dialysate is passed through the spaces between the membrane and the restraining boards. The second and most widely used type is the coil membrane in which two cellophane tubes (each 9 cm in circumference and 108 cm long) are flattened and coiled with an open-mesh spacer material made of nylon. The newest type of kidney is made of hollow fibers. Each fiber has dimensions of 255 and 285 μm inside and outside diameter, respectively, and is 13.5 cm long. Each unit contains up to 11,000 hollow fibers. The blood flows through the fibers while the dialysate is passed through the outside of the fibers. The operational characteristics of the various dialyzers are given in Table 12-3.

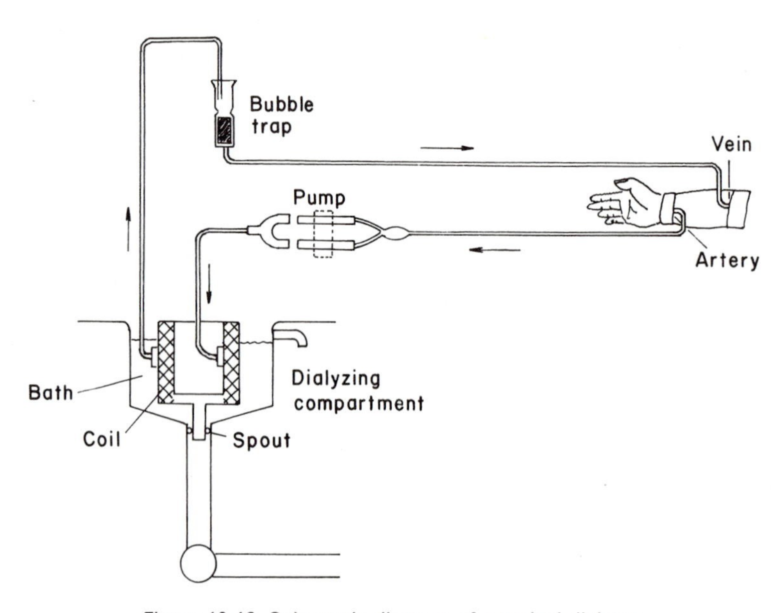

Figure 12-18. Schematic diagram of a typical dialyzer.

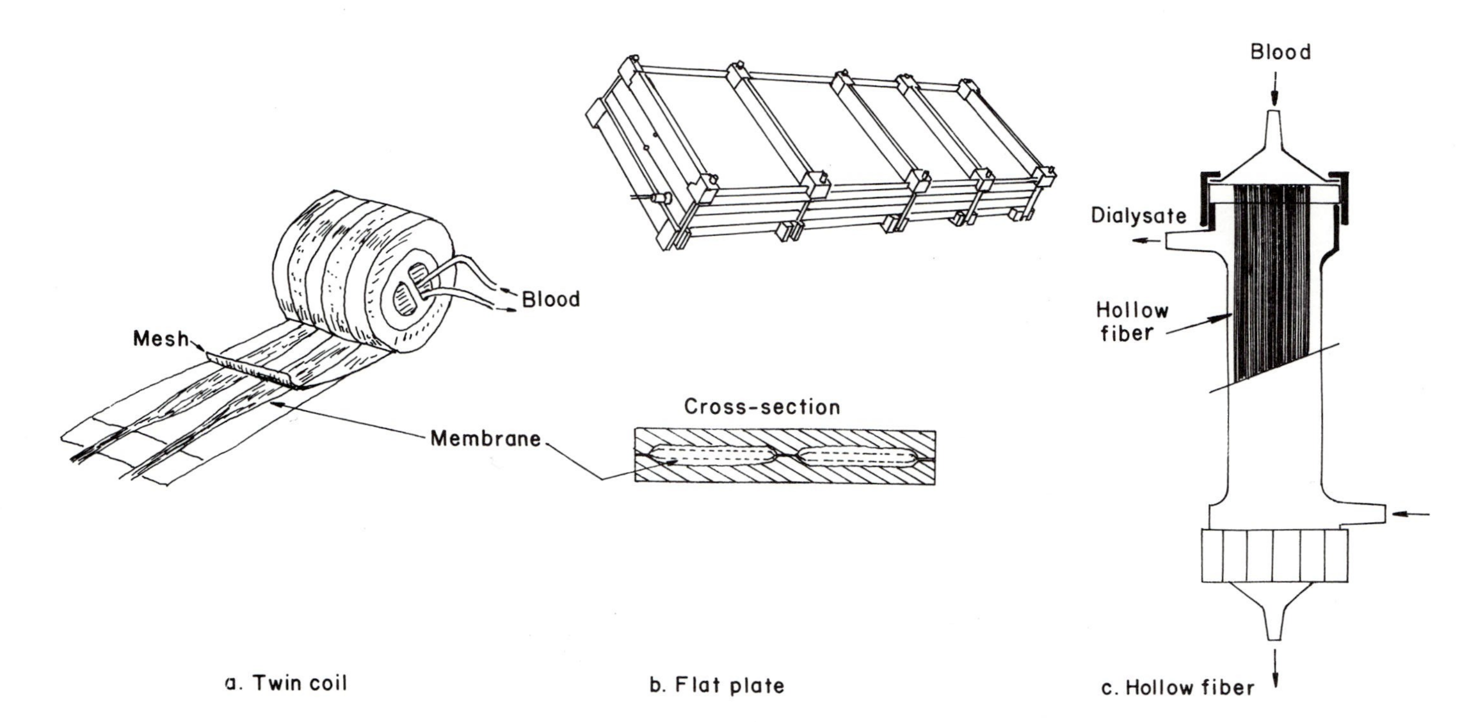

Figure 12-19. Three types of artificial kidney dialyzer; (a) twin coil, (b) flat plate, and (c) hollow fibers.

Table 12-3. Comparison of the Plate and Artificial Kidneys[a]

	Flat plate (2 layers)	Coil (twin)
Membrane area (m^2)	1.15	1.9
Priming volume (liters)	130	1000
Pump needed?	No	Yes
Blood flow rate (ml/min)	140–200	200–300
Dialysate flow rate (liters/min)	2.0	20–30
Blood channel thickness (mm)	0.2	1.2
Treatment time (hr)	6–8	6–8

[a] From D. O. Cooney, *Biomedical Engineering Principles*, Marcel Dekker, New York, 1976.

The fibers can also be made from (soda-lime) glass, which is made porous by phase separation techniques. This type of glass fiber has an advantage over the organic fiber in that it can be reused after cleaning and sterilizing.

Recently there have been some efforts to improve the dialyzers by using charcoals. The blood can be circulated directly over the charcoal or the charcoal can be made into microcapsulates incorporating enzymes or other drugs. One drawback of activated carbon filtering is its ineffective absorption of urea, which is one of the major by-products to be eliminated by the dialysis.

The majority of dialysis membranes are made from cellophane, which is derived from cellulose. Ideally, the membrane should selectively remove all of the metabolic wastes as does the normal kidney. Specifically, the membrane should not selectively sequester materials from dialyzing fluid, should be blood compatible so that an anticoagulant is not needed, and should have sufficient wet strength to permit ultrafiltration without significant dimensional changes. It should allow passages of low-molecular-weight waste products while preventing passage of plasma proteins.

There are two clinical-grade cellophanes available, Cuprophane® (Bemberg Co., Wuppertal, Germany) and Visking® (American Viscose Co., Fredericksburg, Va.). The cellophane films contain 2.5-μm-diameter pores which can filter molecules smaller than 4000 g/mol.

There have been many attempts to improve the cellophane membrane wet strength by cross-linking, copolymerization, and reinforcement with other polymers such as nylon fibers. Also, the surface was coated with heparin in order to prevent clotting. Other membranes such as copolymers of polyethylene glycol and polyethylene terephthalate (PET) can filter selectively due to their alternate hydrophilic and hydrophobic segments. Besides improving the membrane for better dialysis, the main thrust of kidney research is to make the kidney machine more compact (portable or wearable kidney, Figure 12-20a) and less costly (home dialysis, reusable or disposable filters, etc.).

The other important factor in dialysis is the cannula, which is used to gain access to the blood vessels. In order to minimize the repeated trauma on the blood vessels the cannula can be implanted for a long period for chronic kidney

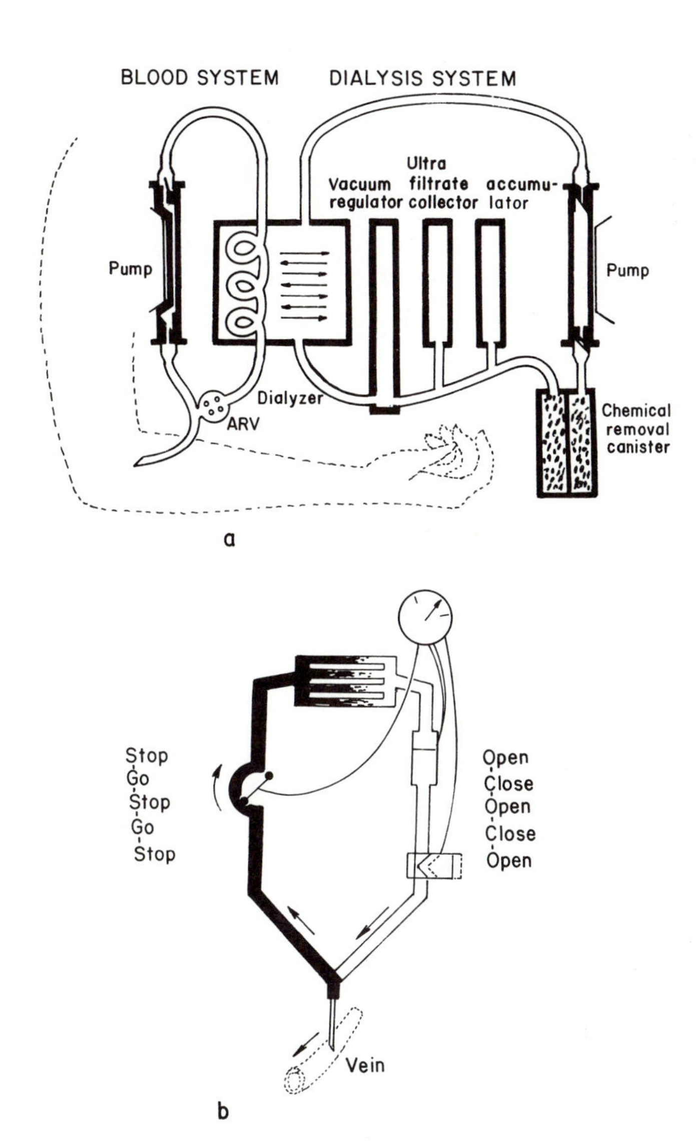

Figure 12-20. (a) Diagram of wearable artificial kidney; (b) schematic arrangement of the single-needle dialysis of Dr. Klaus Kopp. The pump operates continuously or intermittently synchronized with the inflow and outflow of blood. (a) From W. J. Kolff, "Artificial Organs and Their Impact," in *Polymers in Medicine and Surgery*, R. L. Kronenthal, Z. Oser, and E. Martin (eds.), Plenum Press, New York, 1975, pp. 1–28; (b) From W. J. Kolff, *Artificial Organs*, J. Wiley and Sons, New York, 1976.

patients. For the same reason a single-needle dialysis technique has been developed as shown in Figure 12-20b.

In addition to hemodialysis by a machine, dialysis can also be carried out by using the patient's own peritoneum, which is a semipermeable membrane. The blood is brought to the membrane through the microcirculation of the peritoneum while dialysate is introduced into the peritoneal cavity through a catheter, one of which is shown in Figure 11-7 implanted in the abdominal wall. The dialysate is drained through the same catheter after solute exchange takes place and is replaced by a fresh bottle. Glucose is added to the dialysate to increase its osmotic pressure gradient for ultrafiltration since it is impossible to obtain a high hydrostatic pressure gradient.

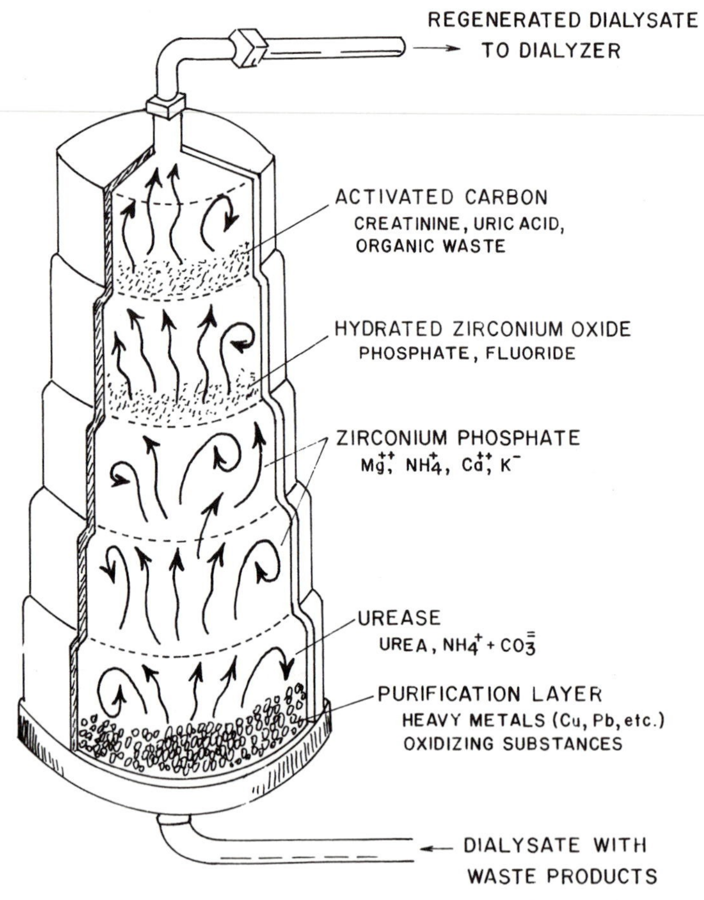

Figure 12-21. Schematic diagram of a sorbent regeneration cartridge. From R. A. Ward, "Investigation of the Risk and Hazards with Devices Associated with Peritoneal Dialysis and Sorbent Regenerated Dialysate Delivery Systems," FDA Contract No. 223–81–5001, revised draft report, 1982.

Recently a sorbent cartridge that regenerates the dialysis fluid for reuse has been advanced. The cartridge was originally developed for hemodialysis and could reduce the water requirement substantially, making dialysis more portable. A schematic diagram of the sorbent regenerated cartridge is shown in Figure 12-21. It is interesting to note that in order to remove urea, enzymolysis is carried out by urease since carbon cannot absorb urea effectively as mentioned before.

Example 12-3

Calculate the number of ions released in a year from a platinum pacemaker tip. Assume that the average current flow is 10 μA and that the surface area is 1 cm^2.

Answer

$Pt \rightarrow Pt^2 + 2e^-$, so that the number of atoms per year is

$$10 \times 10^{-6} \frac{\text{coul}}{\text{sec}} \frac{3.15 \times 10^7 \text{ sec/yr}}{1.6 \times 10^{-19} \text{ coul/electron}} = 1.97 \times 10^{21} \text{ electrons/yr}$$

$$= \underline{9.9 \times 10^{20} \text{ Pt atoms/yr}}$$

PROBLEMS

12-1. What will be the blood urea nitrogen concentration after 5 and 10 hours of dialysis if the initial concentration is 100 mg% (mg% is concentration in mg per 100 ml)? The concentration after dialysis can be expressed exponentially,

$$C_t = C_o \exp\left(\frac{Q_b(b-1)t}{V}\right)$$

in which C_o is the original dialysate concentration, Q_b is the blood flow rate, t is time, V is the volume of body fluid (60% of body weight), and b is a constant determined by mass transfer coefficient (K), Q_b, and surface area (A) of the membrane by $\exp(KA/Q_b)$ according to Cooney (*Biomedical Engineering Principles*, Marcel Dekker, New York, 1976, p. 332).

12-2. Select the most related match.

(a) low coefficient of friction	1. bone ( )
(b) tensile force transmission	2. skin ( )
(c) Laplace equation	3. tendon ( )
(d) Haversian system	4. artery ( )
(e) elastin	5. ligamentum nuchae ( )
(f) Langer's line	6. cartilage ( )

(a) Al_2O_3	1. fluidized bed ( )
(b) hydroxyapatite	2. resorbable ceramic ( )
(c) pyrolytic carbon	3. sapphire ( )
(d) calcium phosphate	4. mineral phase of bone ( )
(e) Bioglass®	5. piezoelectric ( )
(f) barium titanate	6. SiO_2-CaO-Na_2O-P_2O_5 ( )
(g) graphite	7. substrate of heart valve disk ( )

(a) PMMA	1. polyolefin ( )
(b) polyamide	2. styrene–butadiene copolymer ( )
(c) poly-HEMA	3. natural rubber ( )
(d) polyacrylamide	4. Teflon ( )
(e) polytetrafluoroethylene	5. nylon ( )
(f) SBR	6. soft contact lens ( )
(g) polyisoprene	7. bone cement ( )
(h) polypropylene	8. hydrogel ( )

12-3. Answer the following questions:

(a) Calculate the maximum wall tension (N/mm) due to internal pressure developed for a 0.5-cm-diameter artery. Assume the maximum pressure will be 250 mm Hg and the artery is uniform in the longitudinal direction.

(b) If one wishes to replace a 5-cm-long section of the artery, what is the maximum force exerted on the wall?

(c) Can one use silicone rubber for the replacement material if the wall thickness is 1 mm and the safety factor is 10?

12-4. Design a heart valve and give specific materials selected for each part and your reason for each selection.

12-5. Is it practical to use porous glass tubules for the hollow-fiber-type kidney dialysis machine? Suppose that the porous glass has filtering characteristics similar to those of currently used materials, by virtue of appropriate pore size and tubule thickness.

12-6. The surface of a kidney dialysis membrane is coated with poly-HEMA (hydroxyethyl methacrylate). Discuss the advantages and disadvantages.

DEFINITIONS

Anastomosis: Interconnection between two blood vessels.

Aneurysm: Abnormal dilatation or bulging of a segment of a blood vessel, often involving the aorta or pulmonary artery.

Autograft: A transplanted tissue or organ transferred from one part of a body to another part of the same body.

Denier: A unit of weight for measuring the fineness of threads of silk, rayon, nylon, etc., equal to 1 gram per 9000 meters.

Glomerulus: A small tuft or cluster, as of blood vessels or nerve fibers; applied especially to the coils of blood vessels, one projecting into the expanded end or capsule of each of the uriniferous tubules of the kidney.

Incompetence: Incomplete closure of a heart valve.

Neointima: New lining. Intima (blood vessel lining) formed by the ingrowth of tissues through pores of a vascular graft.

Oxygenator: An apparatus by which oxygen is introduced into the blood during circulation outside the body, as during open-heart surgery.

Pacemaker (cardiac): A device designed to stimulate, by electrical impulses, contraction of the heart muscle at a certain rate.

Pseudointima: See neointima.

Regurgitation: Back flow of blood at a heart valve.

Replamineform: Devices or materials made by replicating natural tissues such as sea urchin tentacles. The objective is to achieve better tissue ingrowth into the pores.

Tenacity: A normalized strength of a fiber; strength per unit size (e.g., expressed in terms of denier). Tensile strength = tenacity × density × constant, in appropriate units.

Xenograft: A transplanted tissue or organ transferred from an individual of another species.

BIBLIOGRAPHY

T. Akutzu (ed.), *Artificial Heart 1,* Springer-Verlag, Berlin, 1986.

S. D. Bruck, *Blood Compatible Synthetic Polymers: An Introduction,* Charles C. Thomas, Publ., Springfield, Ill., 1974.

A. H. Bulbulian, *Facial Prosthetics,* Charles C. Thomas, Publ., Springfield, Ill., 1973.

K. B. Chandran, *Cardiovascular Biomechanics,* NYU Press, New York, 1992.

D. O. Cooney, *Biomedical Engineering Principles,* Marcel Dekker, New York, 1976.

H. Dardik (ed.), *Graft Materials in Vascular Surgery,* Year Book Medical Publ., Chicago, 1978.

G. H. Gyers and V. Parsonet, *Engineering in the Heart and Blood Vessels,* J. Wiley and Sons, New York, 1969.

A. D. Haubold, H. S. Shim, and J. C. Bokros, "Carbon Cardiovascular Devices," in *Assisted Circulation,* F. Unger (ed.), Academic Press, New York, 1979, pp. 520–532.

C. A. Homsy and C. D. Armeniades (eds.), Biomaterials for Skeletal and Cardiovascular Applications, *J. Biomed. Mater. Symp.,* No. 3, J. Wiley and Sons, New York, 1972.

H. Lee and K. Neville, *Handbook of Biomedical Plastics,* Chapters 3–5 and 13, Pasadena Technology Press, Pasadena, 1971.

J. E. Liddicoat, S. M. Bekassy, A. C. Beall, D. H. Glaeser, and M. E. DeBakey, "Membrane vs Bubble Oxygenator: Clinical Comparison," *Ann. Surg., 184,* 747–753, 1975.

W. Lynch, *Implants: Reconstructing the Human Body,* Van Nostrand Reinhold Co., Princeton, N.J., 1982.

P. N. Sawyer and M. H. Kaplitt, *Vascular Grafts,* Appleton-Century-Crofts, New York, 1978.

C. P. Sharma and M. Szycher (eds.), *Blood Compatible Materials and Devices,* Technomic, Lancaster, Pennsylvania, 1991.

C. Stanley, W. E. Burkel, S. M. Lindenauer, R. H. Bartlett, and J. G. Turcotte (eds.), *Biologic and Synthetic Vascular Prostheses,* Grune & Stratton, New York, 1972.

Transactions of American Society for Artificial Internal Organs, Published yearly and contains studies related to this chapter.

A. B. Voorhees, A. Jaretski, and A. H. Blackmore, "Use of Tubes Constructed from Vinyon-N Cloth in Bridging Arterial Defects," *Ann. Surg., 135,* 332–336, 1952.

D. F. Williams (ed.), *Systemic Aspects of Blood Compatibility,* CRC Press, Boca Raton, Fla., 1981.

CHAPTER 13

HARD TISSUE REPLACEMENT I: LONG BONE REPAIR

The design principles, selection of materials, and manufacturing criteria for orthopedic implants are the same as for any other engineering products undergoing dynamic loading. Although it is tempting to duplicate the natural tissues with materials having the same strength and shape, this has not been practical or desirable since the natural tissues and organs have one major advantage over the man-made implants, i.e., their ability to adjust to a new set of circumstances by remodeling their micro- and macrostructure. Consequently, the mechanical fatigue of tissues is minimal unless a disease hinders the natural healing processes or unless they are overloaded beyond their ability to heal.

When we try to replace a joint or heal a fractured bone, it is logical that bone repairs should be made according to the best repair course that the tissues themselves follow. Therefore, if the bone heals faster when a compressive force (or strain) is exerted, then we should provide compression through an appropriate implant design. Likewise, if the compression is detrimental for the wound healing, the opposite approach should be taken. Unfortunately, the effects of compressive or tensile forces on the repair of long bones are not fully understood. Moreover, many experimental results thus far are contradictory.

It is believed that the osteogenic and osteoclastic activity is related to the normal activities of the bone *in vivo.* Thus, the equilibrium between *osteogenic* and *osteoclastic* activities can be balanced according to the static and dynamic force applied *in vivo,* i.e., if more load is applied the equilibrium tilts toward more osteogenic activity to counteract the load and vice versa (*Wolff's law*) as shown in Figure 13-1. Of course, excessive load should not be imposed by the implant; too much force can damage the cells rather than enhance their activities.

This cause-and-effect relationship may also be related to the *piezoelectric* phenomenon of bone in which the *strain-generated electric potentials* may trigger the bone remodeling response. This is the basis of the electrically stimulated fracture repair of clinical *nonunions.* The strain-generated potentials in living

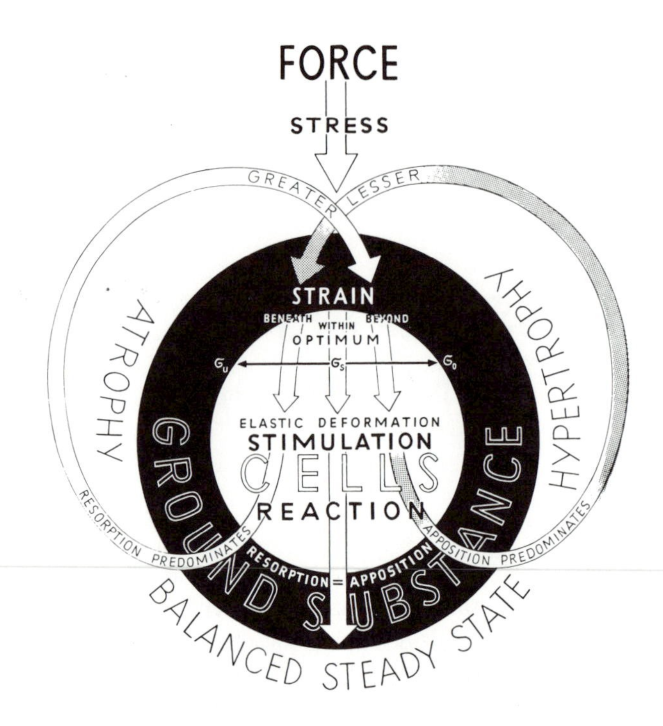

Figure 13-1. A proposed model for feedback mechanism of bone remodeling process due to mechanical energy input. From B. K. F. Kummer, "Biomechanics of Bone," in *Biomechanics: Its Foundations and Objectives,"* Y. C. Fung, N. Perrone, and M. Anliker (eds.), Prentice–Hall, Englewood Cliffs, N.J., 1972, p. 263.

bone are thought to be a result of streaming potentials from fluid flow in the channels in bone.

Historically speaking, until Dr. J. Lister's aseptic surgical technique was developed in the 1860s, various metal devices such as wires and pins constructed of iron, gold, silver, platinum, etc. were not successful largely because of infection after implantation. Most of the modern implant developments have been centered around repairing long bones and joints. Lane of England designed a fracture plate in the early 1900s using steel as shown in Figure 13-2. Sherman of Pittsburgh modified the Lane plate to reduce stress concentration by eliminating sharp

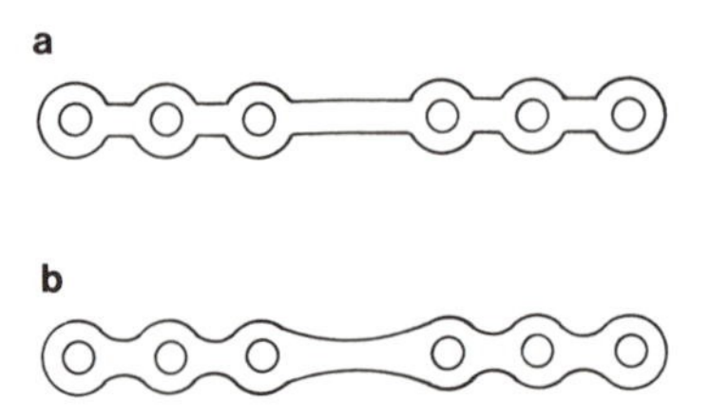

Figure 13-2. Early design of bone fracture plate: (a) Lane; (b) Sherman.

corners (Figure 13-2b). He used vanadium alloy steel for its toughness and ductility. Vanadium steel was used clinically for some years, but problems with *in vivo* corrosion led to its abandonment. Subsequently, Stellite® (Co–Cr based alloy) was found to be the most inert material for skeletal implantation by Zierold in 1924. Soon after, 18-8 (18 wt% Cr, 8 wt% Ni) and 18-8sMo (2–4 wt% Mo) stainless steels were introduced for their corrosion resistance, with 18-8sMo being especially resistant in saline solution. Later, another stainless steel (19 wt% Cr, 9 wt% Ni) named Vitallium® was introduced into medicine. (The name Vitallium® is now used for the Co-based alloy.) Another metal, tantalum, was introduced in 1939 but its poor mechanical properties made it unpopular in orthopedics, yet it found wide use in neurological and plastic surgery. During the post-Lister period, all of the various designs and materials could not be related specifically to the success of an implant despite many studies. It became customary to remove any metal implant as soon as possible after its initial function was served.

13.1. WIRES, PINS, AND SCREWS

Although the exact mechanism of bone fracture repair is not known at this time, stability of the implant with respect to the wound surfaces is clinically known to be an important factor to be considered. Whether the fixation is accomplished by the compressive or tensile force, the reduction (i.e., the placement of the broken bone ends) should be anatomical and the bone ends should be firmly fixed so that the healing processes cannot be disturbed by unnecessary micro- and macromovement. Bone fractures may be conservatively managed (fixed by external means such as a cast), or surgically reduced and fixed. Surgical techniques usually involve the use of metallic fixation devices. These vary in complexity from simple pins to complex multicomponent hip nails. Almost all of these devices are made from metal alloys (discussed in Chapter 5).

13.1.1. Wires

The simplest but most versatile implants are the various metal wires [called Kirschner wires for a diameter less than 3/32 inch (2.38 mm) and Steinman pins for a larger diameter] which can be used to hold fragments of bones together. Wires are also used to reattach the greater trochanter in hip joint replacements or for long oblique or spiral fracture of long bones. The common problems of fatigue combined with corrosion of metals may weaken the wires *in vivo*. The added necessity of twisting and knotting the wires for fastening aggravates the problem since strength can be reduced by 25% or more because of the stress concentration effect. The deformed region of the wire will be more prone to corrosion than the undeformed region as a result of the higher strain energy. The wires are classified as given in Table 13-1.

Table 13-1. Nomenclature and Specifications of Surgical Wires

Suture size	Wire gage No.	American wire gage (diameter)		Standard wire gage (diameter)		Suture diameter (mm)		Knot-pull tensile strength (kg_f) class 3
		in.	mm	in.	mm	Min.	Max.	
10-0						0.013	0.025	0.05
9-0						0.025	0.038	0.06
8-0						0.038	0.051	0.11
7-0						0.051	0.076	0.16
6-0	40	0.0031	0.079	0.0048	0.1222	0.076	0.102	0.27
6-0	38	0.0040	0.102	0.0060	0.152			
5-0	35	0.0056	0.142	0.0084	0.213	0.102	0.152	0.54
4-0	34	0.0063	0.160	0.0092	0.234			
4-0	32	0.0080	0.203	0.0108	0.274	0.152	0.203	0.82
000	30	0.0100	0.254	0.0124	0.315	0.203	0.254	1.36
00	28	0.0126	0.320	0.0148	0.376	0.254	0.330	1.80
0	26	0.0159	0.404	0.0180	0.457	0.330	0.406	3.40
1	25	0.0179	0.455	0.0200	0.508	0.406	0.483	4.76
2	24	0.0201	0.511	0.0220	0.559	0.483	0.559	5.90
3	23	0.0226	0.574	0.0240	0.610	0.559	0.635	7.26
4	22	0.0254	0.643	0.0280	0.712	0.635	0.711	9.11
5	20	0.0320	0.813	0.0360	0.915	0.711	0.813	11.40
6	19	0.0359	0.912	0.0400	1.020	0.814	0.914	13.60
7	18	0.0403	1.061	0.0480	1.220	0.914	1.016	15.90
	10	0.0109	2.590	0.1280	3.260			
	1	0.2893	7.340	0.3000	7.630			
	0	0.3249	8.250	0.3240	8.230			
	6-0	0.5800	14.700	0.4640	11.800			
	7-0			0.5000	12.700			

13.1.2. Pins

The Steinman pin is also a versatile implant and often used for internal fixation in cases when it is difficult to use a plate or when adequate stability cannot be obtained by any other means. The tip of the pin is designed to penetrate bone easily when the pin end is screwed into the bone. The fluting of the pin end differs from that of screws in that the flute angles of the pin are opposite to those of the screws. The reason is, in contrast to a screw or a drill, there is a lack of space between the hole created and the pin. Three types of tip designs are shown in Figure 13-3. The trochar tip is the most efficient in cutting, and hence is often used for cortical bone insertion. The fractured bones can be held together by two or more pins inserted percutaneously away from the fracture site and the pins are fixed by a device such as a Hoffmann external fixator.

13.1.3. Screws

Screws are some of the most widely used devices for fixation of bone fragments to each other or in conjunction with fracture plates. Figure 13-4

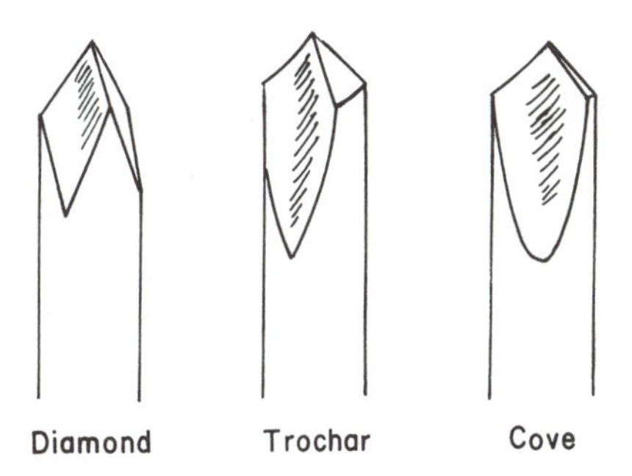

Figure 13-3. Types of Steinman pin tip.

illustrates the various parts of the screw with various head designs. There are basically two types of screws: one is the *self-tapping*, and the other is *non-self-tapping* (Figure 13-5). As the name indicates, the self-tapping screw cuts its own threads as it is screwed in. The extra step of tapping (i.e., cutting threads in the bone) required of the non-self-tapping screws makes them less favorable although the holding power (or pull-out strength) of the two types of screws is about the same. The variations of the thread design (Figure 13-6) do not influence the holding power. However, the radial stress transfer between the screw thread and

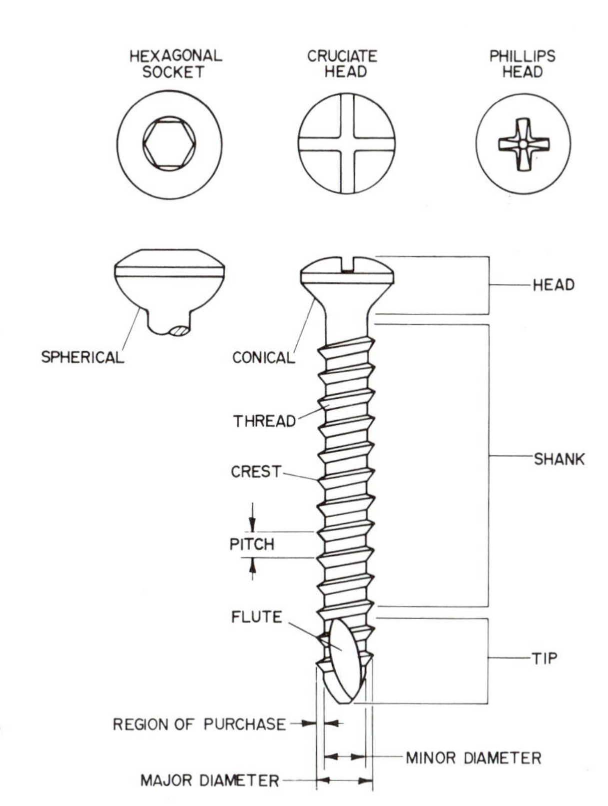

Figure 13-4. Illustration showing the various parts of a self-tapping bone screw, including various head designs. From D. C. Mears, *Materials and Orthopedic Surgery*, Williams and Wilkins, Baltimore, 1979.

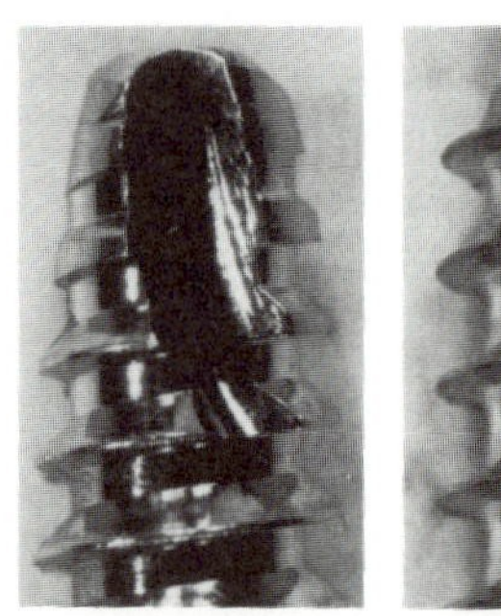

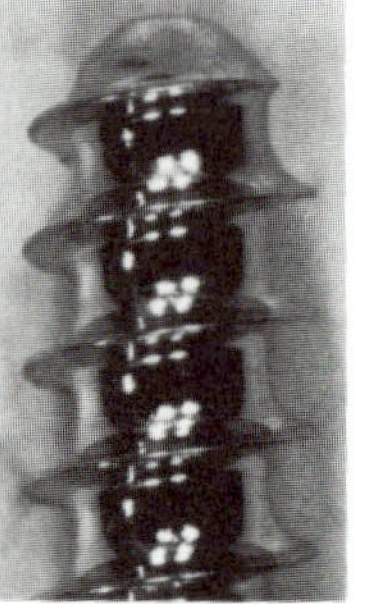

Figure 13-5. Photographs of the points of a self-tapping and a non-self-tapping screw.

bone is slightly less for the V-shaped thread than the buttress thread, indicating the latter can withstand the longitudinal load slightly better.

The rake angle of the cutting edge is also an important factor of a screw design (Figure 13-7). The positive rake angle requires higher cutting force yet results in lower cutting temperature due to less drag while the opposite effect is obtained by the negative rake angle. Almost all bone screws are made with positive rake angles while hard metal cutting drills that can withstand larger cutting loads are made with negative rake angles.

The pull-out strength or holding strength of the screws is an important factor in the selection of a particular screw design. However, regardless of the differences in design, the pull-out strength depends only on the size (diameter) of the screw, as illustrated in Figure 13-8. The larger screw, of course, has a higher pull-out strength.

The tissues immediately adjacent to the screw often necrose and reabsorb initially, but if the screws are firmly fixed then the dead tissues are replaced by

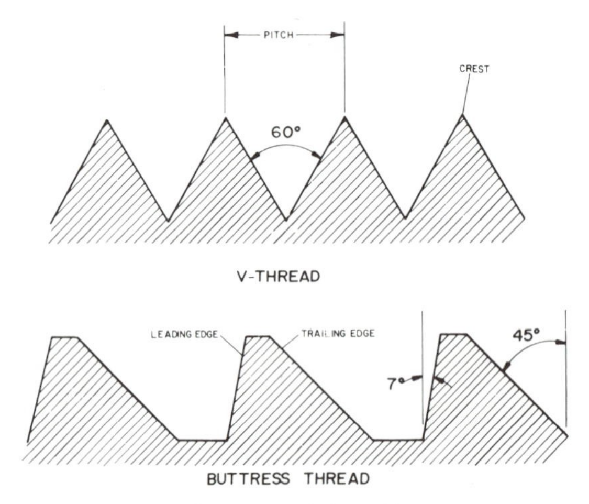

Figure 13-6. Types of thread in screws.

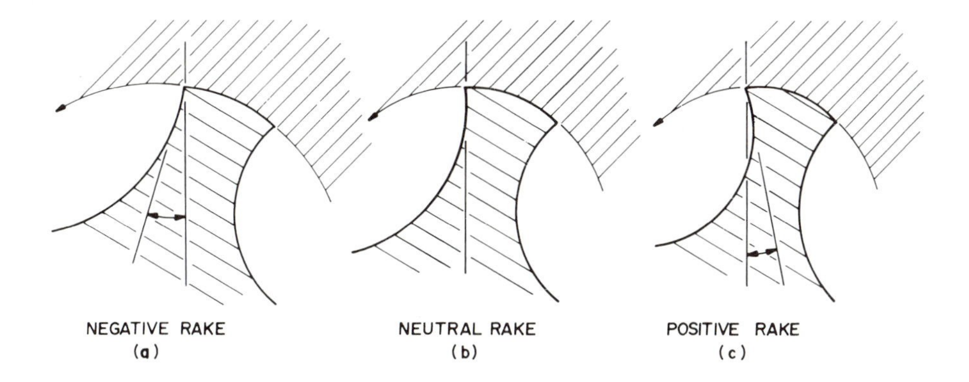

Figure 13-7. Illustration showing the relationship of various rake angles to the outer edge of the cross section of a cutting flute.

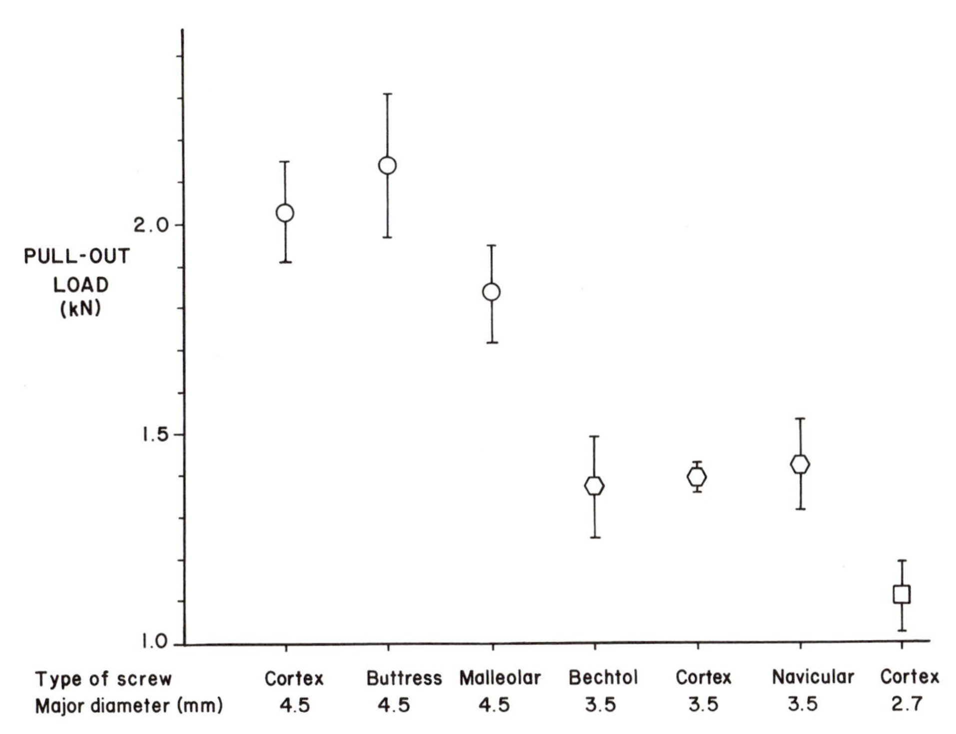

Figure 13-8. Average pull-out strength with 95% confidence limits for various Richards and Richards Osteo bone screws. From Smith & Nephew Richards, Inc., Bone Screw Technical Information, Technical Publ. No. 4167, Memphis, Tenn., 1980.

living tissues. When micro- or macromovement exists, a collagenous fibrous tissue encapsulates the screw. This is the reason why the loading of the repaired bone by the patient should be delayed until a firm fixation takes place between the screw and the bone.

Example 13.1

Calculate the breaking strength of the size 0 suture wire given in Table 13-1. Compare the result with the tensile strength of fully annealed 316L stainless steel (Table 5-2).

Answer

Since the knot pulling strength (3.40 kg_f) is given in Table 13-1,

$$\text{Tensile strength} = \frac{\text{force}}{\text{area}} = \frac{3.4\ kg_f}{(0.404\ mm)^2/4} = 26.52\ kg_f/\ mm^2$$

$$\text{But } 1\ kg_f = 1\ kg \times 9.81\ m/sec^2 = 9.81\ N$$

$$\text{Tensile strength} = 26.52 \times 9.8 \times 10^6\ Pa = \underline{260\ MPa}$$

The ultimate tensile strength of annealed 316L stainless steel is 73,000 psi or 505 MPa, which is almost twice the knot pulling strength. You may recall that other sutures have the same strength reduction after knotting, as a result of stress concentration.

13.2. FRACTURE PLATES

13.2.1. Cortical Bone Plates

There are many different types and sizes of fracture plates, as shown in Figure 13-9. Since the forces generated by the muscles in the limbs are very large generating large bending moments (see Table 13-2), the plates must be strong. This is especially true for the femoral and tibial plates. The bending moment versus bend angle (rotation) of various devices is plotted in Figure 13-10. In comparison with the bending moment at the proximal end of the femur during normal activities (see Table 13-2), one can see that the plates cannot withstand the maximum bending moment applied. Therefore, some type of restriction on the patient's movement is essential in the early stage of healing.

Equally important is adequate fixation of the plate to the bone with the screws, as mentioned previously. However, overtightening may result in necrotic bone as well as deformed screws, which may fail later due to the corrosion process at the deformed region (see Section 5.6.1).

A bone plate device to compress the ends of the fractured bones together is shown in Figure 13-11. A similar effect can be achieved by using a self-compression plate and screw system. The added complexity of the devices and the controversy as to whether the compressive force or strain is beneficial or not had hindered early acceptance of the devices. It is interesting to note that traditionally a large amount of *callus formation* has been considered to be a favorable sign and even

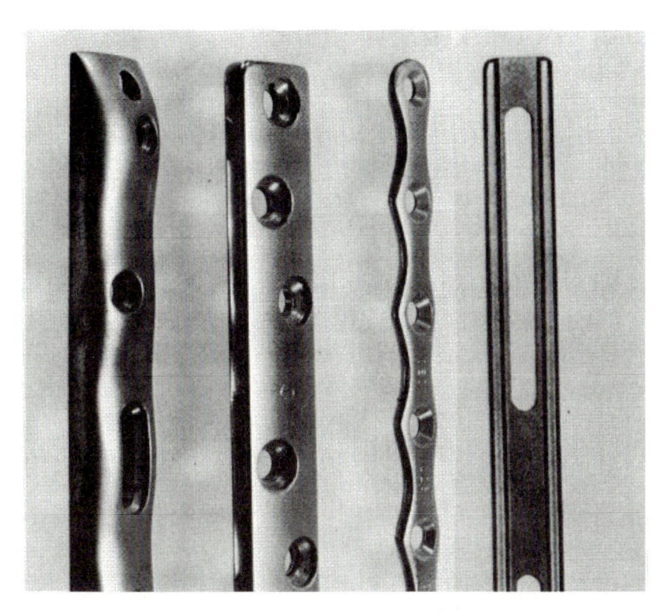

Figure 13-9. Bone plates. From left: Richard-Hirschorn plate (stress is evenly distributed throughout the length of the plate); AO compression plate; Sherman plate (stress is evenly distributed throughout the length of the plate, but the plate is much weaker than the two plates on the left); Egger's plate. From J. A. Albright, T. R. Johnson, and S. Saha, "Principles of Internal Fixation," in *Orthopedic Mechanics: Procedures and Devices,* D. N. Ghista and R. Roaf (eds.), Academic Press, New York, 1978, pp. 123–199.

essential for good healing. However, with the use of the compression plate, the opposite is thought to be a more favorable sign of healing. In this situation the amount of callus formed is proportional to the amount of motion between the plate and the bone. One major drawback of healing by rigid plate fixation is the weakening of the underlying bone such that refracture may occur following

Table 13-2. Greatest Resistable Bending Moment at Proximal End of Femur[a]

	No. of subjects		Bending moment (N m)	
Muscle group	Men	Women	Range	Mean
Hamstring	11		54–93	72
		17	26–54	35
Quadriceps	6		42-60	51
Hip abductors	6		38–108	63
		3	24–48	39
Hip adductors	6		60–126	81
		3	32–40	30

[a] From M. Laurance, M. A. R. Freeman, and S. A. V. Swanson, "Engineering Considerations in the Internal Fixation of Fractures of the Tibial Shaft," *J. Bone Joint Surg.*, *51B*, 754–768, 1969.

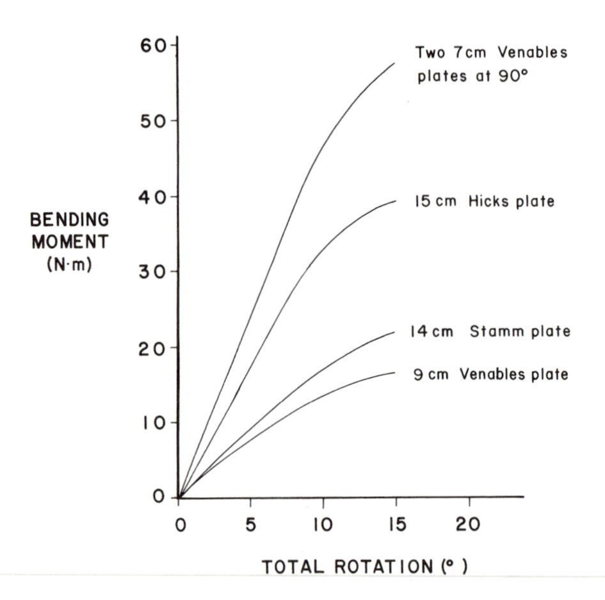

Figure 13-10. Bending moment versus total rotation of various bone plates. From M. Laurance, M. A. R. Freeman, and S. A. V. Swanson, "Engineering Considerations in the Internal Fixation of Fractures of the Tibial Shaft," *J. Bone Joint Surg., 51B*, 754–768, 1969.

removal of the plate. This is largely due to the *stress-shield effect* on the bone underneath the rigid plate. The stiff plate can carry so much of the load that the bone is understressed, so that it is reabsorbed by the body according to Wolff's law. In addition, the tightening of the screws that hold the bone plate creates a concentrated stress on the bone; this can have a deleterious effect if the screws are in cancellous bone. In order to alleviate osteopenia due to stress shielding, resorbable bone plates made of polyglycolic acid (PGA) or polylactic acid (PLA) have been tried experimentally. Problems associated with implant uniformity and in achieving the correct rate of resorption have hindered progress. In addition, it has been difficult to make sufficiently strong screws from this type of material. If conventional metal screws are used instead, they must be removed surgically after healing as is the case with conventional metal bone plates. Moreover, manipulation of the resorbable plate *in situ* by the surgeon is difficult if not impossible; additional equipment such as an oven may be required to heat and shape these thermoplastic materials to fit the bone.

13.2.2. Cancellous Bone Plate

A considerable amount of care must be exercised when fixing cancellous bone since this kind of bone has lower density and much lower stiffness and strength than cortical bone. An example of the fixation of the ends of a long bone is shown in Figure 13-12, in which the fractured bones are fixed with a combination of screws, plates, bolts, and nuts. The bulk necessary for the adequate stabilization of the fracture increases the chance of infection near the site of implants.

Sometimes one can fix a cancellous bone fracture by using a simple nail as shown in Figure 13-13. This is a special case since the patient was a young child

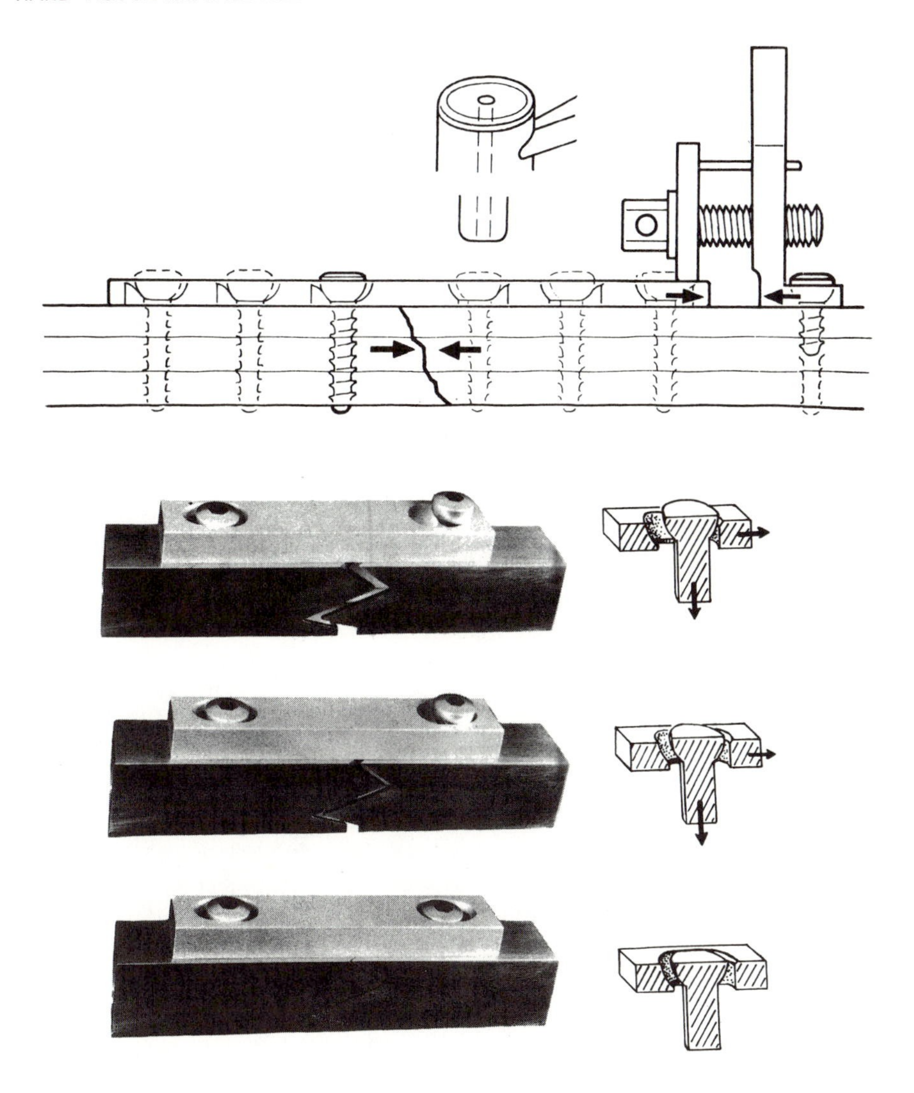

Figure 13-11. Principle of a dynamic compression plate (DCP): method of compression with a device (upper) and illustration of the principle by a model (lower). From M. Allgower, P. Matter, S. M. Perren, and T. Ruedi, *The Dynamic Compression Plate, DCP*, Springer-Verlag, Berlin, 1973, p. 18.

(who incidentally has two to three times the trabecular bone mass normally present in the adult cancellous femoral head and neck region) and since the epiphyseal plate lies close to the hip joint the loading is essentially normal to the fracture surface. Also the freely mobile hip joint relieves stress except in the compression cycle. Obviously, a wide range of choices is available. The choice is largely determined by the surgeon(s), not by the patients and bioengineers.

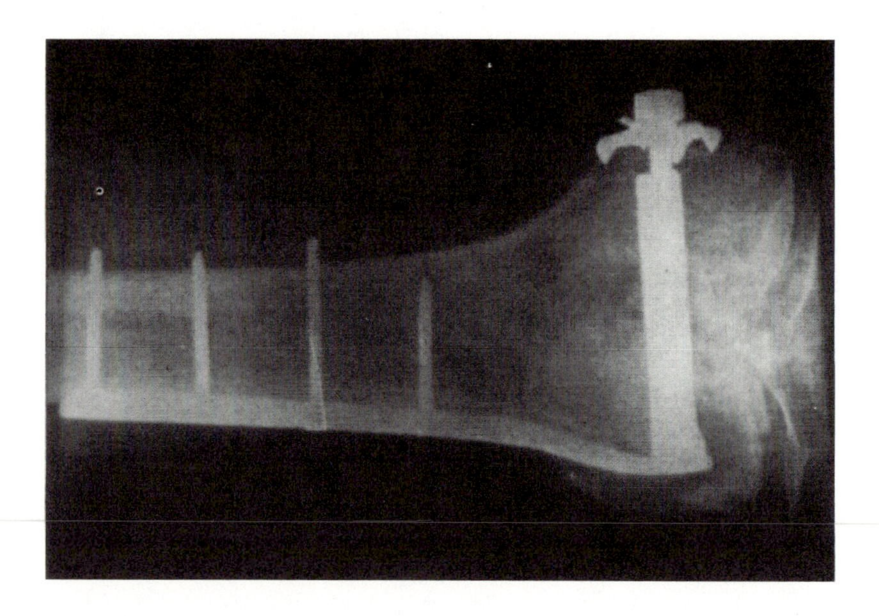

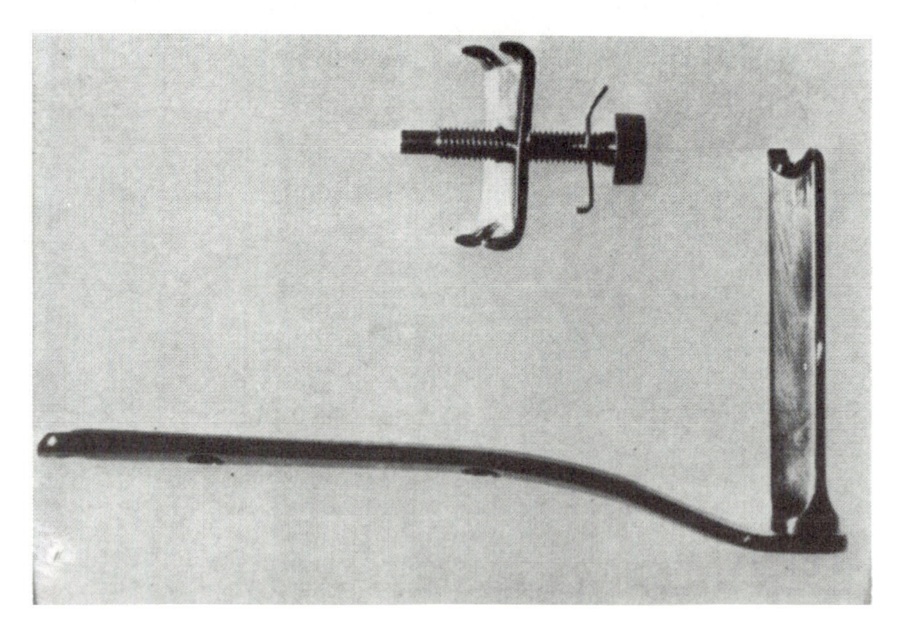

Figure 13-12. Devices to fix cancellous bone in a supracondylar fracture of the femur. From A. Brown and J. C. D'Arcy, "Internal Fixation for Supra-condylar Fractures of the Femur in Elderly Patients," *J. Bone Joint Surg.*, *53B*, 420–424, 1971.

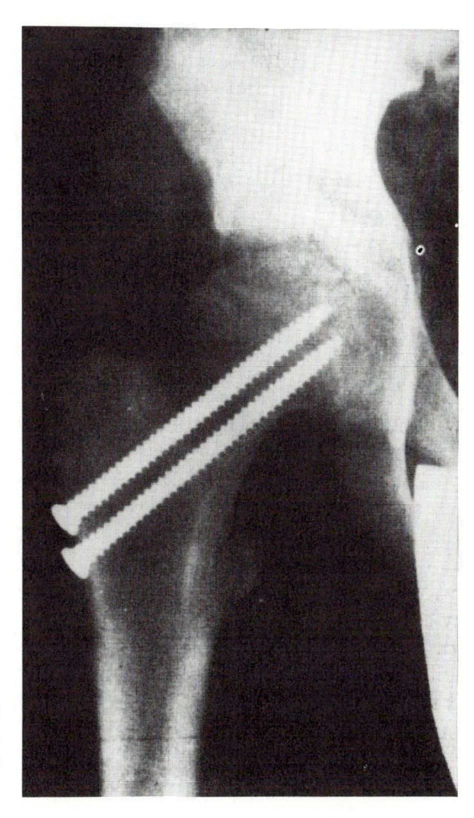

Figure 13-13. An example of a simple fracture fixation of cancellous bone. From H. M. Frost, *Orthopedic Biomechanics*, Charles C. Thomas Publ., Springfield, Ill., 1973, p. 444.

Example 13-2

A bioengineer is asked to make a composite material from carbon fiber and PMMA resin (as a matrix) for a bone fracture plate. He is given the following data:

Material	E (GPa)	Density (g/cm^3)	Strength (MPa)
Carbon fiber	250	1.95	5000
PMMA	3	1.20	70

(a) What is the amount of carbon fiber required to make the modulus of the plate 100 GPa? Assume the fibers are aligned in the direction of the test.
(b) How many fibers are required if the diameter of the fibers is 10 μm and the cross-sectional area of the specimen is 1 cm^2?
(c) How much stress in the fiber direction can the composite take?

Answer

(a) Using a simple mixture rule,

$$E_c = E_f V_f + E_m V_m \tag{1}$$

$$V_f + V_m = 1 \tag{2}$$

(2) ⇒ (1) and rearranging

$$V_f = \frac{E_c - E_m}{E_f - E_m} = \frac{97}{247} = \underline{0.39\ (39\ \text{vol\%})}$$

(b) Volume of fibers = $\pi r^2 NL$
where N is the total number of fibers and L is the length of composite.
Volume of composite = 1 $\text{cm}^2 L$

$$\frac{V_f}{F_c} = \frac{\pi r^2 N \quad L}{1 \qquad L} = 0.39$$

$$N = \frac{0.39\ 1}{\pi r^2} = \frac{0.39}{\pi (10 \times 10^{-4})^2} = \underline{1.24 \times 10^5}$$

(c) $$\sigma_c = \sigma_f V_f + \sigma_m V_m$$

$$\sigma_c = 5000 \times 0.39 + 3000 \times 0.02 \times 0.61 = \underline{1.99\ \text{GPa}}$$

The fiber can only stretch 2% (5/250) at its maximum load while the PMMA can stretch 2.3% (70/300) at its maximum load. Actually, in this geometry both the fibers and matrix undergo the same strain when the composite is loaded along the fibers, so that the actual strength will be somewhat less than the value predicted by the simple rule of mixtures.

13.3. INTRAMEDULLARY DEVICES

Intramedullary devices are used to fix fractures of long bones. The devices are snugly inserted into the medullary cavity (Figure 13-14). This type of implant should have some spring in it to exert some elastic force inside the bone cavity to prevent rotation of the device and to fix the fracture firmly.

Compared to the plate fixation, the intramedullary device is better positioned to resist bending since it is located in the center of the bone. However, its torsional resistance is much less than that of the plate. It is also believed that the intramedullary device destroys the intramedullary blood supply although it does not disturb the periosteal blood supply in contrast to the case of plate fixation. Another advantage of the intramedullary device is that it does not require opening of a large area to operate, and the device can be nailed through a small incision.

Many studies have examined the medullary blood supply and fracture healing in view of the extensive damage to the medullary canal by the insertion of the device. The long bone blood supply comes from three sources: the nutrient arteries and their intramedullary branches, the metaphyseal arteries, and periosteal arteries. If fracture occurs, the extraosseous circulation from the surrounding soft tissues becomes active and forms the fourth source of blood supply. When the intramedullary canal is reamed and a device is inserted, the nutrient artery

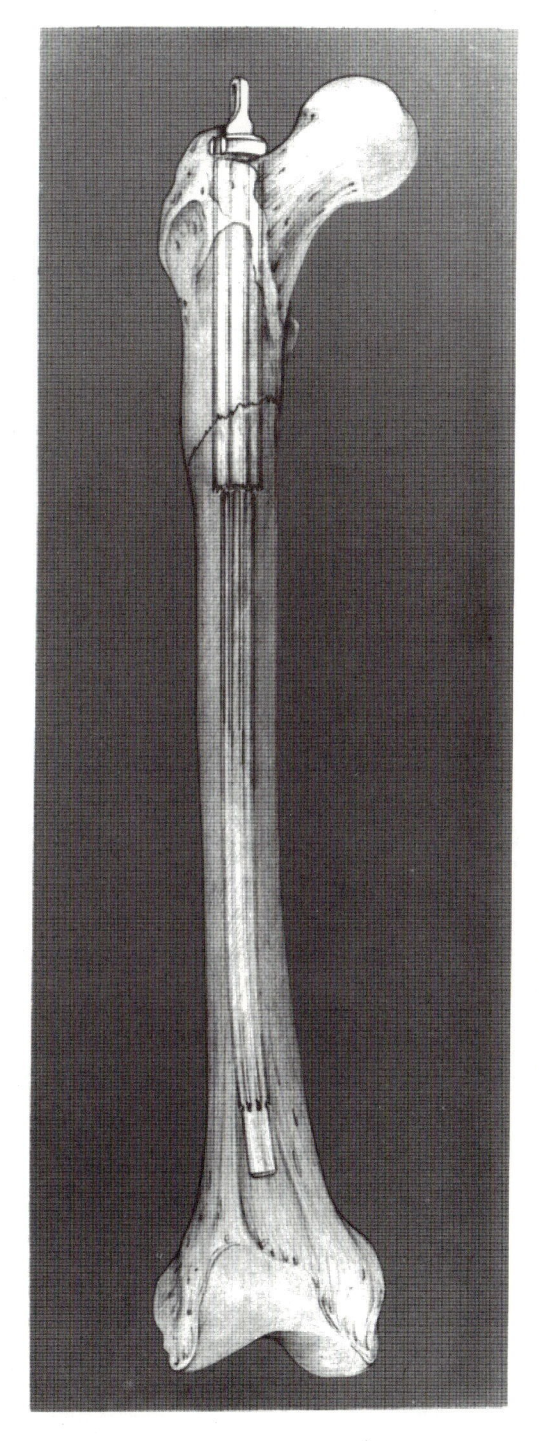

Figure 13-14. Illustration of an intramedullary device used in the femur.

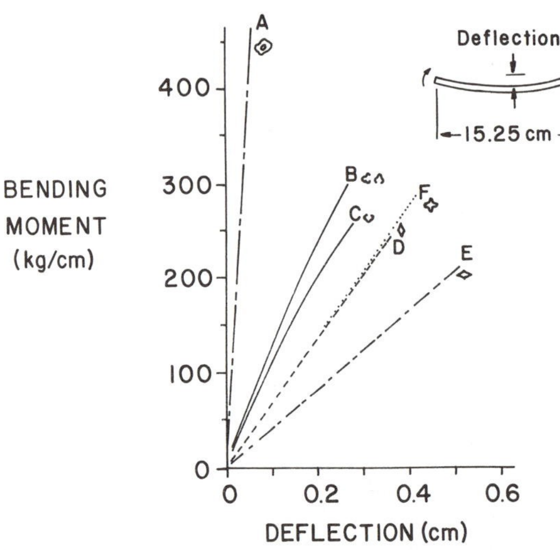

Figure 13-15. Bending deflections of the femur and of three intramedullary nails (9-mm cloverleaf, Schneider, and diamond-shaped) are compared as they would appear under identical loading. The length of each structure is taken as 15.25 cm. Curve A for the femur shows the bone to be more rigid than any of the nails. The cloverleaf nail is stiffer with the slot in tension (curve B) than in compression (curve C). The diamond-shaped nail is 50% more rigid when bent in its major plane (curve D) than in its minor plane (curve E). The Schneider nail has the same rigidity as the diamond-shaped nail (curve F), but even in its most unfavorable orientation the Schneider nail has a higher ultimate bending strength (not shown). From R. Soto-Hall and N. P. McCloy, "Cause and Treatment of Angulation of Femoral Intramedullary Nails," *Clin. Orthop. Relat. Res.*, *2*, 66–74, 1953.

and its branches are destroyed. Nevertheless, this procedure does not significantly damage the viability of the bone. Thus, a solid reunion can be achieved with this method of treatment. Other studies have demonstrated that the tight-fitting nail delays healing because of the time necessary to reestablish an intracortical (or intramedullary) circulation.

There are many different types of intramedullary devices varying to a large extent only in their cross-sectional shapes. For a given size of device, the resistance

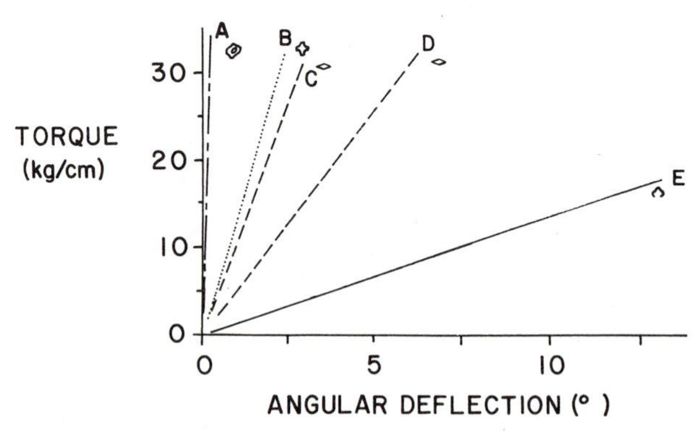

Figure 13-16. Curves of torque versus angular deflection for the femur and three intramedullary nails. The femur (curve A) is more rigid than any of the nails. The Schneider nail (curve B) is about one tenth as rigid as the femur. As the length of the diamond shape is doubled, the rigidity is halved (curves C and D). The cloverleaf nail is the least rigid (curve E). The length of each nail is 20.25 cm except for curve D. From R. Soto-Hall and N. P. McCloy, "Cause and Treatment of Angulation of Femoral Intramedullary Nails," *Clin. Orthop. Relat. Res.*, *2*, 66–74, 1953.

to bending and torque is different for different devices, as shown in Figures 13-15 and 13-16. The closed (solid) designs of the four-flanged nail (Schneider) and the diamond-shaped nail showed higher resistance to torsion than the open cloverleaf nail. However, in bending, the cloverleaf nail showed the highest resistance due to its larger bending moment of inertia.

The intramedullary nail is usually plated to the proximal femur for the fixation of a femoral neck and intertrochanteric bone fracture as shown in Figure 13-17. There are many different types of cross-sectional area designs to prevent rotation after fixation, as shown in Figure 13-18. Note that all of them except the V-shaped devices have a guide hole in the center. The femoral neck

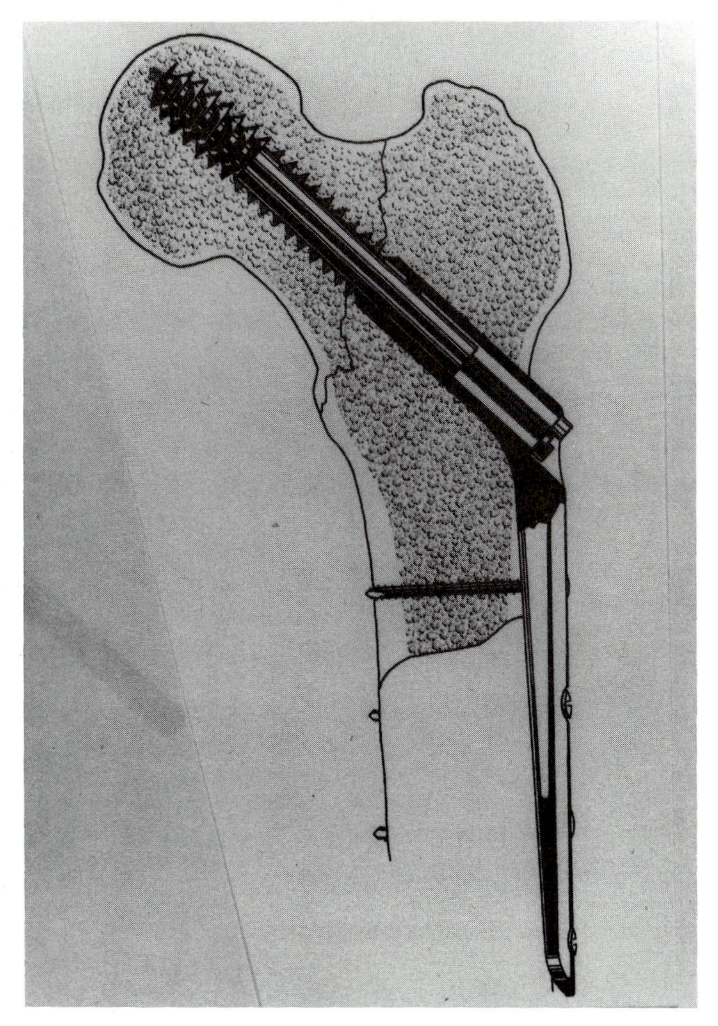

Figure 13-17. An illustration of hip fixation nail on a fractured femoral head. (Courtesy of DePuy, Division of Bio-Dynamics, Inc., Warsaw, Ind.)

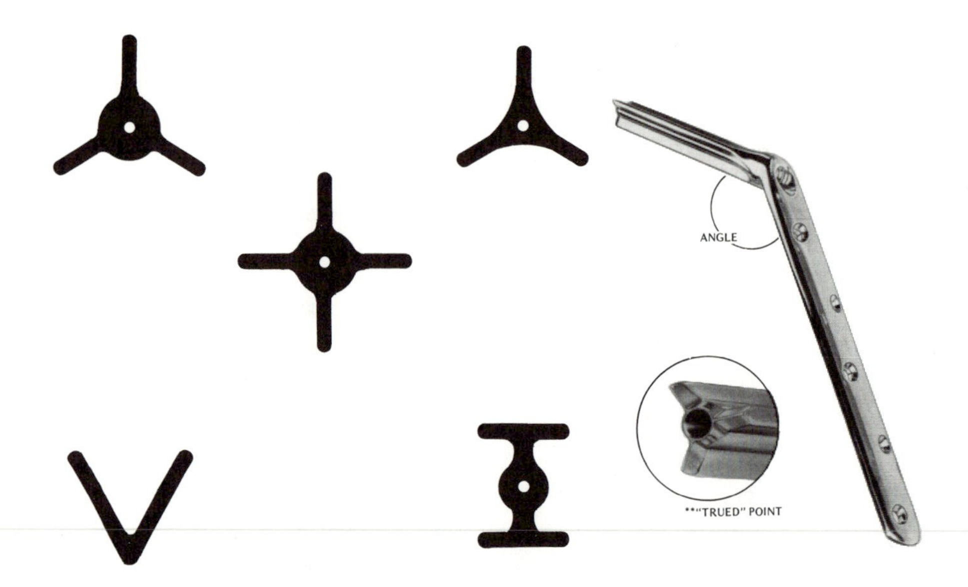

Figure 13-18. Cross-section of various hip nails and a typical implant and its insertion tip. (Courtesy of Smith & Nephew Richards, Inc., Memphis, Tenn.)

fracture fixation is usually made to compress the broken bones together by tightening a screw, which also helps to stabilize the fracture.

13.4. SPINAL FIXATION DEVICES

When the spinous elements of the backbone are deformed in such a manner that the length of the anterior elements is longer than that of the posterior ones, the resulting structure is bent backward and is called *lordosis.* The opposite condition is called *kyphosis.* There are forward and backward curvatures in the normal spine. A lateral curvature of the spine is always abnormal and is known as *scoliosis.* In severe cases of such spinal deformities, internal and external fixation is called upon to correct the situation. There are several designs to stabilize or straighten the curvatures, one of which is shown in Figure 13-19. Other designs include plates that are attached to the spinous processes by using bolts and expanding spinal jacks that are hooked to the spine through the articular process so that the distraction (tension force) can be adjusted during implantation.

The main problems with these devices are the fatigue failure due to the sharp notches for the hooks, the adjustment or extension of the device as the spine is being straightened, and the necrosis of the bones where the fixation device is attached due to the concentrated load. This necrosis results from the tremendous moment, of more than 100 N m, exerted by the trunk muscles. As the spine is straightened, it is harder to distract without hooks since the leverage on the spine becomes smaller. Thus, multiple hooks are sometimes attached to obviate this problem.

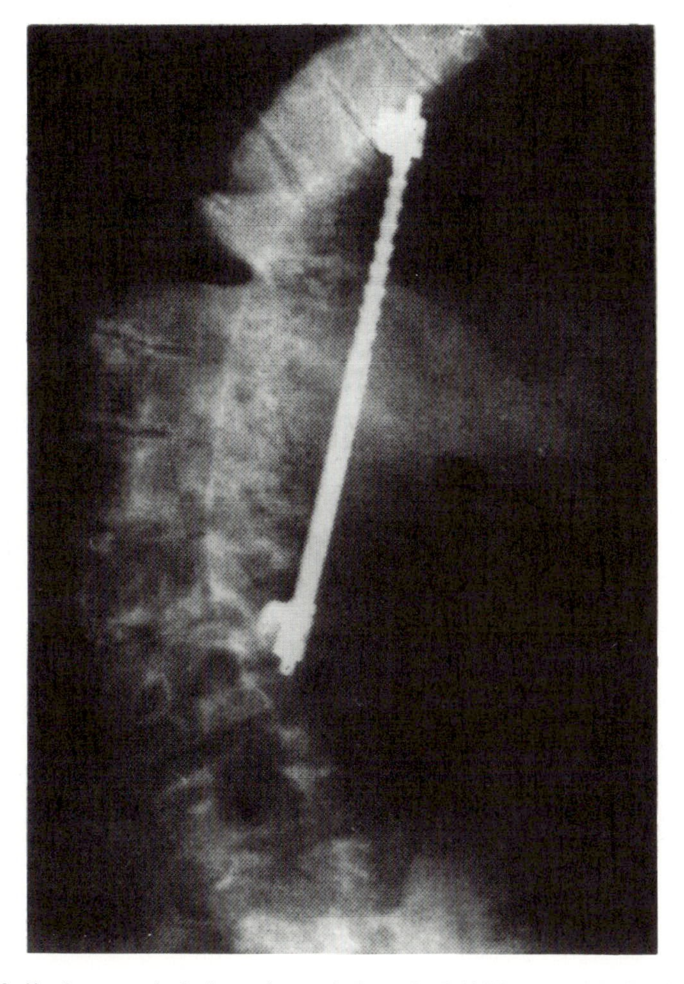

Figure 13-19. Harrington spinal distraction rod. From D. F. Williams and R. Roaf, *Implants in Surgery*, W. B. Saunders Co., Philadelphia, 1973.

13.5. FRACTURE HEALING BY ELECTRICAL AND ELECTROMAGNETIC STIMULATION

It has been recognized that some of the most challenging problems in orthopedic surgery are the clinical nonunions or delayed unions of fractures and congenital and acquired pseudoarthroses. Until recently, the usual methods of treatment were grafting, plating, and nailing. When one or a combination of any of the methods failed to repair the gap in the bone, especially in the lower extremities, amputation was the only alternative. Recently, many investigators demonstrated that the application of electrical energy by means of *direct* (or *alternating*) *current* or an *electromagnetic field* can enhance and stimulate the osteogenic activities. Although the exact mechanism of the stimulation is not fully elucidated, the electrical stimulation may be related to the highly electronegative nature of the fracture site resulting from the increases in ionic and metabolic activities (Figure 10-6). The extra electrical energy input into the wound area seems to trigger more osteogenic cellular activities. Bone tissue responds favorably only over appropriate ranges of input power, current, and voltage (10–40 μW, 5–20 μA, and less than 1 V). The response to the stimulation is also closely related to the nature of electrode material, surface area, and location.

One commercially available electrical stimulator is shown in Figure 13-20. The four electrodes (cathodes) are inserted into the fracture site transcutaneously after drilling the bone and the positive electrode (anode) is placed over the skin using a conducting pad. The electrode can be used with or without internal fixation devices, as shown in Figure 13-21.

The magnetic stimulators use a pair of Helmholtz coils, which are aligned across the wound site. Electrical current in the coils generates a magnetic field that penetrates the tissues; the changing magnetic field induces an electric field, hence also an electric current in the bone. The coils are excited with a monophasic waveform with a 150-msec period, and with a repetition rate of 75 Hz. The stimulation is applied as shown in Figure 13-22. This pulse input induces 1–1.6 mV/cm of potential gradient in the bone. The magnetic stimulation has

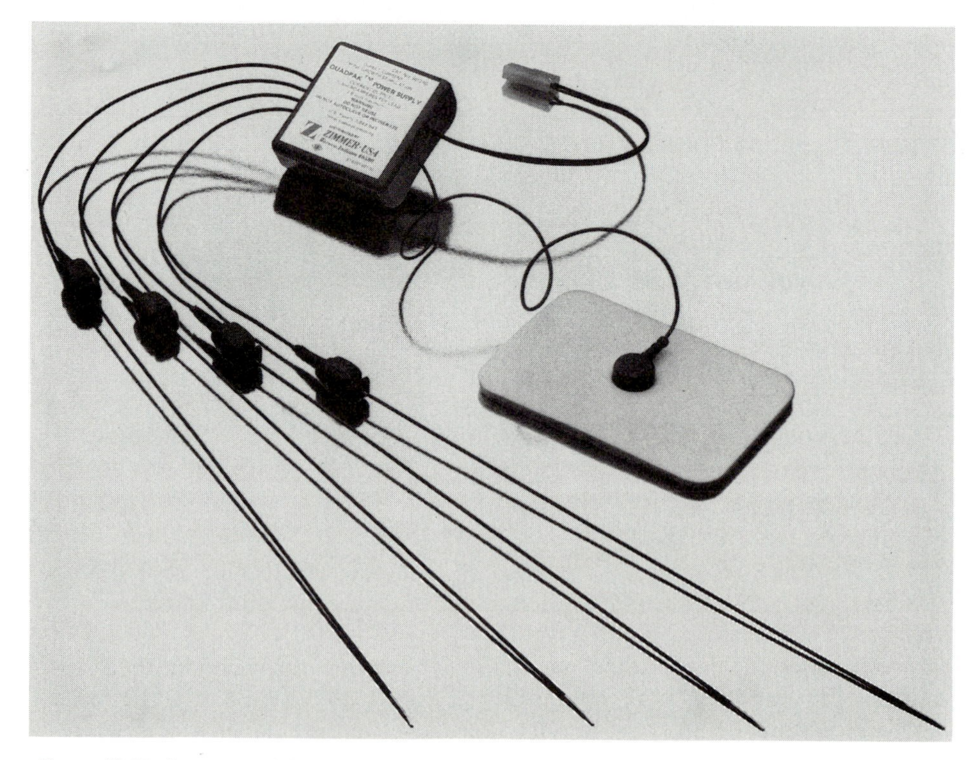

Figure 13-20. A commercial electrical stimulator for bone fracture repair. (Courtesy of Zimmer: USA, Warsaw, Ind.)

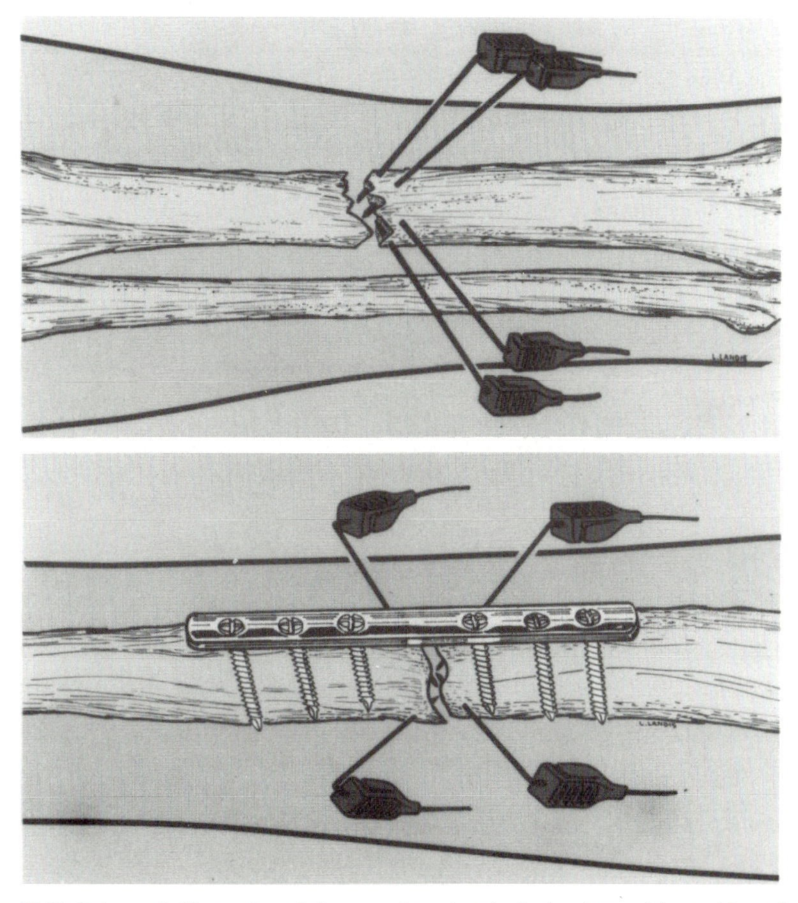

Figure 13-21. Schematic illustration of the use of an electrical stimulator with or without fracture fixation devices.

one big advantage over the direct current stimulation; namely, it is a noninvasive technique. The efficacy of both types of stimulation is about the same, over a 70% success rate.

Example 13-3

A bioengineer is studying the effect of electrical stimulation on the healing of bone fractures. In an experimental model, a platinum wire is inserted into a fracture site and a small electric current passed through it. As a control, an identical platinum wire is inserted into a similar fracture but no current is applied. The rationale is to isolate the effect of the electric current from the effect of the wire alone.

(a) What are the variables that will affect the results of this study?
(b) Write a critique of the experimental procedure.

Answer

(a) Any study of this nature will be influenced by many factors including the following:

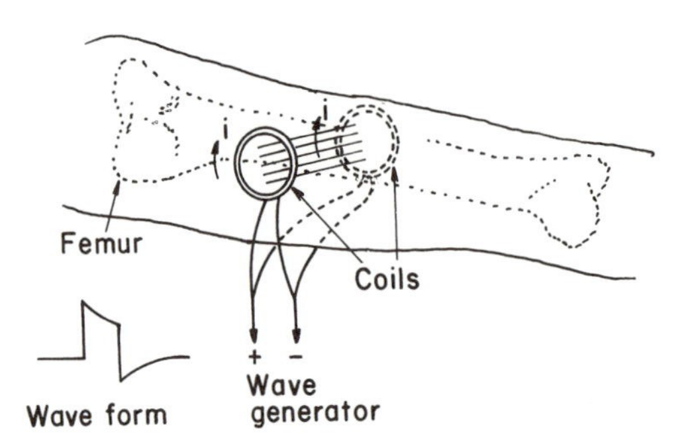

Figure 13-22. Representation of pulsed electromagnetic field bone stimulation.

(1) The nature of the injury that occurs during fracture and during surgery, the location of the fracture in the skeleton, and the type of bone (cortical or cancellous).
(2) The patient's physical condition, age, sex, and state of health.
(3) The type of metal of which the electrode is made, its geometry and surface area.
(4) The electrical power source voltage, current, frequency content, DC signal content, and duty cycle.
(5) The length of the observation period. The effect may occur only during a limited time interval following stimulation.

(b) To some extent the experiment can discriminate the active from the inactive electrodes provided the fracture sites are matched for factors listed under (1)-(3) above. However, it is difficult to isolate the effect of electric current alone since *ions* are released from both electrodes. The inactive electrode releases ions via corrosion. The active electrode provides an electron current and an ion current. Additional experiments such as trace metal analysis of tissues and urine would be needed to evaluate how much of each is present.

The issue of metal ion currents is circumvented by the use of pulsing electromagnetic fields (PEMF) for stimulation. This technique is noninvasive. A possible confounding variable in the study of this method is the heating of the tissue caused by the induced currents as well as the heat from the coils used to generate the fields. This heating effect is small. Nevertheless, a proper scientific study should include collateral experiments to isolate the effect of heating from the effect of the electromagnetic fields.

PROBLEMS

13-1. A bioengineer is asked by his supervisor to construct a bone fracture plate from the following materials: 316L SS, Ti6Al4V, cast CoCrMo, wrought CoNiCrMo, Al_2O_3, carbon fiber-carbon composite.

(a) Which two would be chosen for their biocompatibility?
(b) Which one would be chosen for its strength?
(c) Which one would be chosen for its specific strength (σ/ρ)?
(d) Which one(s) should be chosen for its FDA approval reason alone?
(e) If Ti6Al4V were chosen, discuss its advantages and disadvantages for making the fracture plate.

13-2. Explain possible degradation mechanisms of polymers *in vitro* and *in vivo*. Read Section 7.5.)

13-3. What is the Oppenheimer effect? (See B. S. Oppenheimer *et al.*, *Cancer Res.*, *21*, 137, 1961.)

13-4. After bone screws have been removed from the healed bone, holes remain that act as stress concentrators. Estimate the value of the stress concentration factor. Discuss the short-term consequences for the patient. What do you think will happen over a period of years? (Refer to M. Laurance, M. A. R. Freeman, and S. A. V. Swanson, "Engineering Considerations in the Internal Fixation of Fractures of the Tibial Shaft," *J. Bone Joint Surg.*, *51B*, 754–768, 1969; also D. B. Brooks, A. H. Burstein, and V. H. Frankel, "The Biomechanics of Torsional Fractures: The Stress Concentration Effect of a Drill Hole," *J. Bone Joint Surg.*, *52A*, 507–514, 1970; A. H. Burstein, J. D. Currey, V. H. Frankel, K. G. Heiple, P. Lunseth, and J. C. Vessely, "Bone Strength: The Effect of Screw Holes," *J. Bone Joint Surg.*, *54A*, 1143–1156, 1972.)

13-5. A bioengineer has designed a bone plate that has a rubber washer between plate and screws. What advantages and disadvantages would this insert have in comparison with conventional screws and plate? What would be the effect of such a washer in conjunction with a compression plate?

13-6. An intramedullary fixation device is designed to be of tubular shape and made of homogeneous metal. Discuss the advantages and disadvantages. Suppose the design included two orthogonal holes on each end, perpendicular to the tube axis. Would this modification be advantageous?

13-7. Design a self-compression bone plate that forces the bone fragments together as the screws are tightened. Illustrate its principle in the form of a two-hole plate.

DEFINITIONS

Compression plate: Bone plate designed to give compression on the fractured site of a broken bone for fast healing. A minimum amount of callus is observed; this is considered to be a better sign of healing in comparison with the conventional plate.

18-8sMo: One of the stainless steels made of 18 wt% Cr, 8 wt% Ni, and a small (2–4 wt%) amount of Mo for salt water corrosion resistance.

Electrical stimulation: Tissues can be stimulated by applying a small amount of electrical energy in either alternating or direct current at low electrical potential so as not to hydrolyze the water medium. Dosage-response relationships are not fully established yet for various tissues.

Fracture plate: Plate used to fix broken bones by open (surgical) reduction. Screws are used to fix it to the bone.

Helmholtz coil: A coil wound with wires for generating a (pulsed) electromagnetic field to induce electric current in the tissues to stimulate their growth.

Hoffmann external fixator: A mechanical device used to reattach broken long bones using percutaneously inserted pins to compress the bones together for faster healing.

Intramedullary device: A rodlike device inserted into the intramedullary marrow cavity to promote healing of long bone (spiral, noncomminutive) fractures.

Kirschner wire: Metal surgical wires with diameters less than 3/32 in (2.38 mm).

Kyphosis: Abnormally increased convexity in the curvature of the lumbar spine.

Lane plate: First fracture plate used extensively for repair of broken long bones designed by Dr. Lane in the 1900s.

Lordosis: Abnormally increased concavity in the curvature of the lumbar spine.

Schneider nail: One of the intramedullary rods used to fix broken long bones.

Scoliosis: Lateral curvature of vertebral column; it is always abnormal.
Steinman pin: Metal pins with diameters more than 3/32 in (2.38 mm).
Stellite®: Co–Cr alloy.
Strain-generated electric potential: Electric potential generated by the strains during deformation of the bone. They may be related to the remodeling processes of bone.
Stress-shield effect: Prolonged reduction of stress on a bone may result in porotic bone (osteoporosis) which may weaken it. The process can be reversed if the natural state of stress can be restored to its original state.
Vitallium®: Originally the name given to the 19-9 stainless steel but later used to designate a Co–Cr alloy.

BIBLIOGRAPHY

M. Allgower, P. Matter, S. M. Perren, and T. Ruedi, *The Dynamic Compression Plate, DCP*, Springer-Verlag, Berlin, 1973.

C. A. L. Bassett, "The Development and Application of Pulsed Electromagnetic (PEMFs) for Ununited Fractures and Arthrodeses," *Orthop. Clin. North Am.*, *15*, 61–88, 1984.

C. O. Bechtol, A. B. Ferguson, and P. G. Laing, *Metals and Engineering in Bone and Joint Surgery*, Balliere, Tindall and Cox, London, 1959.

C. T. Brighton, Z. B. Friedenberg, E. I. Mitchell, and R. E. Booth, "Treatment of Nonunion with Constant Direct Current," *Clin. Orthop. Relat. Res.*, *124*, 106–123, 1977.

J. H. Dumbleton and J. Black, *An Introduction to Orthopedic Materials*, Charles C. Thomas Publ., Springfield, Ill., 1975.

C. R. Hassler, E. F. Rybicki, R. B. Diegle, and L. C. Clark, "Studies of Enhanced Bone Healing via Electrical Stimuli," *Clin. Orthop. Relat. Res.*, *124*, 9–19, 1977.

G. Kuntscher, *The Practice of Intramedullary Nailing*, Charles C. Thomas Publ., Springfield, Ill., 1947.

D. C. Mears, *Materials and Orthopedic Surgery*, Williams & Wilkins, Baltimore, 1979.

H. K. Uhthoff (ed.), *Current Concepts of Internal Fixation of Fractures*, Springer-Verlag, Berlin, 1980.

C. S. Venable and W. C. Stuck, *The Internal Fixation of Fractures*, Charles C. Thomas Publ., Springfield, Ill., 1947.

D. F. Williams (ed.), *Compatibility of Implant Materials*, Sector Publ. Ltd., London, 1976.

D. F. Williams (ed.), *Fundamental Aspects of Biocompatibility*, Vols. 1 and 2, CRC Press, Boca Raton, Fla., 1981.

D. F. Williams and R. Roaf, *Implants in Surgery*, W. B. Saunders Co., Philadelphia, 1973.

CHAPTER 14

HARD TISSUE REPLACEMENT II: JOINTS AND TEETH

The articulation of joints poses some additional problems as compared with long bone fracture repairs. These include wear and corrosion and their products, as well as complicated load transfer dynamics. In addition, the massive nature of the (total) joint replacements such as the knee and the elbow and their proximity to the skin also renders the greater possibility of infection. More importantly, if the replacement fails for any reason, it is much more difficult to replace the joint a second time since a large portion of the natural tissue has already been destroyed.

For these reasons orthopedic surgeons try to salvage the existing joint if possible and make the use of implants a last resort. However, the hip prosthesis has shown favorable acceptance in recent years by older patients. The four primary indicators for any joint replacement are (1) pain, (2) instability, (3) stiffness, and (4) deformity.

The use of dental implants for missing or extracted teeth has not been as successful as the use of joint replacements; the difference is due mainly to the *percutaneous* nature of the dental implants. Exposure of the interface between implants and tissues to the hostile oral environment makes them easily infected. In fact, it is very difficult if not impossible to get rid of a low-grade infection even for a successful and functional dental implant. In addition, the extremely large and varying direction of the forces applied during mastication limit the selection of implant materials.

14.1. JOINT REPLACEMENTS

The hip and shoulder joints have a *ball-and-socket* articulation while other joints such as the knee and the elbow have a *hinge-type* articulation. However, they all possess two opposing smooth cartilaginous articular surfaces that are lubricated by viscous synovial fluid. This fluid is made of polysaccharides that adhere to the cartilage and upon loading can be permeated out onto the surface to reduce friction. The cartilage is not vascularized, and the nutrition of the tissues appears to be by a diffusional process.

For the hip and knee joints, nature provided the surface of the joint with a large area to distribute the load as shown in Figure 14-1. The shock of loading can be absorbed further by the trabecular subchondral bone underlying the cartilaginous tissue, which also transfers the load gradually due to its viscoelastic properties.

The actual articulation of the joint is performed by coordinating the ligaments, tendons, and muscles. The analysis of the forces acting on the various tendons and ligaments is very complicated. Even the center of the knee joint cannot be determined with any great precision; in fact, it shifts position with each movement. The eccentric joint movement helps to distribute the load throughout the entire joint surface.

Some joints such as the knee have fibrous, cartilaginous menisci shaped like wedges located between the sliding surfaces (Figure 14-1). It is believed that the main function of the menisci is to transfer the load over a larger area than is possible without them.

The joint forces applied during a range of activities are given in Table 14-1. Of course, the forces applied during walking vary considerably with each motion,

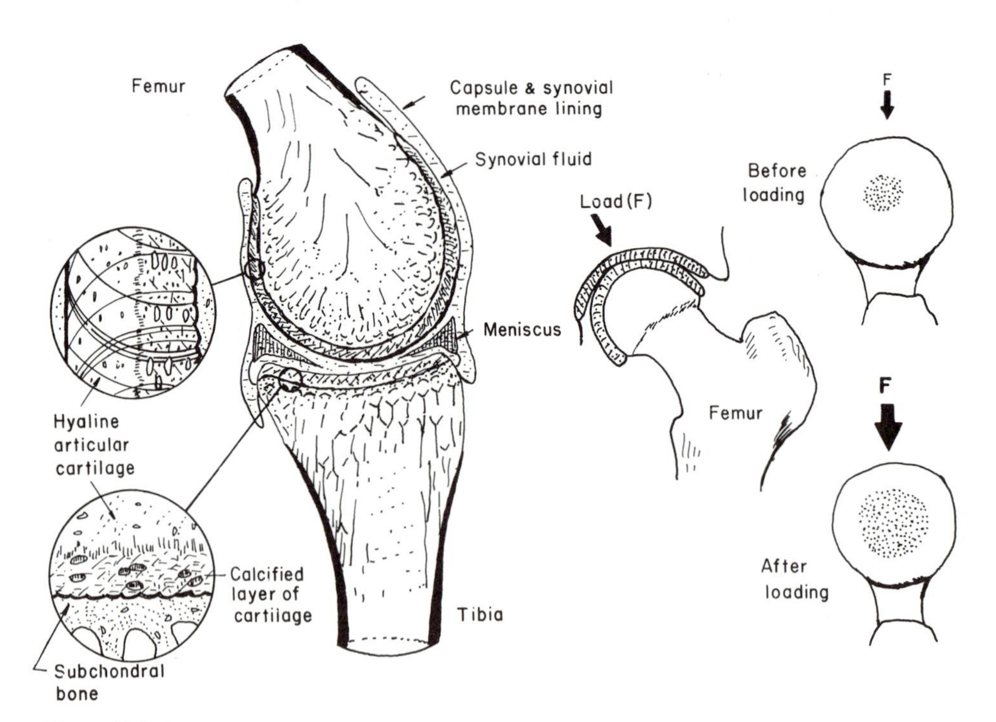

Figure 14-1. Structural arrangement of a knee joint and hip joint before and after loading. From H. M. Frost, *Orthopedic Biomechanics*, Charles C. Thomas Publ., Springfield, Ill., 1973, p. 444.

Table 14-1. Average Maximum Values of Forces at Hip and Tibio-Femoral Joints during a Range of Activities[a]

	Maximum joint force Multiples of body weight	
Activity	Hip	Knee
Level walking		
Slow	4.9	2.7
Normal	4.9	2.8
Fast	7.6	4.3
Up stairs	7.2	4.4
Down stairs	7.1	4.4
Up ramp	5.9	3.7
Down ramp	5.1	4.4

[a] From J. P. Paul, "Loading on Normal Hip and Knee Joints and Joint Replacements," in *Advances in Hip and Knee Joint Technology*, M. Schaldach and D. Hohmann (eds.), Springer-Verlag, Berlin, 1976, pp. 53–70.

as shown in Figure 14-2. It should not be surprising that the forces are up to 8 times the body weight as a result of the leverage geometry of the muscles and the dynamic nature of human activity. Biomechanical analysis can be of use in designing a better implant since it is necessary to know the load that will be applied to the implant in order to design sufficient strength and stiffness into it.

14.1.1. Lower Extremity Implants

14.1.1.1. Hip Joint Replacements. The early methods of correcting hip joint malfunctions involved only the acetabular cup or femoral head. One technique of restoring the hip joint function is to place a cup over the femoral head while the surface of the acetabulum is also resected to fit the cup. The implant served as a *mold* interposing between the articulating surfaces, which eventually adapt according to the function of the joint. Today, both the acetabulum and femoral head surfaces are replaced as shown in Figure 14-3.

Some surgeons have tried to replace the femoral head with implants of various designs after resection as shown in Figure 14-4. The wide variety of implants reflect the limited knowledge of the function of joints and the ability of the joint to accommodate insult imposed on it by the various implants. Most femoral head replacements are made with the installation of an acetabular cup. This is the so-called *total hip joint replacement* (arthroplasty). Dr. J. Charnley introduced the use of bone cement in the fixation of artificial hip joints on the advice of Dr. D. Smith in the late 1950s and this helped to popularize the procedure throughout the world. The initial success of the procedure has been tempered by the problems related not only to the bone cement but also the implants *per se*, surgical techniques, patient selection, etc.

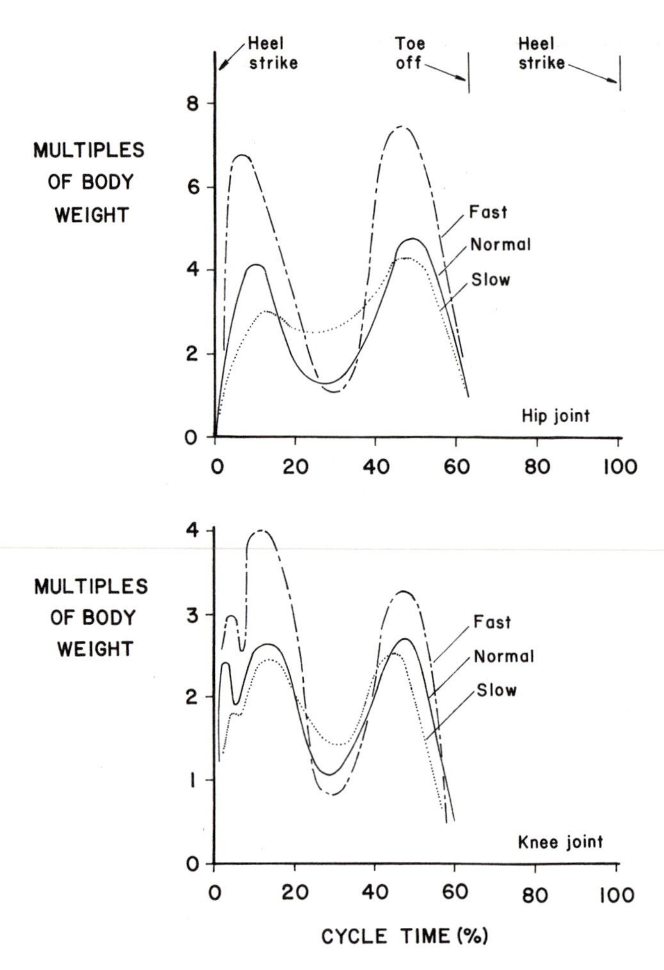

Figure 14.2. Variation of forces with time of hip and knee joint in walking. From J. P. Paul, "Loading on Normal Hip and Knee Joints and Joint Replacements," in *Advances in Hip and Knee Joint Technology*, M. Schaldach and D. Hohmann (eds.), Springer-Verlag, Berlin, 1976, pp. 53–70.

Surgical insertion of a total hip replacement is done as follows. The diseased femoral head is cut off, and the medullary canal of the femur is drilled and reamed to prepare it for the stem of the prosthesis. The cartilage of the acetabulum is also reamed. The PMMA bone cement is prepared from polymer powder and monomer liquid. When the cement reaches the correct "doughy" consistency, it is packed into the medullary canal of the femur and the femoral stem is inserted. The acetabular component is similarly cemented. The alignment and articulation of the artificial ball-and-socket joint are then verified. In cases where both joints are severely arthritic, the operations are performed bilaterally. The various types of hip implants can be grouped into ball and socket, retained ball and socket, trunnion bearing, floating acetabulum, and double cup, as illustrated in Figure 14-5.

Figure 14-3. An example of modern double cup arthroplasty. In the early days, only the femoral cup was placed (mold arthroplasty) in order to obtain a movable joint. (Courtesy of Howmedica, Inc., Rutherford, N.J.)

The single most difficult problem of the hip joint as well as other joint replacements is the fixation of the implants. This is because the implant lies on the cancellous bone, which is much weaker than compact bone, and there may be insufficient trabeculae to support the increased load imposed. Also, the stress concentration of the implant at points of sharp contact, such as the *calcar region* and the (lateral) end of the femoral stem, causes the already weakened bone to resorb. In fact, the first wide acceptance of the total hip replacement was achieved by providing an acceptable fixation using an acrylic bone cement (see Section 7.3.4).

The cement serves not only the initial attachment of the implant with bone; it also acts as a shock absorber since it is a viscoelastic polymer. The bone cement also helps to spread the load more evenly over a large area and reduces the stress concentration on the bone by the prosthesis. However, the bone stress in the region of the distal stem is much higher than normal. In contrast, the bone stress in the proximal region is reduced by the presence of the implant, as shown in Figure 14-6. This stress shielding effect causes bone resorption of the proximal region due to the reduced stress on this region, which, in turn, will lead to either

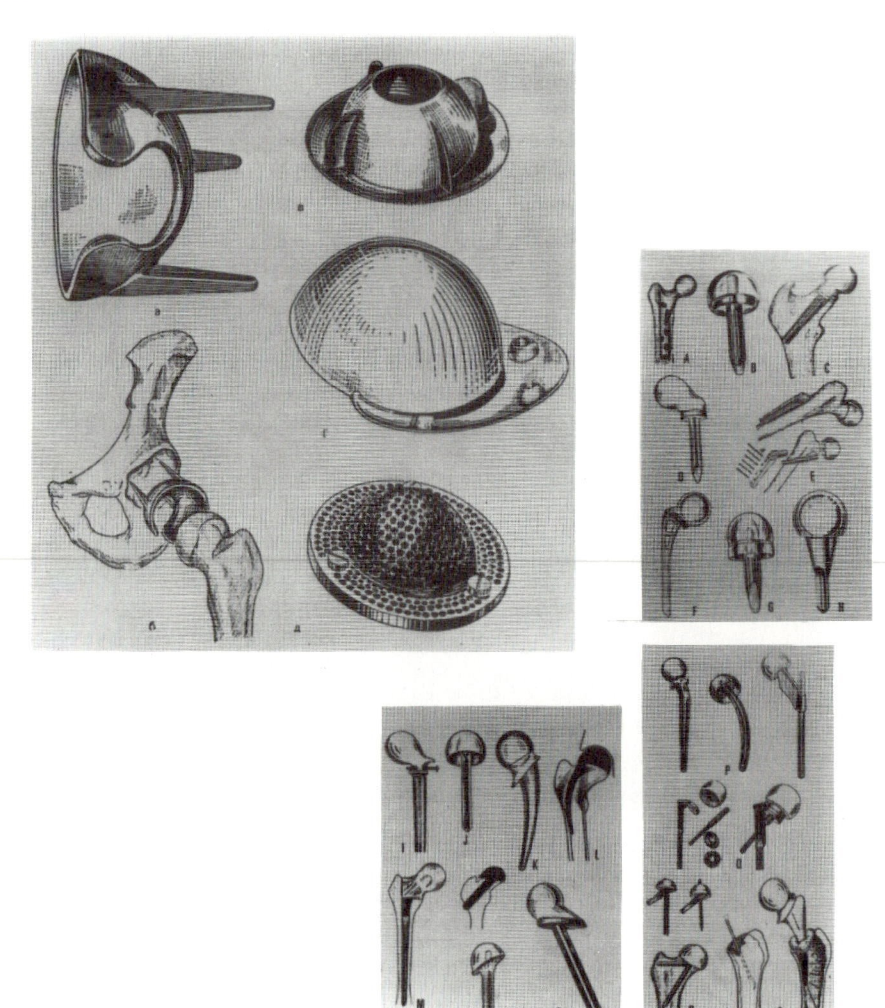

Figure 14-4. Various designs of acetabular and femoral head components of hip prostheses. From D. F. Williams and R. Roaf, *Implants in Surgery*, W. B. Saunders Co., Philadelphia, 1973.

loosening or fracture of the stem. Nevertheless, total hip joint prostheses commonly last 10 years in older patients.

To obviate the problem of calcar resorption, one can beneficially increase the load in the proximal calcar region by making the neck portion of the stem longer. However, this arrangement increases the moment applied in the midstem, predisposing it to fracture.

The cement itself sometimes creates problems such as monomer vapors interfering with the body's systemic function, thereby decreasing the blood pressure. The highly exothermic polymerization reaction can cause a local temperature rise which can result in cell necrosis as mentioned in Section 7.3.4. Also,

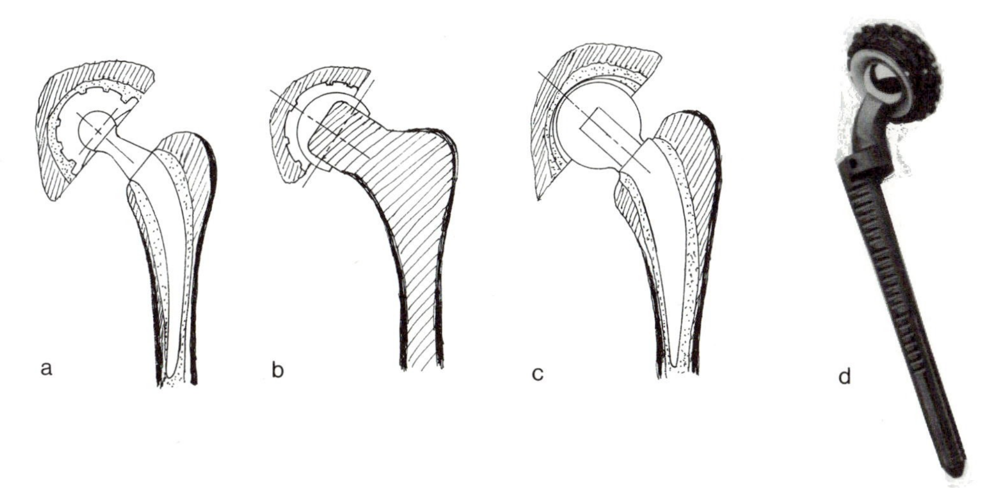

Figure 14-5. Different types of total hip implants. (a) Ball and socket; (b) double cup; (c) trunnion; and (d) retained ball and socket. (Sivash design, courtesy of United States Surgical Corp., New York, N.Y.)

the extensive intramedullary cavity preparation for the cement space can block the bone sinusoids, causing tissue necrosis and fat embolism.

Another problem is the difficulty of removal and the extent of tissues destroyed if the implant has to be removed for any reason. Thus, the *replaceability* of the implant is an important aspect of the design. In this regard, the original

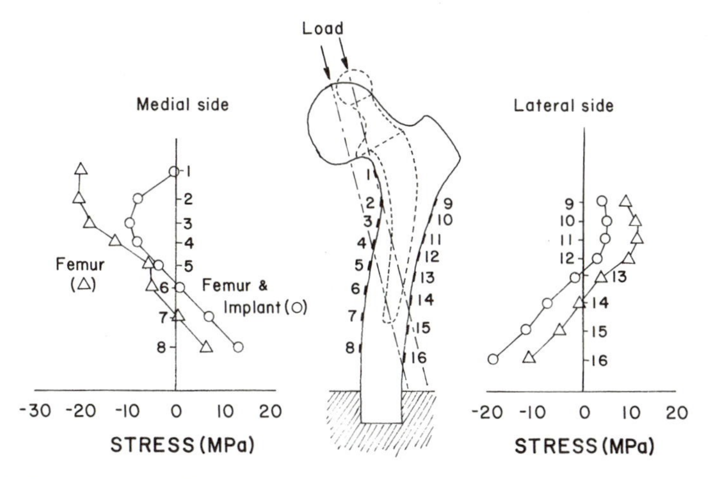

Figure 14-6. The stresses on the surface of the femoral stem by a load of 4000 newtons. The numbers indicate the location of strain gages to measure the deformations. Note that there is no stress in position 1 (calcar region) after insertion of the implant. From S. A. V. Swanson and M. A. R. Freeman (eds.), *The Scientific Basis of Joint Replacement*, J. Wiley and Sons, New York, 1977.

Sivash implant has an inherent weakness due to its design, that is, the whole prosthesis has to be replaced even if only one component has failed. It is now designed so that it is possible to replace one part without removing the others.

The friction between the ball and cup of the hip joint is problematical when it creates excessive torque. Especially for large loads, the frictional torque becomes very significant for the cobalt–chromium alloy hip joint, as shown in Figure 14-7. The stainless steel–polyethylene and cobalt–chromium alloy–polyethylene combinations are better for reducing the frictional torque and wear than the all-metal system. The high frictional torque of the all-metal system may also be due to the larger surface contact area since the femoral head is much larger than in the metal–polymer prosthesis. In actual use, the all-metal system works well without exerting high frictional torque. This is mainly due to the lubrication of the surfaces by tissue fluids.

The problems caused by cement/prosthesis interface can be diminished by precoating the prosthesis with bone cement at the factory; such implants are commercially available. Precoating not only increases the interfacial strength but also helps to eliminate bare metal exposure, and reduces the amount of bone cement used at the time of surgery, thus reducing the amount of heat and monomers released at the time of surgery, etc. The pre-coating technique can be used for any orthopedic implants.

The prosthesis/bone interface generates the most significant problems in joint replacements. Specifically, the failure of hip joint replacements is usually the result of *loosening* of the acetabular and femoral components. Loosening can be largely divided into mechanical and radiological loosening. Loosening may

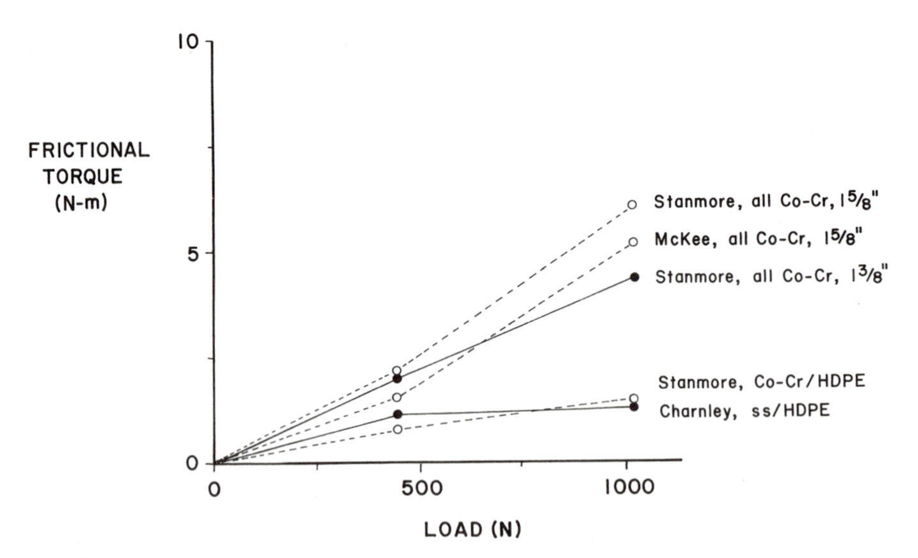

Figure 14-7. Frictional torque versus applied load for various hip prostheses. From J. N. Wilson and J. T. Scales, "Loosening of the Total Hip Replacements with Cement Fixation," *Clin. Orthop. Relat. Res.*, *72*, 145–160, 1970.

cause pain or it may not be related to any clinical symptoms. Inadequate fit of implants, and surgical and cementing techniques were thought to be the main factors in loosening. Some workers attributed the loosening to the blood clots interposed at the time of surgery and shrinkage of bone cement during polymerization. Also, bone remodeling and resorption in the proximal femur due to the stress-shielding by the stiff prosthesis could contribute to the loosening in long-term implants.

Interface problems in orthopedic implants are discussed more fully in Section 14.3.

14.1.1.2. Knee Joint Replacements. The development and acceptance of knee joint prostheses have been slower than for the hip joint due to the knee's more complicated geometry and biomechanics of movements, and lesser stability in comparison with the hip. The incidence of knee joint degeneration is higher than that of any other joints, as shown in Figure 14-8.

The knee joint implants can be classified into hinged and nonhinged types. The latter is further divided into uni- and bicompartmental. A cross section of a natural knee is shown in Figure 14-1 and typical artificial knee joints are shown in Figure 14-9. The knee surgery with bone cement requires a complete cleanup of the cement and bone chip debris, which can cause severe damage to the articulating surfaces, especially the tibial plateau. The selection of a particular implant depends on the health of the knee and the types of disease and range of activities required. As in the case of hip joint replacement, the major problems are loosening and infection. Sinking of the tibial plateaus can occur, the result of crushing of the trabecular bone under the implant. This problem can be corrected by making the prosthesis larger. A metallic backing is made under the polymer (ultra-high-molecular-weight polyethylene), as shown in Figure 14-10; the metal in this case is in contact with cortical bone, which is much stronger than trabecular bone.

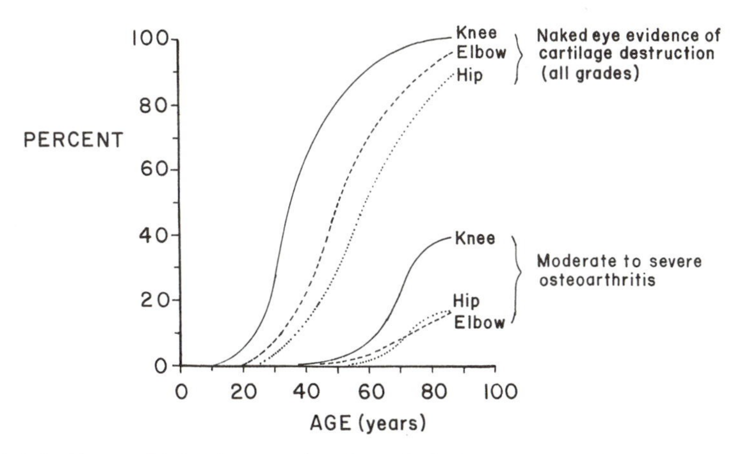

Figure 14-8. Incidence of joint degeneration. From A. S. Greenwald and M. B. Matejczyk, "Knee Joint Mechanics and Implant Evaluation," in *Total Knee Replacement*, A. A. Savastano (ed.), Appleton–Century–Crofts, New York, 1980, p. 1130.

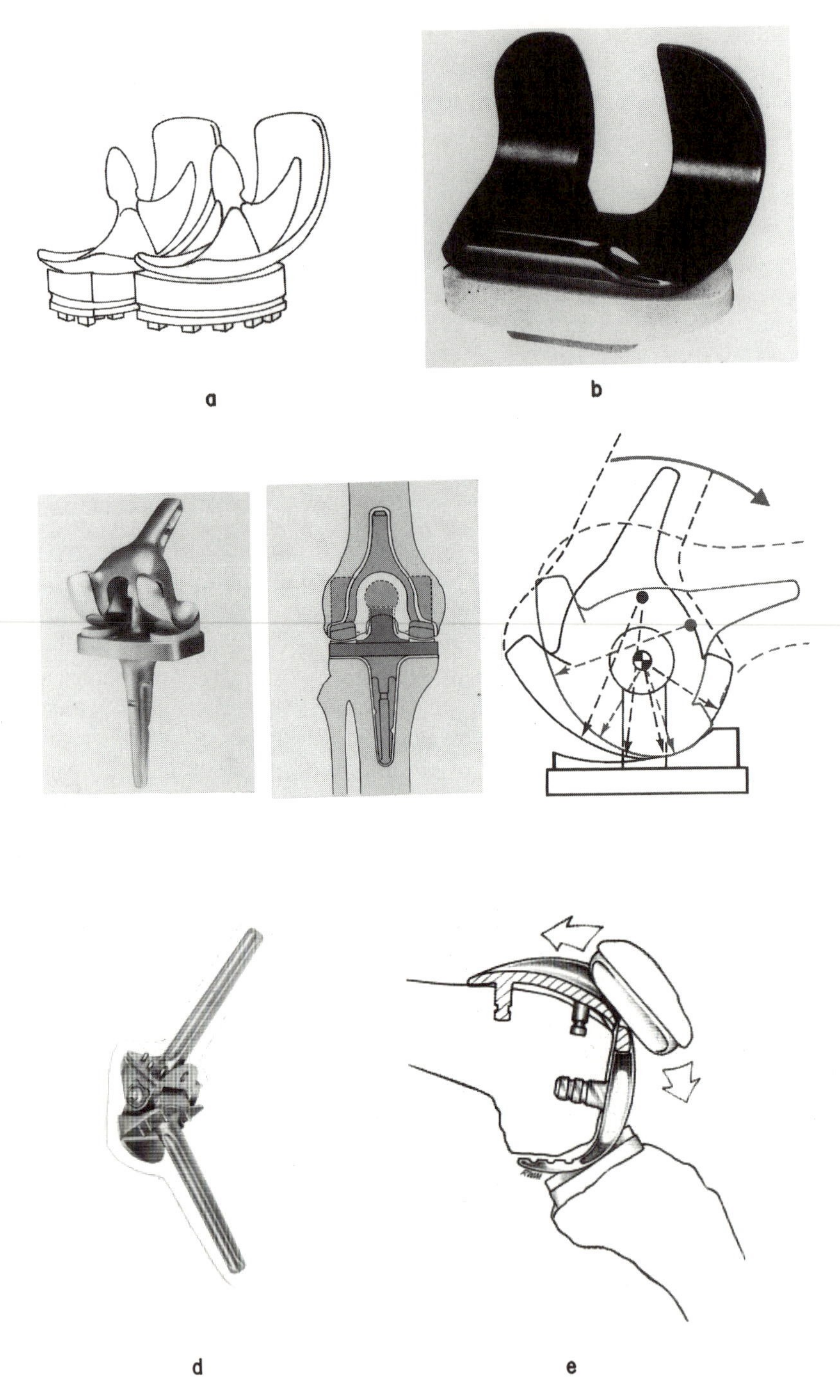

Figure 14-9. Examples of various types of knee replacements. (a) Marmor (L. Marmor, in *Total Knee Replacement*, A. A. Savastano (ed.), Appleton–Century–Crofts, New York, 1980, pp. 107–123); (b) Freeman–Swanson ICLH (M. A. R. Freeman, in *Total Knee Replacement*, pp. 59–82); (c) spherocentric (D. A. Sonstegard *et al.*, *J. Bone Joint Surg.*, *59A*, 602–616, 1978); (d) Walldius (B. Walldius, in *Total Knee Replacement*, pp. 195–216); (e) Bechtol (C. O. Bechtol, Bechtol Patello-Femoral Joint Replacement System, Smith & Nephew, Inc., Memphis, Tenn.).

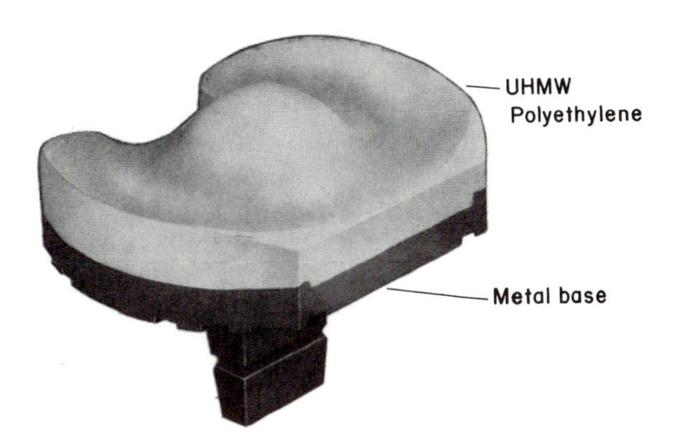

Figure 14-10. Typical metal base tibial implant designed to prevent sinking of the component.

Porous coated implants have been developed to avoid the problems associated with cemented prostheses. A porous surface layer permits bony tissue ingrowth to achieve a dynamic interface of bone and implant (Figure 14-11). Of course, this implant does not require the use of bone cement as most other

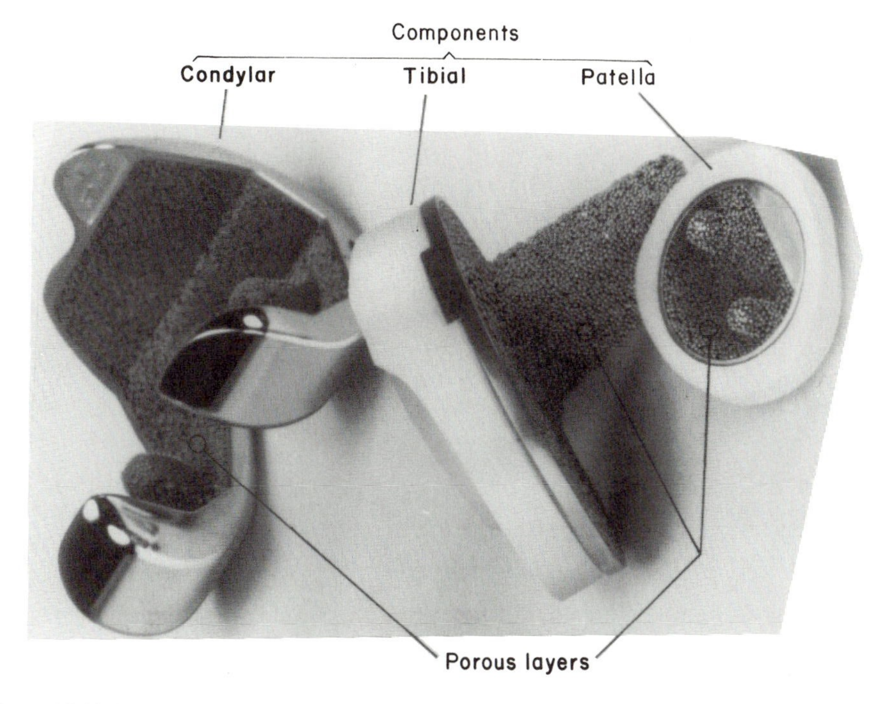

Figure 14-11. Porous metal-coated total knee implant. From *The Porous Coated Anatomic* (*PCA*) *Total Knee System*, Orthopaedic Division, Howmedica, Inc., Rutherford, N.J., 1981.

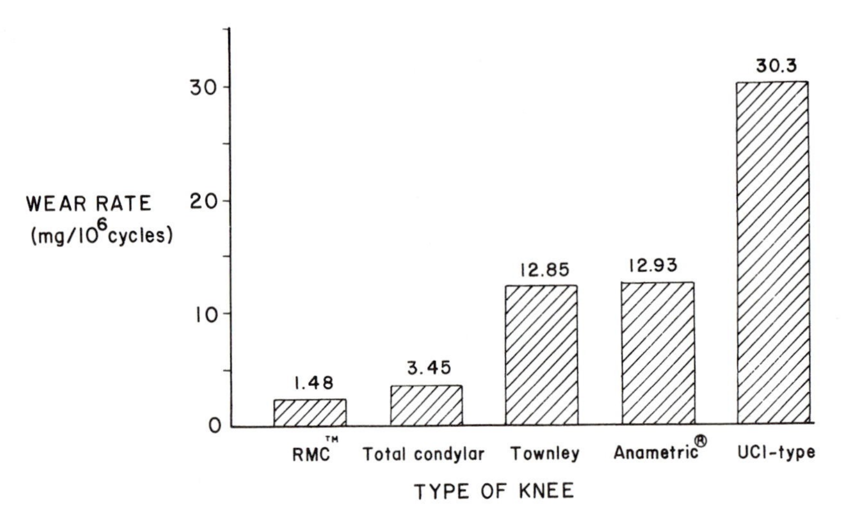

Figure 14-12. Bar graph of the wear rate results. From *In Vitro* Testing of the RMC Total Knee, R & D Technical Monograph No. 3468, Smith & Nephew, Richards Medical Co., Memphis, Tenn., 1978.

orthopedic implants do. The porous coated implants should be used for relatively healthy knees since their stability is entirely dependent on the ingrown tissues. It is also expected that the time required before the patient can walk (ambulate) will be much longer than for the cement-fixed case since it will take some time for the tissues to grow into the pores and premature loading may be detrimental for the ingrowth process.

The wear of the surface of the tibial plateaus of knee implants can be significant, as shown in Figure 14-12, which is an in vitro measurement. Note the unusually high wear rate of the UCI-type knee. The in vivo performance may be quite different.

14.1.1.3. Other Joint Replacements. Other joint replacements such as ankle and toe have not enjoyed as much success as those of hip and knee joints. This is mainly because of the extremely complicated motion of the ankle joints in addition to their many (three) articulating surfaces. The toe joints have very high stress imposed on them and no satisfactory implants have been found yet.

The materials used to construct ankle joints are usually Co–Cr alloy and ultra-high-molecular-weight polyethylene (UHMWPE). Recently, a carbon fiber-reinforced UHMWPE was tried in the fabrication of a tibial component to provide higher strength and creep resistance. Although these implants have been used for a relatively short time and little clinical experience is available, the usual problems were prosthetic loosening, limited motion, and pain at the ankle joints.

Example 14-1

A bioengineer is trying to determine the amount of gap developed between bone and cement when a femoral hip replacement arthroplasty is performed. It is assumed that the system is modeled as a set of concentric cylinders. Calculate the gap developed between bone and cement if the temperatures of cement, implant, and bone reached 55, 50, and 45°C, respectively throughout each component uniformly. (Assume the thermal expansion coefficient (α) of the implant is $17 \times 10^{-6}/°C$ and no direct adhesion takes place between bone and cement.)

Answer

From Table 3-1 the linear coefficients of thermal expansion are 8.3 and $81 \times 10^{-6}/°C$ for bone and cement; therefore, the shrinkage after equilibration with body temperature for each component will be (in a one-dimensional approximation)

Implant: $\Delta l = 17 \times 10^{-6}/(37\text{–}50)°C = -3.32\ \mu m$
Cement: $\Delta l = 0.6\ cm \times 81.0 \times 10^{-6}/°C = -8.75\ \mu m$
Bone: $\Delta l = 8.3 \times 10^{-6}/(37\text{–}45)°C = -0.53\ \mu m$

Since the metal implant shrinks, the diameter is only 3.32 μm but the cement cannot shrink the full 8.75 μm because of the stiff implant. Therefore, the real gap between bone and cement is (1/2)(3.32–0.53) μm, which is 1.4 μm. The cement will impose a *hoop* stress around the implant. If the hoop stress becomes large enough, the cement may break by its own shrinkage stress. (Cement breakage and the microgap between the implant and bone have been observed in a clinical situation, although the former is a rare case.) The linear shrinkage of 0.145% (8.75 μm/0.6 cm × 100) is close to the reported value for the cement (0.12–0.27% for Simplex-P), although the α is used for solid glassy PMMA.

Example 14-2

Consider a total hip replacement prosthesis, a simplified diagram of which is shown below. The proximal portion of the supporting bone cement has crumbled, so that there is an

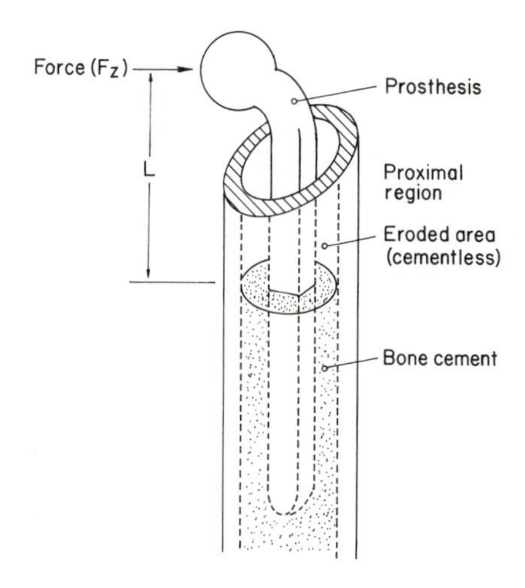

8-in (203 mm) distance (defined as L) between the point of application of the force at the head of the prosthesis and the undamaged distal bone cement. Assume that the stem of the prosthesis is $b = 0.8$ in (20.3 mm) wide and $h = 0.31$ in (7.87 mm) deep in the direction of bending. The patient weighs 100 lb (45.5 kg), and walks slowly with the damaged hip replacement in place. Suppose that for the hip geometry in question, the bending component of force transverse to the implant stem is three times the body weight, or 300 lb. This is realistic since the forces in walking can be from 3 to 5 times body weight. Neglect the bending moment due to the vertical component of load. The stem is made of a cobalt chrome with a yield strength $\sigma_y = 122{,}000$ psi and an ultimate tensile strength $\sigma_{ult} = 185{,}000$ psi. Determine whether the implant will break.

Answer

Consider cantilever bending with the load horizontal and the stem (considered as a beam) vertical. Then $\sigma_{zz} = My/I = Fzy/I = 6FL/bh^2$ since $y = h$ at the surface where the stress is greatest, and the area moment of inertia $I = bh^3/12$, so

$$\sigma_{zz} = (300\text{ lb} \times 6 \times 8\text{ in})/(0.1\text{ in}^2 \times 0.8\text{ in}) = \underline{180{,}000\text{ psi (1.24 GPa)}}$$

which exceeds the yield strength and is nearly as large as the ultimate strength. The stem will yield at the first step and fracture due to fatigue after a few steps. The stress is so

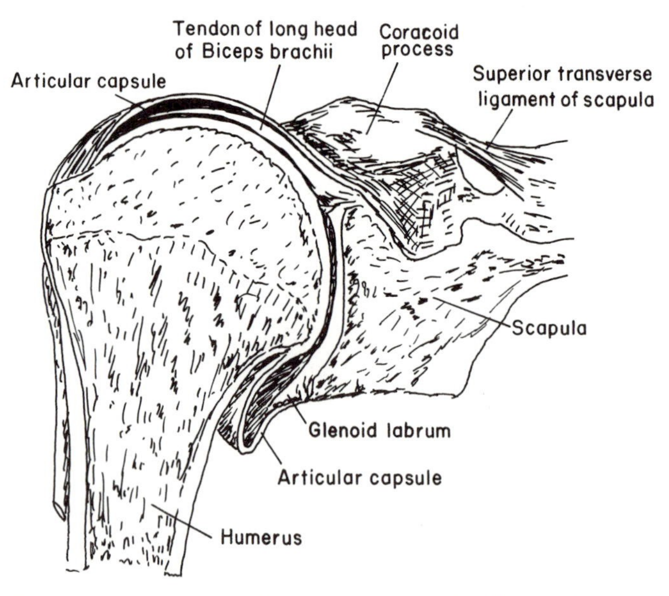

Figure 14-13. The anatomy of the glenohumeral joint. From C. M. Goss (ed.), Gray's Anatomy, Lea & Febiger, Philadelphia, 1975, p. 316.

large as a result of lack of proximal support of the stem ($z = 8$ in). This example is an extreme example of a continuing clinical problem. Breakdown of the proximal cement causes excessive stress to occur in the stem, which can fracture in fatigue after many cycles of walking. Loosening of the implant can also cause pain to the patient.

14.1.2. Upper Extremity Implants

14.1.2.1. Shoulder Joint Replacements. The major shoulder joint motion originates from the ball-and-socket articulation of the glenohumeral joint as shown in Figure 14-13. The hemispherical, incongruent joint provides the largest motion in the body. As in the hip joint replacements, the first shoulder joint replacement by Neer was attempted by merely replacing the humeral head, as shown in Figure 14-14. Note the large articulating surface and two holes, which provide better fixation and resistance to rotation. The implant can be fixed with or without bone cement. Figure 14-15 shows some of the shoulder implant designs, most of which are a cup-and-head type as in hip joint prostheses.

14.1.2.2. Finger Joint Replacements. A finger contains distal, middle, and proximal joints and is provided with adequate controls by ligaments and tendons

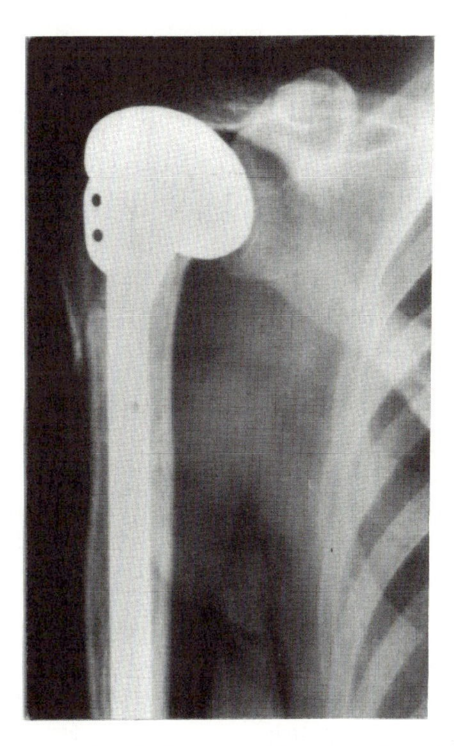

Figure 14-14. Neer's humeral implant (C. S. Neer, *J. Bone Joint Surg.*, *56A*, 1–13, 1974). (Courtesy of Smith & Nephew, Richards Medical Co., Memphis, Tenn.)

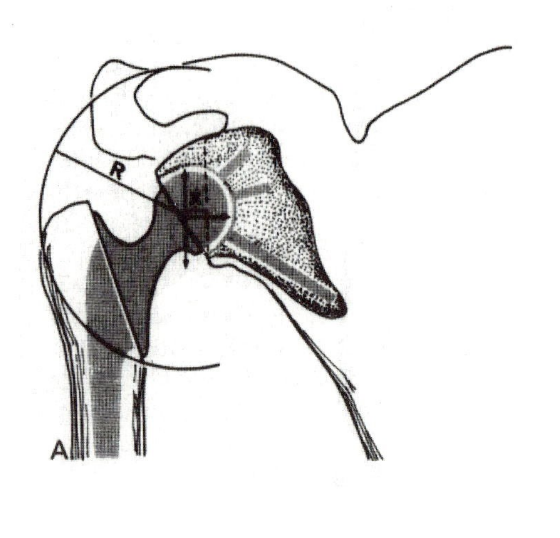

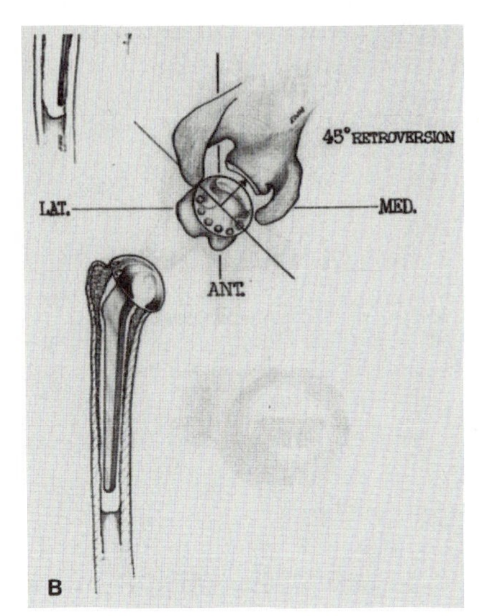

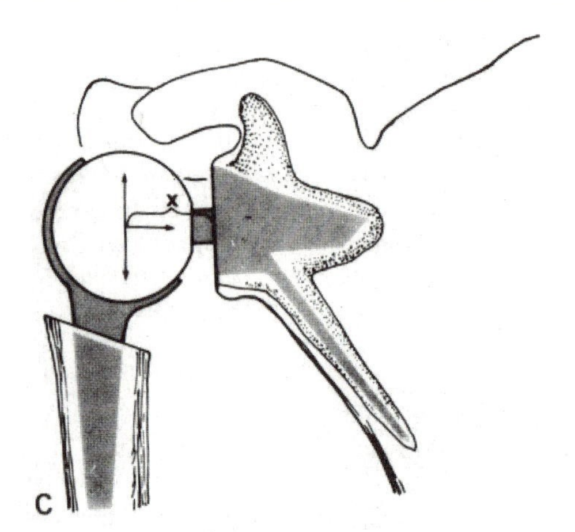

Figure 14-15. Types of shoulder joint prostheses: (A) Stanmore, (B) Bechtol, (C) Fenlin. From J. M. Fenlin, Jr., "Total Glenohumeral Joint Replacement," *Orthop. Clin. North Am.*, *6*, 565–583, 1975.

so as not to collapse under a compressive load. The exact mechanics of the joint movements are rather complex. The traditional treatment for arthritic degeneration is the resection of the joint, which usually relieves pain and corrects deformity but results in a loss of stability and strength. One can divide the various types of resectional and implant arthroplasties into five major categories as shown in Figure 14-16. Alternate approaches of joint replacement were divided into four categories: hinge type, polycentric type, space-filler type, and combination of space filler and hemiresection arthroplasty. Some actual finger joint implants are shown in Figure 14-17.

The concept of fixing the implant with an encapsulated collagenous membrane, which is created around the implant, is believed to be a clinically sound

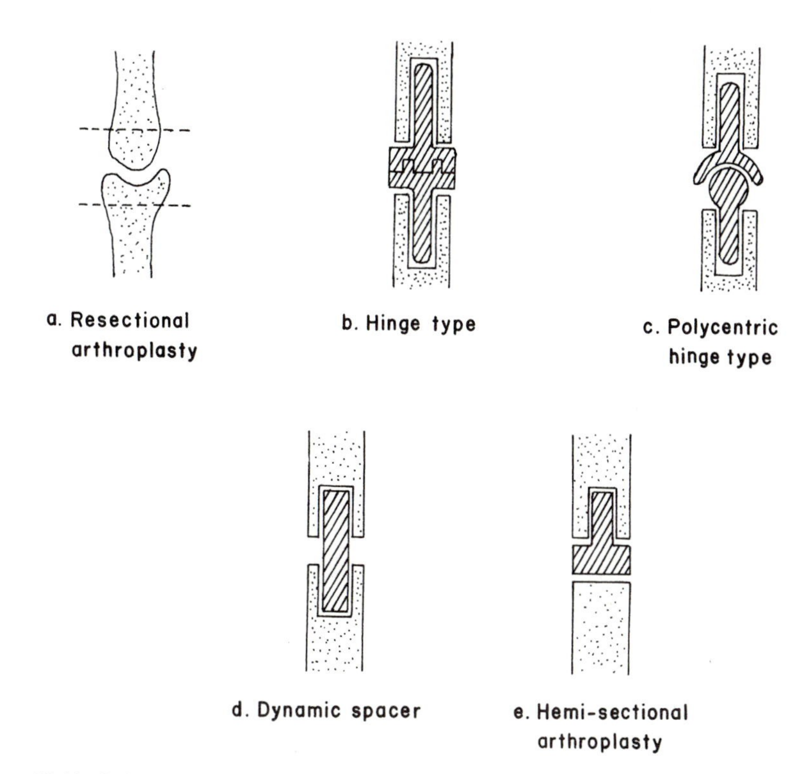

Figure 14-16. Schematic diagrams indicating the various types of resectional and implant arthroplasties available for treatment of joints in the hand. From R. I. Burton, "Implant Arthroplasty in the Hand: An Introduction," *Orthop. Clin. North Am.*, *4*, 313–316, 1973.

one. This primarily takes advantage of nature's response to any implant in the body, especially a moving implant, that is, encapsulation. The finger joint is best suited for this means of fixation since cortices of finger bones are often too thin to allow ingrowth of adequate amount of bone tissues around a metal prosthesis.

14.1.2.3. Other Joint Replacements. The *elbow joint* is a hinge-type joint allowing mostly flexion and extension, but having a polycentric motion. Most elbow joint implants are either hinge or surface replacement types. The major problems are the loosening of implants and the limited soft tissue coverage of the implant making it vulnerable to infection. One of the experimental elbow prostheses *in situ* is shown in Figure 14-18.

The *wrist joint* allows flexion, extension, adduction, and abduction primarily through the radiocarpal joint. Since the wrist joint arthroplasty for prosthetic replacement includes the removal of the capitate, where the anatomical instantaneous center or axis of motion of a radiocarpal complex is located, it is very difficult to place the prosthesis in the correct position. The unnatural position of the implant will constrain its movement, causing an excessive generation of bending moments in adduction or abduction. This is also a major cause of complications of total wrist replacements.

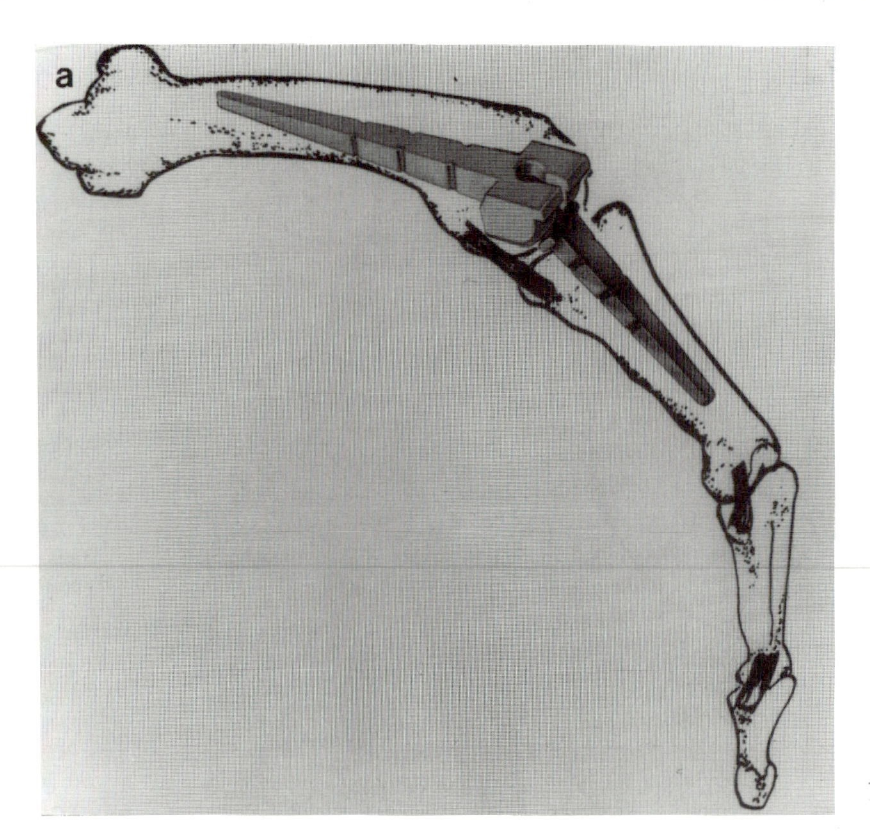

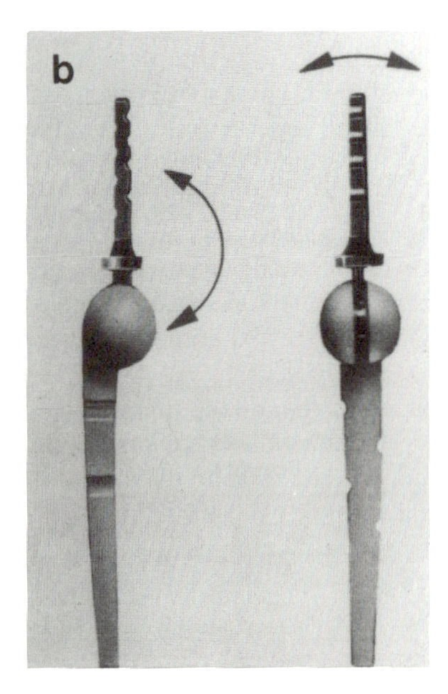

Figure 14-17. Various types of finger joint prostheses. (a) Schultz; (b) St. Georg; (c) Stefee; (d) Swanson; (e) Calnan-Nicolle; (f) Niebauer-Cutter; (g) Lord's Bonded Bion elastomer–titanium joint.

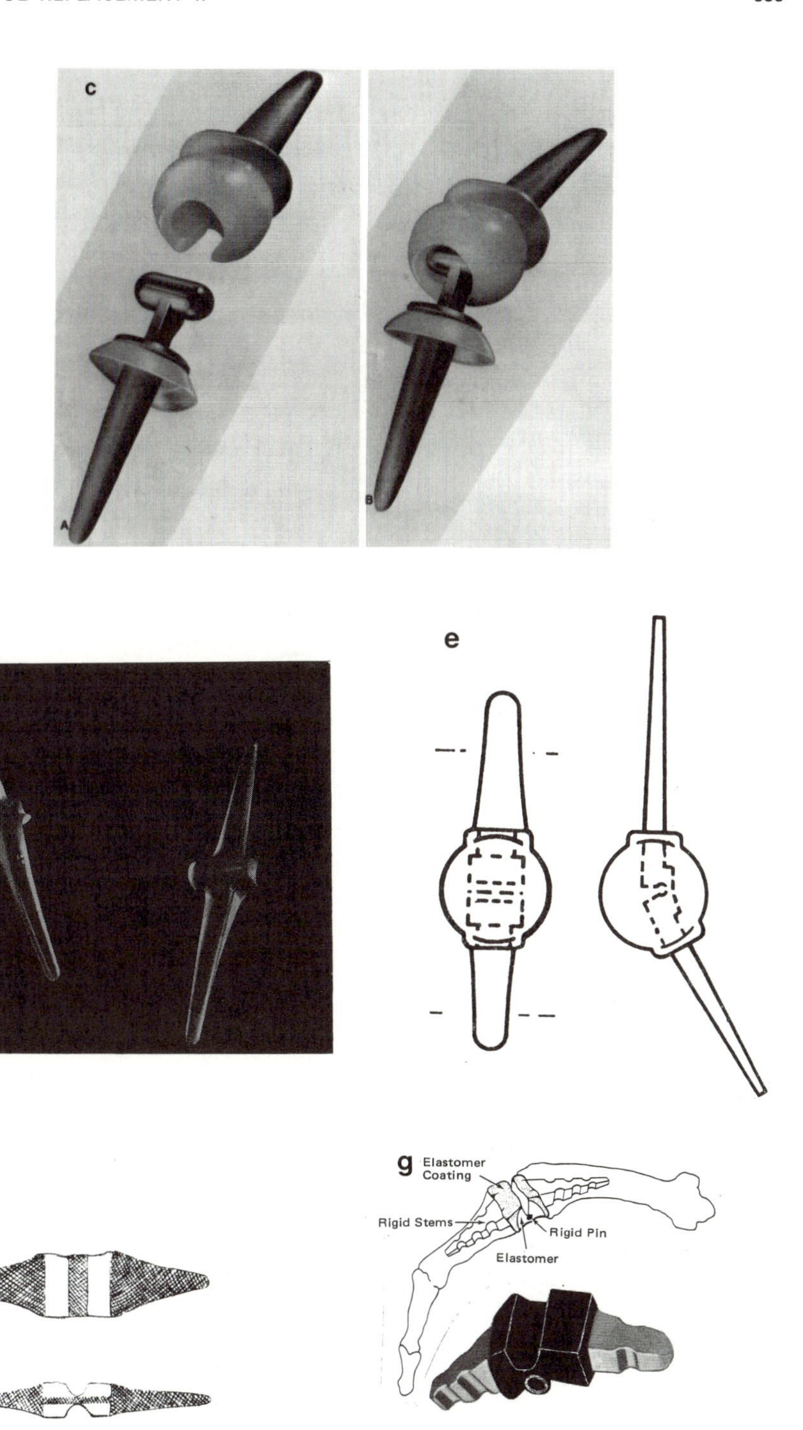

Figure 14-17. (*Continued*)

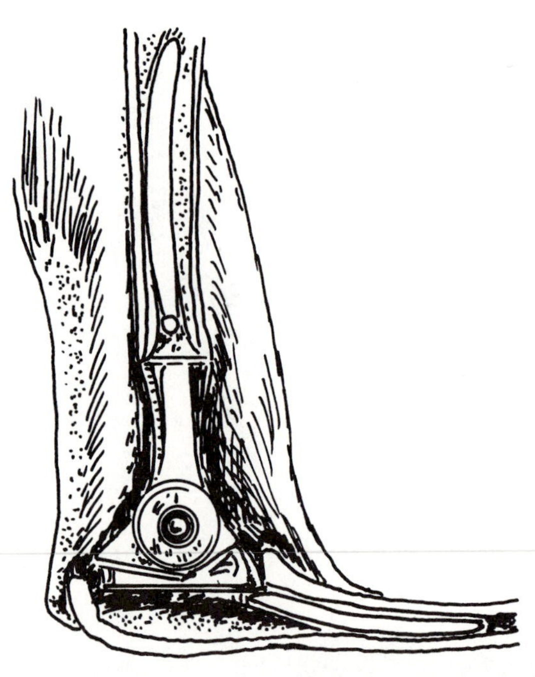

Figure 14-18. A schematic lateral view of a Dee elbow prosthesis. Apart from a portion of the humeral intramedullary stem, the remainder of the implant is buried within the bone. From Institution of Mech. Eng. Report, Joint Replacement in the Upper Limb, *Eng. Med.*, *6*, 90–93, 1977.

Figure 14-19 illustrates some of the wrist implants now in use. The Meuli and Volz implants are ball-in-socket while Swanson's is the space-filler type made of silicone rubber, like his finger joint prosthesis. The clinical results are poorer than other joint replacements because of the more complex motion of the wrist and surgical inconsistency caused by lack of landmarks to rely on as mentioned before.

Example 14-3

Bioglass® is coated on the surface of stainless steel for making implants that are more biocompatible. The 10-mm-diameter stainless steel is coated with 1-mm-thick Bioglass. Use the following data to answer (a)–(d).

Material	Young's modulus (GPa)	Strength (MPa)	Density (g/cm^3)
Stainless steel	200	300 (yield)	7.9
Bioglass	300	300 (fracture)	4.5

(a) What is the maximum load the composite can carry in the axial direction?
(b) What is the Young's modulus of the composite in the axial direction?
(c) What is the density of the composite?
(d) Give two reasons why you would not use this composite to make orthopedic implants over plain stainless steel.

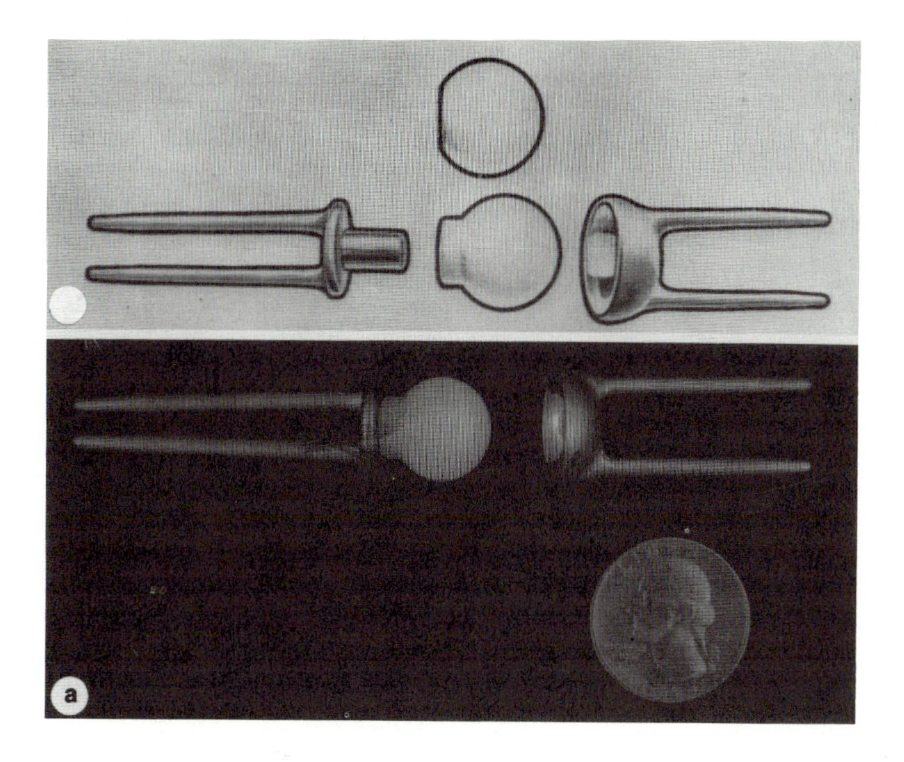

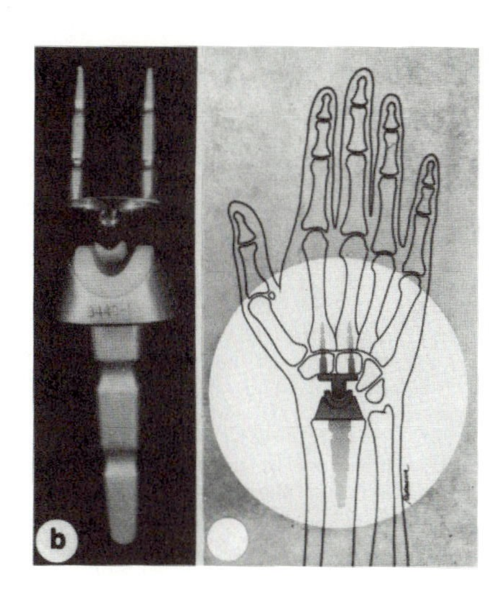

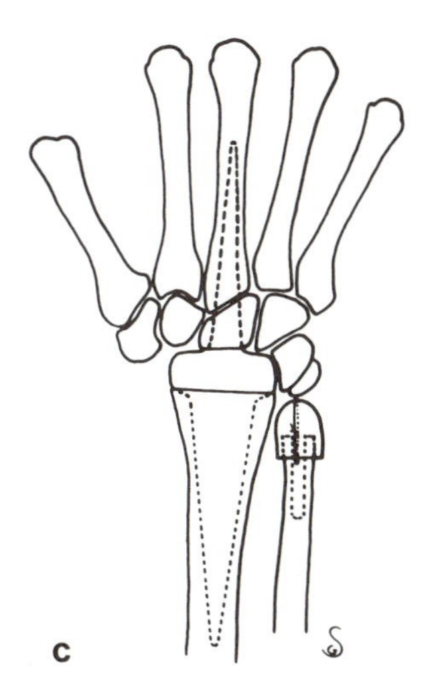

Figure 14-19. Various types of wrist prostheses: (a) Meuli; (b) Volz; (c) Swanson. (a) From H. C. Meuli, "Alloarthropstik des Handgelenks," *A. Orthop. Ihre Grenzgeb.*, *113*, 476–478, 1975; (b, c) from A. B. Swanson, *Flexible Implant Resection Arthroplasty in the Hand and Extremities*, C. V. Mosby Co., St. Louis, 1973.

Answer

(a) In this composite system we calculate the maximum strain for each component by using Hooke's law, hence

$$\text{Max. strain for stainless steel} = \frac{300 \times 10^6 \text{ N/m}^2}{200 \times 10^9 \text{ N/m}^2} = 1.5 \times 10^{-3}$$

$$\text{Max. strain for Bioglass} = \frac{300 \times 10^6 \text{ N/m}^2}{300 \times 10^9 \text{ N/m}^2} = 1.0 \times 10^{-3}$$

Therefore, the maximum strain for the composite is the smaller of the two, that is, $\underline{1 \times 10^{-3}\ (0.1\%)}$.

(b) The cross-sectional areas of each component are

$$A_{SS} = \pi(5 \times 10^{-3} \text{ m})^2 = 25\pi \times 10^{-6} \text{ m}^2$$

$$A_{BG} = \pi[(6 \times 10^{-3} \text{ m})^2 - (5 \times 10^{-3} \text{ m})^2] = 11\pi \times 10^{-6} \text{ m}^2$$

Therefore,

$$F_{comp} = F_{SS} + F_{BG}$$

$$= 200 \times 106 \text{ N/m}^2 \times 25 \times 10^{-6} \text{ m}^2 + 300 \times 10^6 \text{ N/m}^2 \times 11\pi \times 11^{-6} \text{ m}^2$$

$$= 8300\pi \text{ N}$$

$$= \underline{26.1 \text{ kN}}$$

(c) Since,

$$E_{comp} = E_1 V_1 + E_2 V_2 + \cdots$$

$$= E_1 A_1 + E_2 A_2 + \cdots$$

where V and A are volume and area fractions of each component, therefore

$$E_{comp} = 200 \times 25\pi/26\pi + 300 \times 11\pi/36\pi$$

$$= \underline{231\ \text{GPa}}$$

$$\rho_{\text{comp}} = \rho_1 V_1 + \rho_2 V_2 + \cdots$$

$$= \rho_1 A_1 + \rho_2 A_2 + \cdots$$

$$= 7.9 \times 25\pi/36\pi + 4.5 \times 11\pi/36\pi$$

$$= \underline{6.85\ \text{g/cm}^3}$$

(d) (1) It is difficult to obtain perfect bonding between the stainless steel and Bioglass because of the differences in thermal expansion coefficient. In addition, Bioglass is brittle; consequently, a moderate or severe bending load on the implant causes the Bioglass to develop cracks and microcracks. (2) The bonding between the (hard) tissue and Bioglass® cannot be maintained for a prolonged time without further weakening of the Bioglass since a continuous dissolution of the surface film is essential for the bonding to be maintained (remember the tissue cells renew themselves).

14.2. DENTAL IMPLANTS

In this section, we examine total tooth replacements or alveolar bone augmentation with man-made materials. The use of artificial materials for restoring portions of living teeth (i.e., filling cavities) is dealt with elsewhere in this book (see Sections 5.4, 6.3, 6.4, 8.1, 8.3.1).

Replacement of whole teeth is a very challenging task in view of the transcutaneous (or percutaneous) nature of the implant in the hostile oral environment, which continually changes its chemical composition, pH, temperature, etc. Teeth experience large forces (up to 850 N) on a small area; consequently, they undergo the most severe compressive stress in the body. Satisfactory materials or techniques have not yet been found that withstand not only the compressive stress but also the added torque and shear stresses during mastication.

In a recent comprehensive conference, dental implants were classified largely into two categories: subperiosteal/staple/transosteal and endosseous tooth implants. The former support dentures and the latter restore the function of teeth with or without a supporting bridge framework.

14.2.1. Endosseous Tooth Implants

The endosseous dental implant is inserted into the site of missing or extracted teeth to restore the original function. The ideal implant would be the tooth itself, replaced in the same socket from which it was lost. Teeth that have been knocked out traumatically can sometimes be replanted. In most cases of tooth loss,

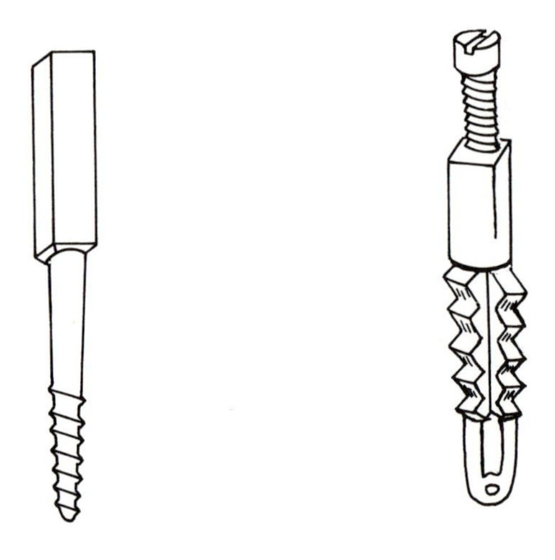

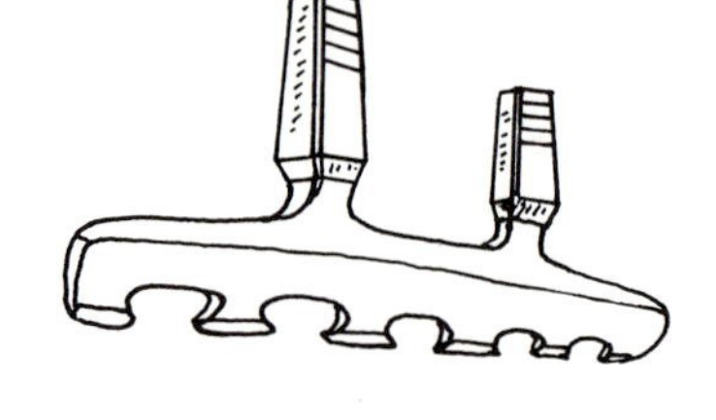

Figure 14-20. Various designs of self-tapping dental (root) implants. From D. E. Grenoble and D. Voss, "Materials and Designs for Implant Dentistry," *Biomed. Mater. Devices Artif. Organs, 4,* 133–169, 1976.

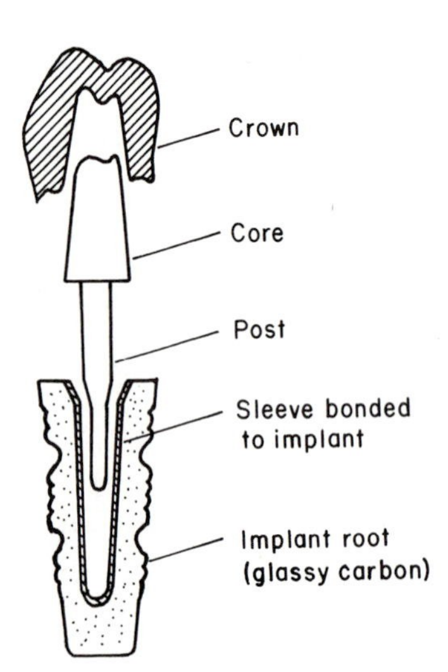

Figure 14-21. The tooth root implant fabricated from glassy carbon. From D. E. Grenoble and D. Voss, "Materials and Designs for Implant Dentistry," *Biomed. Mater. Devices Artif. Organs, 4,* 133–169, 1976.

however, the teeth or their supporting structures are diseased. Artificial teeth supported by the gums (false teeth) are a partial solution to the problem of tooth loss; however, they present problems of their own: lack of stability, poor esthetics, and resorption of the jaw bone. Artificial teeth fixed to the jawbone represent an attempt to achieve more natural replacements; these are known as endosseous implants. There are many different types of designs for endosseous implants as shown in Figure 14-20. The main idea behind the various root portions of self-tapping screws—spiral, screw-vent, and blade-vent implants—is to achieve immediate stabilization as well as long-term viable fixation. The post will be covered with an appropriate crown after the implant has been fixed firmly for about 14 months. Some implants have used a more complicated system of fixation as shown in Figure 14-21, in which the implant root is first implanted in the tooth extraction site (preferably after complete healing of the site) and completely buried, then the post is installed using a cement through a punctured hole on the mucosa, and finally the crown is made. Despite the more elaborate work and design of the implants, the success rate of this implant system did not improve over other implants such as blade vents (Linkow types). The major problems were (1) initially the quality control of the vitreous carbon (Vitrident®) implants were not stringent (almost 20% of one batch was deemed defective by one investigator); (2) the brittle nature of the vitreous carbon makes the implant bulky, consequently the oral surgeon has to remove a larger amount of alveolar bone to make room for the root; and (3) the implants were distributed indiscriminately to general dentists, resulting in unqualified people implanting a rather complicated implant. A high level of failure was the result.

The survival rates of the popular blade-vent implants vary for various investigators because of many factors: surgical techniques, patient selection, evaluation criteria, etc. Figure 14-22 shows the unadjusted and adjusted survival

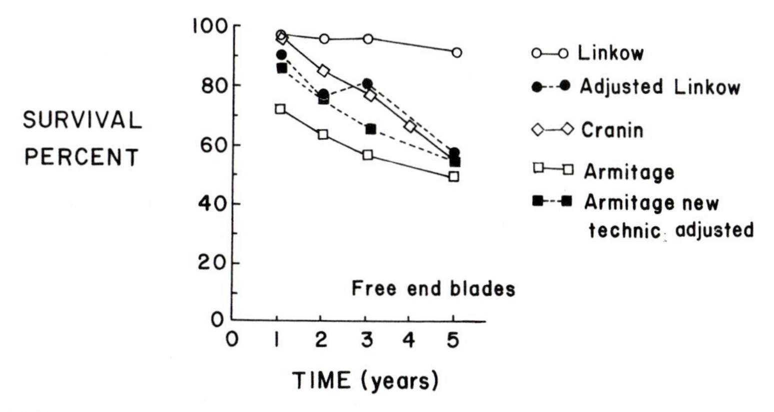

Figure 14-22. Unadjusted and adjusted survival rates for all free-end blade implants. The midpoint between Grades 2 and 3 of radiographic bony defect was used as the criterion for adjusting the survival rates in the Linkow and Armitage data. From K. K. Kapur, "Benefit and Risk of Blade Implants: A Critique," in *Dental Implants: Benefits and Risk*, NIH Publ., No. 81–1531, U.S. Department of Health and Human Services, Bethesda, 1980.

rate for free-end blade implants as determined by various investigators. The variations largely reflect differences in the evaluation criteria used by different investigators. Because of these large differences in reported results, the NIH–Harvard Consensus Development Conference participants stated that "no specific survival estimates can be quoted because of either insufficient sample size or conflicting results" for the endosseous dental implants including vitreous carbon implants.

Most of the blade-vent endosseous implants are made from stainless steel, Co–Cr alloy, Ti, and Ti6Al4V alloy. There have been efforts to coat the surfaces of the implants with ceramics (alumina, zironia, and hydroxyapatite) and polytetrafluoroethylene composite (Proplast®) with little significant improvement in their performance. Others used pyrolytic carbon, polycrystalline alumina, and single-crystal alumina. Recently, surface-textured implants and porous implants with electrical stimulation have been tried. Some researchers have even tried to use anorganic bone/acrylic polymer (PMMA) composite material to induce bony tissue to grow into the spaces originally filled with anorganic bone. This results in ingrown tissue and fixation of the bone to the artificial tooth root.

Example 14-4

During an experiment involving tissue growth into the pores of a porous-surfaced tooth implant, a push-out test was performed. The experimenter sectioned into 2-mm thick disks (two for each implant) and one of the curves is given on the facing page (the diameter is 4.6 mm).

(a) Calculate the maximum interfacial shear stress between the bone and the implant.
(b) What is the stiffness of the interface?
(c) Is the interfacial strength adequate for fixation of the implant?

Answer

(a) The "nominal" shear stress is

$$\sigma = \frac{F}{A} = \frac{F}{\pi \times D \times H} = \frac{11 \text{ N}}{28.9 \text{ mm}^2} = \underline{0.38 \text{ MPa}}$$

(b) The stiffness can be calculated:

$$S = \frac{\sigma}{\varepsilon} = \frac{5.8 \text{ N}/28.9 \text{ mm}^2}{0.05} = \underline{4 \text{ MPa}}$$

Note that the strain is the "shear" strain.

(c) The interfacial strength may not substantially contribute to the total mastication load (because the natural tooth has a conical shape, the applied load is distributed into two major components, shear and compression); therefore, the interfacial shear strength calculated in (a) is a high enough value. (The direct attachment of artificial tooth by tissue growth into the pores of an implant has certain advantages. However,

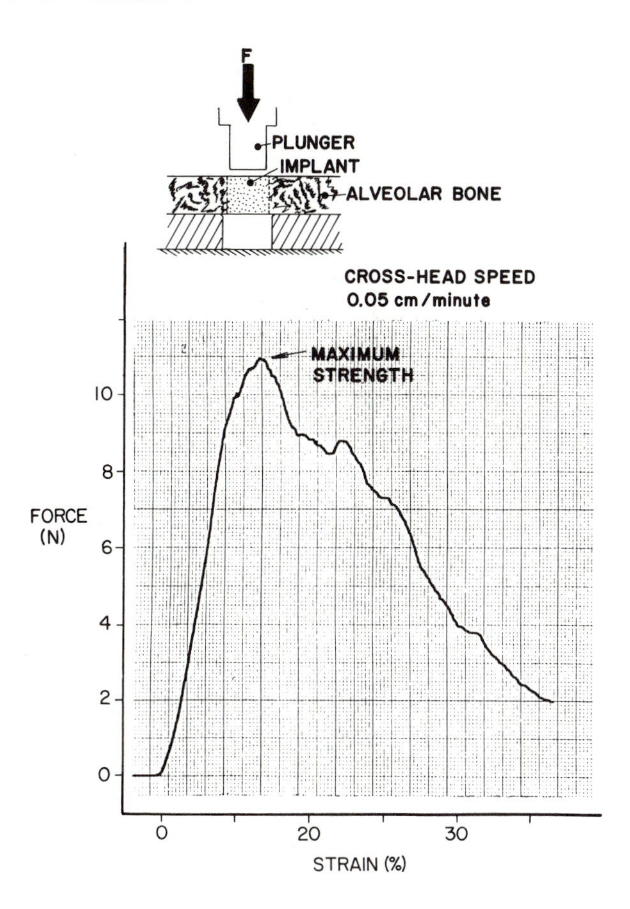

the types of tissues, whether hard or soft, to be grown into to achieve a viable fixation and give the best results have not been established.)

14.2.2. Subperiosteal and Staple/Transosteal Implants

Implants have been successfully used to provide a framework for dentures for the edentulous alveolar ridge, as shown in Figure 14-23. Although similar functions can be duplicated by osseous dental implants, the periosteal and/or staple/transosteal implants have been developed to compensate for the weakness of the thin alveolar ridge in many edentulous patients. The rationale is to provide better support for dentures or other bridgework. Unfortunately, these implants thus far have experienced problems similar to those associated with individual dental implants.

Materials used for these implants are primarily metals, stainless steel, Co–Cr alloy, and Ti alloy, owing to their ease of fabrication in a conventional dental laboratory. Some advocate coating of metals with other inert materials, such as

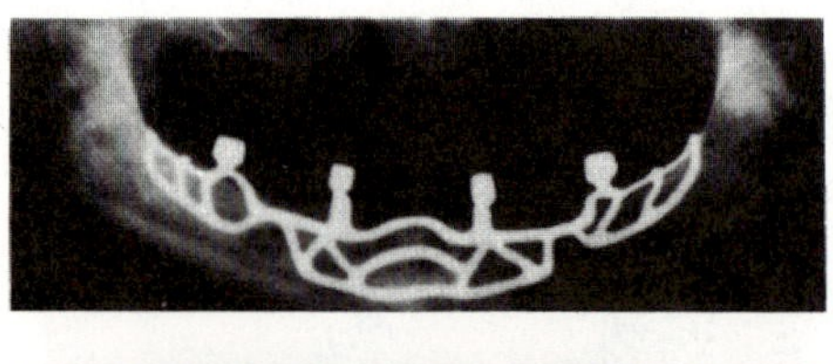

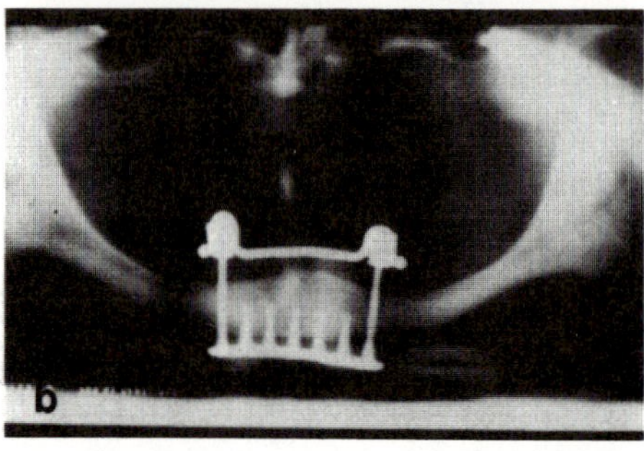

Figure 14-23. (a) A mandibular subperiosteal implant. From P. A. Schnitman and L. B. Schulman (eds.), *Dental Implants: Benefits and Risk,* NIH Publ. 85–1531, U.S. Department of Health and Human Services, Bethesda, 1980. (b) Seven-pin mandibular staple. From I. S. Small, "Benefit and Risk of Mandibular Staple Bone Plates," in *Dental Implants: Benefits and Risk,* pp. 139–151.

carbon and ceramics. It is suspected that these coatings will result in a marginal improvement, as in endosseous dental implants.

14.3. INTERFACE PROBLEMS IN ORTHOPEDIC IMPLANTS

The fixation of orthopedic implants has been one of the most difficult and challenging problems. The fixation can be (1) active mechanical fixation by using screws, pins, wires, and bone cement, (2) passive mechanical fixation by allowing relative motion between implant and tissue surfaces, (3) biological fixation by allowing tissues to grow into the interstices of pores or the textured surface of implants, (4) direct chemical bonding between implant and tissues.

This section is concerned with the pros and cons of the various fixation techniques. Also considered are newer techniques such as the use of electrical and pulsed electromagnetic field stimulation, chemical stimulation by using calcium phosphates, direct bonding with bone by glass-ceramics and resorbable particle-impregnated bone cement. Such techniques take advantage of both the immediate fixation of bone cement and the long-term fixation of tissue ingrowth.

One of the inherent problems of the orthopedic implantation is the fixation of the devices and the maintenance of the interface between the device and host tissue. The fixation can be classified into several categories (see also Figure 14-24):

1. Direct interference of (passive) noninterference fit
2. Mechanical fixation using screws, bolts, nuts, wires, etc.

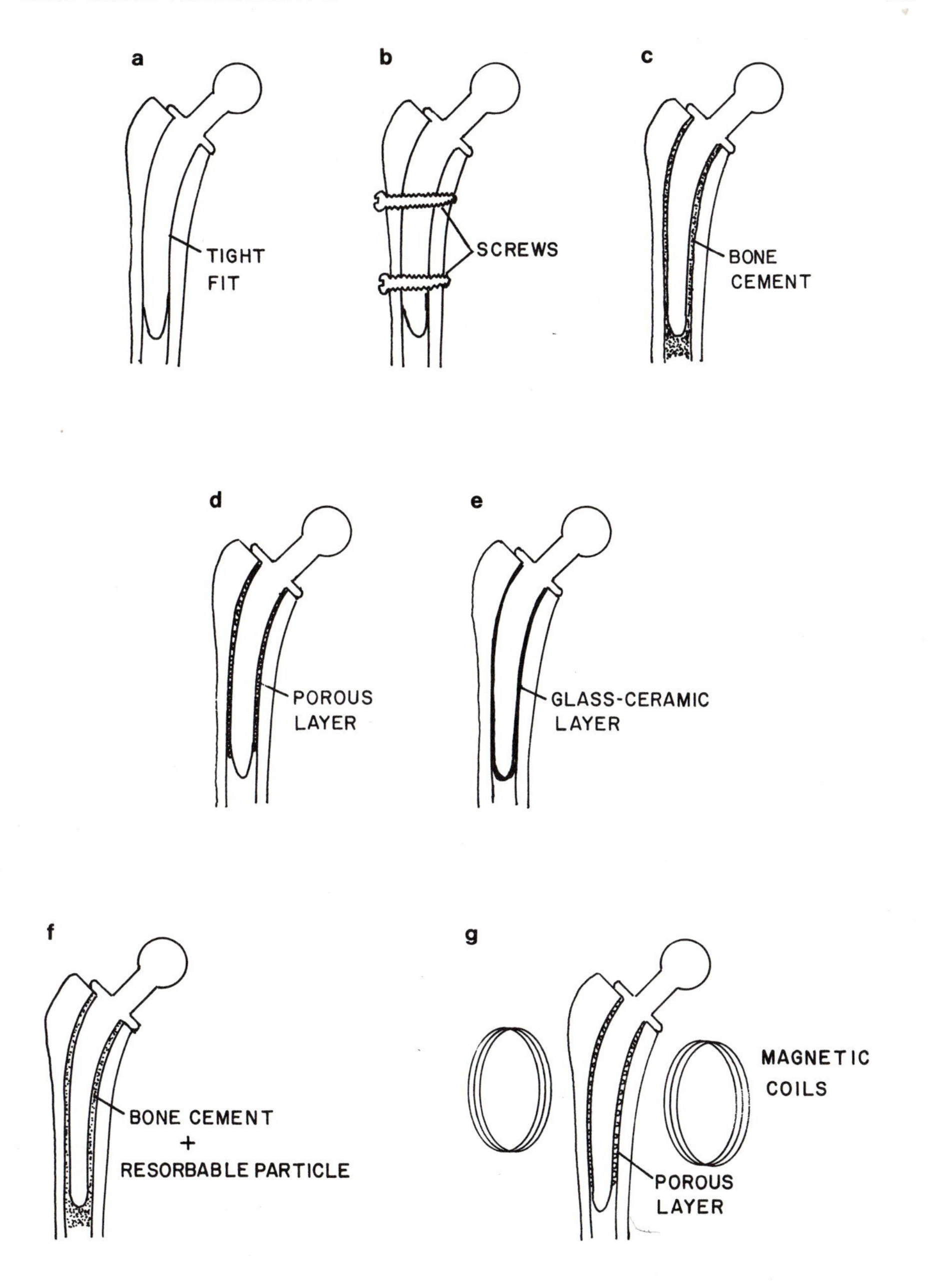

Figure 14-24. Schematic illustration of the various fixation methods. (a) Direct interference or (passive) noninterference fit. (b) Mechanical fixation using screws, bolts, nuts, wires, etc. (c) Bone cement or grouting. (d) Porous ingrowth (biological) fixation. (e) Direct chemical bonding using adhesives or after coating with direct bonding material layer. (f) Bone cement with resorbable particles. (g) Porous ingrowth controlled by using electrical or electromagnetic stimulation. From J. B. Park, Interface Problems in Orthopedic Implants, Invited Lecture, Korean Orthopedic Association, Pusan, Korea, April 1989.

3. Bone cement interdigitation
4. Porous ingrowth (biological) fixation
5. Direct chemical bonding using adhesives or after coating with direct bonding layer
6. Combination of categories 3 and 4 by incorporating resorbable particles into bone cement
7. Porous ingrowth controlled by using electrical or electromagnetic stimulation

The major failure mode of modern orthopedic implants is loosening. Loosening is an interface problem; however, the underlying causes may include bone resorption around the implant, fatigue of the cement, biocompatibility problems, surgical technique, and others. We will review some of these fixation techniques and discuss some possible solutions related to the total joint prostheses.

14.3.1. Bone Cement Fixation

The bone cement fixation creates two interfaces: (1) bone/cement and (2) cement/implant. According to an earlier study, the incidence of problems (usually loosening) related to both interfaces for the femoral prostheses were evenly divided (about 10 and 11%) for cement/implant and bone/cement interface, respectively. The cement/implant interface problems can be minimized by precoating the metal with bone cement or PMMA polymer to which the new bone cement can adhere well. The precoating can achieve a good bonding between the cement and prostheses during the manufacturing process as mentioned earlier.

The problem of the bone/cement interface cannot be so easily overcome since this problem derives from the intrinsic properties of the bone cement and of the bone as well as extrinsic factors such as cementing technique. The toxicity of the monomer, inherent weakness of the cement as a material (see Table 14-2), and inevitable inclusion of pores can contribute to the problem of loosening at the bone/cement interface.

The problem of the bone/cement interface may be solved by utilizing the concept of bone ingrowth. Specifically, the bone cement can be used for the initial fixation medium and yet provide tissue ingrowth space later; this is done by incorporating resorbable particles such as inorganic bone as shown in Figure 14-25. Recent studies in rabbits and dogs indicate that the concept can be effective. In one experiment the bone particle-impregnated bone cement was used to fix femoral stem prostheses in femora of dogs (one side was experimental, another was control). After a predetermined time the femora were harvested and sectioned into disks. Push-out test measured the interfacial strength for the bone/cement interface. Figure 14-26 shows that the experimental side exhibited increasing strength up to 5 months while the control side had a decrease in strength but stabilized after that time. The histology studies also disclosed integration of tissues into the spaces left by the dissolved particles. It was found that about 30% of bone particles can provide continuous porosity for the bone to grow and still give a reasonable compromise to other parameters. If too high

Table 14-2. Mechanical Properties of Materials Used for Joint Prosthesis and Bone

Material	Young's modulus (GPa)	UTS[a] (MPa)	Elongation (%)	Density (g/cm^3)
Metals				
316L SS (wrought)	200	1000	9	7.9
CoCrMo (cast)	230	660	8	8.3
CoNiCrMo (wrought)	230	1800	8	9.2
Ti6Al4V	110	900	10	4.5
Ceramics				
Alumina (Al_2O_3, polycrystalline)	400	260	0.065	3.9
Glass-ceramic (Bioglass®)	200[b]	200	0.1	2.5[b]
Hydroxyapatite	120	200	0.17	3.2
Polymers				
PMMA (solid)	3	65	5	1.18
PMMA bone cement	2	30	3	1.1
UHMW polyethylene	1	30	200	0.94
Bone				
Femur (compact), long axis	17	130	3	2.0
Femur (compact), tangential	12	60	1	2.0
Spongy bone	0.1	2	2.5	1.0

[a] UTS, ultimate tensile strength.
[b] Estimated values.

a concentration of bone particles were used, viscosity would be excessive during the polymerization/dough stage, and the strength of the cement might be reduced. At this time the concept of bone particle-reinforced PMMA has not been demonstrated in human patients.

14.3.2. Porous Ingrowth (Biological) Fixation

An effort to develop a viable interface between tissue and implant has been ongoing over the three decades since Smith (1963) tried to develop a bone substitute with porous aluminate ceramic impregnated with epoxy resin (called

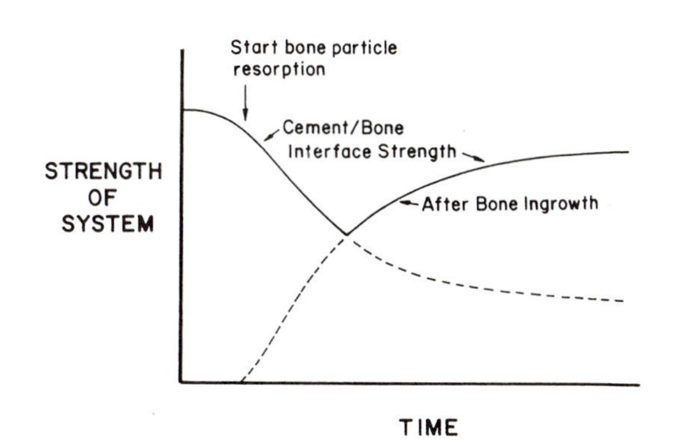

Figure 14-25. Basic concept of fixation by bone cement with resorbable particles. Immediate fixation is achieved as in ordinary bone cement. Particles are resorbed by the body, and bone grows into the resulting voids. From Y. K. Liu, J. B. Park, G. O. Njus, and D. Stienstra, "Bone Particle Impregnated Bone Cement I. In Vitro Study," *J. Biomed. Mater. Res.*, *21*, 247–261, 1987.

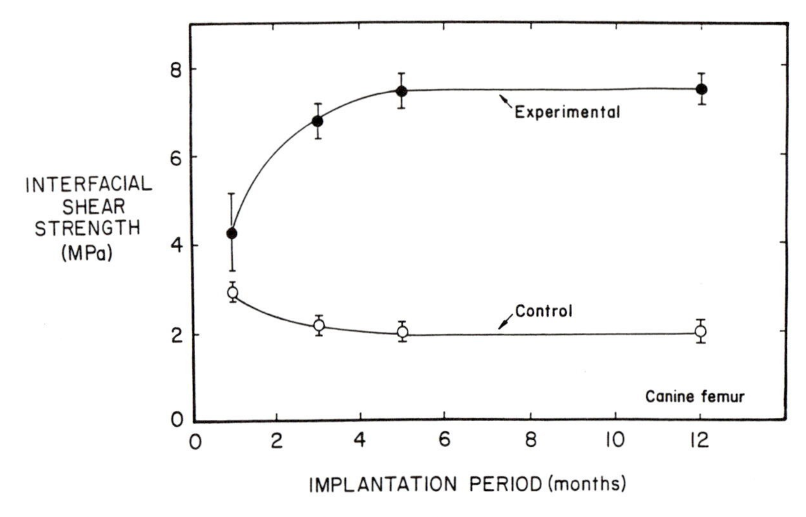

Figure 14-26. Maximum interfacial shear strength between bone and bone cement versus implant period. The femoral stems were implanted with ordinary bone cement and cement with bone particles. In both cases the interfacial strength stabilized after 5 months for this canine model. From K. R. Dai, Y. K. Liu, J. B. Park, and Z. K. Zhang, "Bone-Particle-Impregnated Bone Cement: An in Vivo Weight-Bearing Study," *J. Biomed. Mater. Res.*, *25,* 141–156, 1991.

Cerocium®). Although the material showed a good adherence to the tissues, the pore size (average 18 μm diameter) was too small to allow any bony tissue ingrowth. Later, ceramics, metals, and polymers were used to test the ingrowth ideas. Basically, any inert material will allow bony tissues to grow into any spaces large enough to accommodate osteons. However, for the ingrown bone to be continuously *viable* the pore space must be large enough (more than 75 μm diameter for the bony tissues), contiguous to the surface of the implant, and in contact with the bone. In addition, Wolff's law dictates that the ingrown tissues should be subjected to bodily loading in order to prevent resorption after the initial ingrowth has taken place.

Additional difficulties with the bony ingrowth concept include the fact that it is surgically unforgiving, a long immobilization time is required for the tissue to grow for the initial fixation, the uncertainty of the time until the patient can walk, and the difficulty in eradicating any infection that might occur. Moreover, once the interface is destroyed by accidental overloading, it cannot be reattached with certainty. Furthermore, the porous coating may weaken the underlying prosthesis itself. In metallic implants there is an increased danger of corrosion due to the increased surface area. Because of these many problems and the poorer-than-expected clinical results of porous fixation, some have insisted that the cement technique is still the better choice at this time.

In order to alleviate these problems, the following modifications have been tried.

1. *Precoating the metallic porous surface with ceramics or carbons.* This method has had some limited success. The problem of coating deep in the pores

and the thermal expansion difference between the metal and ceramic materials make a uniform and good adherent coating very difficult. Another attractive material for coating is hydroxyapatite ceramic, which is similar to bone mineral. It is not yet known whether this material is superior to other ceramics. Some preliminary studies in our laboratory indicate that in the early period of fixation the bioactive apatite coating may be more beneficial but the effect may diminish later as shown in Figure 14-27. In these studies, a simple cortical bone plug was replaced by an implant in the canine femur.

2. *Precoating with porous polymeric materials on metal stem.* Theoretically, this method is a better solution to the long-term implant problem since this method has two distinctive advantages over method 1 discussed above. First, the low-modulus polymeric material could transfer the load from the implant to the bone more gradually and evenly than metal. Moreover, it can prevent a stress-shielding effect on the ingrown tissues. Second, this method would reduce corrosion of the metal. One major problem with this technique is the weak interface strength between the polymer and metal stem especially in a dynamic loading condition *in vivo*.

3. *Porous ingrowth with electrical or electromagnetic stimulation.* This technique combines the porous ingrowth with the stimulating effect of electrical or electromagnetic fields. Direct current stimulation can indeed accelerate tissue ingrowth as shown in Figure 14-28. The effect is strongest in the early stages of the healing and diminishes later. The direct current stimulation has one distinctive problem, that is, the invasive nature of the implanted electrodes. The pulsed electromagnetic field stimulation method is better since the stimulation can be carried out extracorporeally. A preliminary study using canine femur indicates that it can be effective as shown in Figure 14-29. More studies are needed to verify this result.

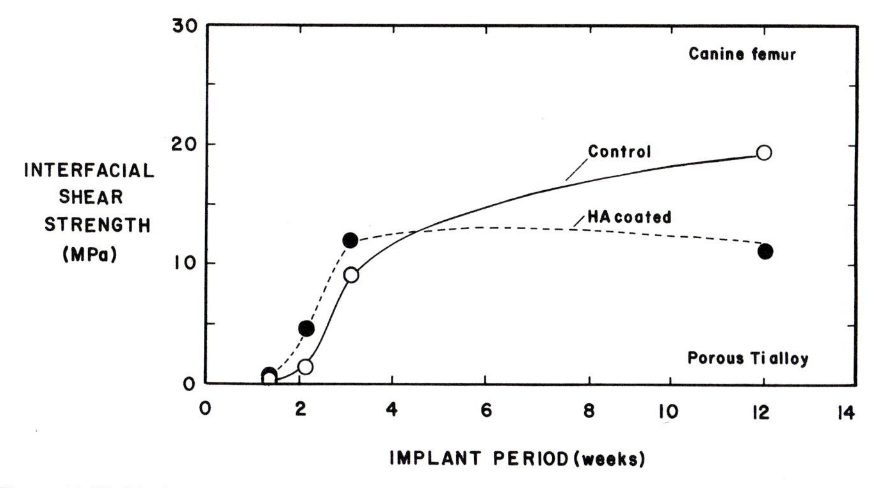

Figure 14-27. Maximum interfacial shear strength between bone and bioactive ceramic-coated porous plug implants versus implant period. Plugs with and without (control) coating were implanted in the cortices of canine femurs. (S. H. Park and J. B. Park, unpublished data, University of Iowa, 1988.)

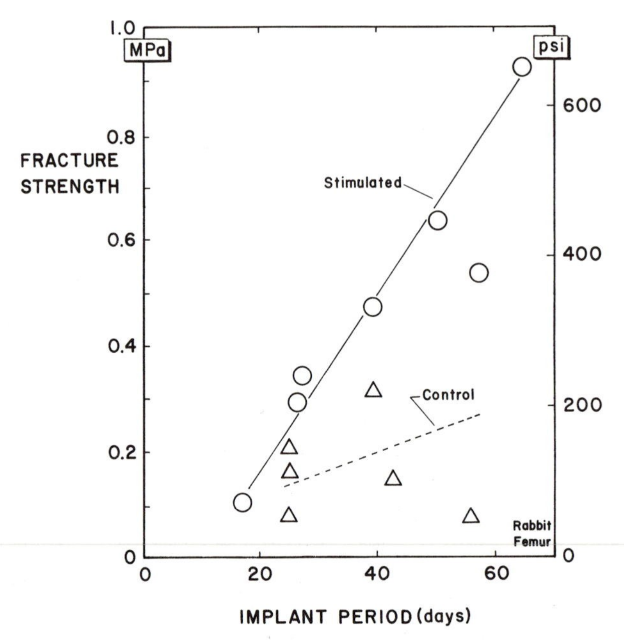

Figure 14-28. Maximum interfacial shear strength between bone and porous ceramic plug implants versus implant period with and without (control) direct current stimulation. Plugs were implanted in the cortices of canine femurs. From J. B. Park and G. H. Kenner, "Effect of Electrical Stimulation on the Tensile Strength of the Porous Implant and Bone Interface," *Biomater. Med. Devices Artif. Organs*, *3*, 233–243, 1975.

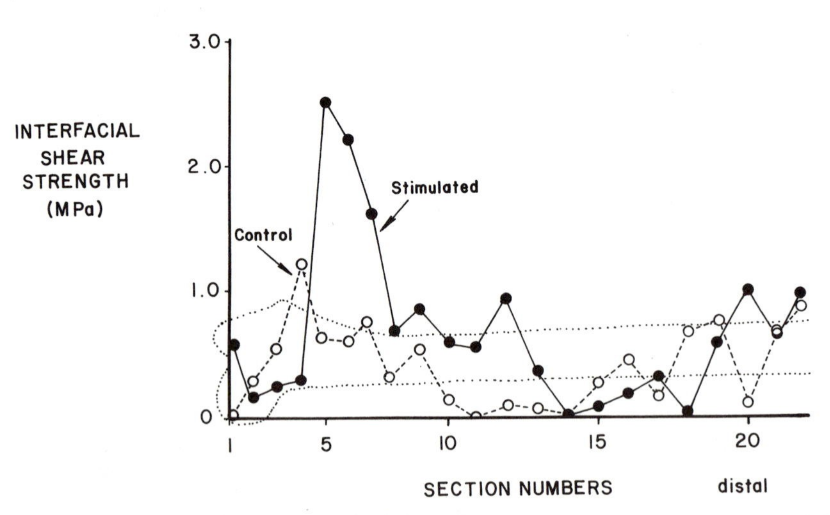

Figure 14-29. Maximum interfacial shear strength between bone and porous metallic intramedullary implant with and without (control) pulsed electromagnetic field stimulation. The porous implants were implanted in the medullary canals of canine femurs (one side was the control, the other experimental). (J. B. Park, unpublished data, University of Iowa, 1988.)

4. *Porous ingrowth with use of filling materials.* There has been some effort to use filling materials around the porous implant since it is very difficult to prepare the tissue bed with the exact shape of the prosthesis as is required to eliminate the micromovement of the prosthesis after implantation. Bone matrix proteins and hydroxyapatite crystals can be used as filling materials. The success of this technique is not fully documented.

14.3.3. Direct Bonding between Bone and Implant

Some glass-ceramics showed direct bonding with the bone via a selective dissolution of the surface film. A glass-ceramic has not yet been successfully coated on a metal surface; moreover, a glass-ceramic is too brittle to use in implants.

14.3.4. Interference and Passive Fixation

These techniques have been applied in limited circumstances such as hip and finger joint prosthesis fixation. Because of the nature of the loading (mostly compressive) and (wedge) shape, an interference fit can be used to fix the femoral stem of a hip joint. This technique is largely used for ceramic stems in view of the large size of the stem. The stem is large because of unpredictability of the failure of brittle materials such as alumina ceramic. The stem can be seated distributing the stress over a large area, thus diminishing stress necrosis of bone. Most likely this fixation method will induce collagenous membrane formation at the interface between bone and implant unless relative motion between them can be eliminated. It is also conceivable that the sinking of the implant under load may continue throughout its life.

The passive fixation of finger joints such as Swanson's prosthesis is based on an entirely different concept of fixation. It depends on the development of a collagenous membrane between implant and bone in which the prosthesis glides in and out. This fixation can provide minimal rigidity of the joint, making it sufficient to hold a cup of coffee but not for putting screws into a wooden block.

In summary, one would like to have firm fixation of an implant for its usefulness and longevity. However, if the implant fails, a revision arthroplasty has to be performed and the removal of the failed implant sometimes becomes a very difficult job. This is a rather catch-22 situation since we need to make the interface strong in order to prevent failure at least related to the loosening of the implant.

Another problem is related to the longevity of the implants. Because of the longer life of humans we have to provide a long-lasting system especially for the young people who receive joint replacements for trauma more often than for arthritis. The original optimism regarding the porous ingrowth fixation method has been tempered somewhat and we need to explore a better method of fixation. Certainly we should inform patients of the limitations of the implant and advise them accordingly so that they not abuse it but use it properly. Lastly, the orthopedic doctors, biomedical engineers, and biologists should work together in order to solve the complicated problems of implant fixation.

PROBLEMS

14-1. Why are the average maximum values of force at the hip and knee several times the body weight? Propose a method of measuring the forces transmitted through the joints in vivo.

14-2. Describe the sequence of events by which the production of wear debris in an artificial joint can lead to joint pain and loosening of the prosthesis.

14-3. A simple way to prevent the stem of a hip prosthesis from breaking through the cortex of a weak femur would be to use a metal band that circumscribed the weak area and distributed the load around the cortex. Describe the advantages and disadvantages of this approach.

14-4. Assuming that you wished to attach a femoral head prosthesis by using a porous coating on a metallic stem, would you choose a ceramic, metallic, or plastic coating material? Discuss all aspects of the problem including fabrication, compatibility, stress distribution, etc.

14-5. What are some of the advantages and disadvantages of porous materials (for the attachment of joint replacements) in comparison with PMMA bone cement fixation?

14-6. It was reported that a maximum of only 15% of the available space in a porous implant was occupied by ingrown tissue for knee joints (especially the tibial component) retrieved after a few years. Discuss the consequences and possible reasons for such a small amount of tissue present.

14-7. Explain how you would control the rate of resorbable particle dissolution for particle-reinforced bone cement. Consider the possibility of a rapid early dissolution followed by a slower dissolution over a period of several years.

14-8. Discuss the idea of precoating radiopacifying agents such as barium sulfate ($BaSO_4$) with PMMA to increase the strength of PMMA bone cement. Consider the fact that 10% by weight of barium sulfate inclusions decreases the strength of the cement by 10%.

DEFINITIONS

Acetabulum: The socket portion of the hip joint.
Arthroplasty: Plastic repair of a joint.
Edentulous: Without teeth.
Endosseous: In the bone, referring to dental implants fixed to the jawbone.
Proplast®: A composite material made of carbon and polytetrafluoroethylene. It has a moderate porosity and strength.
Subchondral: Bone lying under the cartilage of a joint.
Vitrident®: A defunct dental root implant made of glassy carbon.

BIBLIOGRAPHY

H. Amstutz, K. L. Markolf, G. M. McNeice, and T. A. Gruen, "Loosening of Total Hip Components: Cause and Prevention," *The Hip*, C. V. Mosby Co., St. Louis, 1976, pp. 102–116.

W. Barb, J. B. Park, A. F. von Recum, and G. H. Kenner, "Intramedullary Fixation of Artificial Hip Joints with Bone Cement Precoated Implants: I. Interfacial Strengths," *J. Biomed. Mater. Res.*, *16*, 447–458, 1982.

J. Black and J. H. Dumbleton (eds.), *Clinical Biomechanics: A Case History Approach*, Churchill Livingstone, Edinburgh, 1981.

B. A. Blencke, H. Bromer, and K. K. Deutscher, "Compatibility and Long-Term Stability of Glass Ceramic Implants," *J. Biomed. Mater. Res.*, *12*, 307-316, 1978.

R. A. Brand (ed.), *The Hip: Non-Cemented Hip Implants*, C. V. Mosby Co., St. Louis, 1987, pp. 213-358.

J. Charnley, *Acrylic Cement in Orthopedic Surgery*, Livingstone, Edinburgh, 1970.

J. Charnley, *Acrylic Cement in Orthopedic Surgery*, Williams & Wilkins, Baltimore, 1970.

J. Charnley, *Low Friction Arthroplasty of the Hip*, Springer-Verlag, Berlin, 1979.

A. N. Cranin (ed.), *Oral Implantology*, Charles C. Thomas Publ., Springfield, Ill., 1970.

J. H. Dumbleton, *Tribology of Natural and Artificial Joints*, Elsevier Scientific Publ., Amsterdam, 1981.

N. S. Eftekhar, *Principles of Total Hip Arthroplasty*, C. V. Mosby Co., St. Louis, 1978.

A. E. Flatt (ed.), *The Care of Minor Hand Injuries*, C. V. Mosby Co., St. Louis, 1972.

V. H. Frankel and A. H. Burstein, *Orthopedic Biomechanics*, Lea & Febiger, Philadelphia, 1971.

H. M. Frost, *Orthopedic Biomechanics*, Charles C. Thomas Publ., Springfield, Ill., 1973.

D. N. Ghista and R. Roaf (eds.), *Orthopedic Mechanics: Procedures and Devices*, Academic Press, New York, 1978.

N. Gschwend, "Design Criteria, Present Indication, and Implantation Techniques for Artificial Knee Joints, in *Advances in Artificial Hip and Knee Joint Technology*, M. Schaldach and D. Hohmann (eds.), Springer-Verlag, Berlin, 1976, pp. 90-114.

L. L. Hench and H. A. Paschall, "Direct Chemical Bond of Bioactive Glass-Ceramic Materials to Bone and Muscle," *J. Biomed. Mater. Res. Symp.*, *4*, 25-43, 1973.

J. S. Hirshhorn, A. A. McBeath, and M. R. Dustoor, "Porous Titanium Surgical Implant Materials," *J. Biomed. Mater. Res. Symp.*, *2*, 49-69, 1972.

C. A. Homsy, T. E. Cain, F. B. Kessler, M. S. Anderson, and J. W. King, "Porous Implant Systems for Prosthetic Stabilization," *Clin. Orthop. Relat. Res.*, *89*, 220-231, 1972.

R. C. Johnston, "The Case for Cemented Hips," in *The Hip*, R. A. Brand (ed.), C. V. Mosby Co., St. Louis, 1987, pp. 351-358.

J. J. Klawitter and S. F. Hulbert, "Application of Porous Ceramics for the Attachment of Load Bearing Internal Orthopedic Applications," *J. Biomed. Mater. Res. Symp.*, *2* (1), 161-229, 1972.

D. C. Mears, *Materials and Orthopedic Surgery*, Williams & Wilkins, Baltimore, 1979.

E. Morscher (ed.), *The Cementless Fixation of Hip Endoprosthesis*, Springer-Verlag, Berlin, 1984.

J. B. Park, "Acrylic Bone Cement; In vitro and in vivo Property-Structure Relationship—A Selective Review," *Ann. Biomed. Eng.*, *11*, 297-312, 1983.

J. B. Park, W. W. Choi, Y. K. Liu, and T. W. Haugen, "Bone Particle Impregnated Polymethylmethacrylate: In vitro and in vivo Study," in *Tissue Integration in Oral and Facial Reconstruction*, D. Van Steenberghe (ed.), Excerpta Medica, Amsterdam, 1986, pp. 118-124.

P. Predecki, J. E. Stephan, B. E. Auslander, V. L. Mooney, and K. Kirkland, "Kinetics of Bone Growth into Cylindrical Channels in Alumina Oxide and Titanium," *J. Biomed. Mater. Res.*, *6*, 375-400, 1972.

S. Raab, A. M. Ahmed, and J. W. Provan, "Thin Film PMMA Precoating for Improved Implant Bone-Cement Fixation," *J. Biomed. Mater. Res.*, *16*, 679-704, 1982.

A. A. Savastano (ed.), *Total Knee Replacement*, Appleton-Century-Crofts, New York, 1980.

M. Schaldach and D. Hohmann (eds.), *Dental Implants: Benefits and Risk*, NIH Publ. No. 81-1531, U.S. Department of Health and Human Services, Bethesda, 1980.

P. A. Schnitman and L. B. Schulman (eds.), *Dental Implants: Benefits and Risk*, NIH Publ. No. 81-1531, U.S. Department of Health and Human Services, Bethesda, 1980.

H. S. Shim, "The Strength of LTI Carbon Dental Implants," *J. Biomed. Mater. Res.*, *11*, 435-445, 1977.

L. Smith, "Ceramic-Plastic Material as a Bone Substitute," *Arch. Surg.*, *87*, 653-661, 1963.

M. Spector, "Bone Ingrowth into Porous Polymers, in *Biocompatibility of Orthopedic Implants*, Vol. II, D. F. Williams (ed.), CRC Press, Boca Raton, Fla., 1982, pp. 55-88.

A. B. Swanson, *Flexible Implant Resection Arthroplasty in the Hand and Extremities*, C. V. Mosby, St. Louis, 1973.

S. A. V. Swanson and M. A. R. Freeman (eds.), *The Scientific Basis of Joint Replacement*, J. Wiley and Sons, New York, 1977.

A. R. Taylor, *Endosseous Dental Implants*, Butterworths, London, 1970.

A. M. Weinstein, J. J. Klawitter, F. W. Cleveland, and D. C. Amoss, "Electrical Stimulation of Bone Growth into Porous Al_2O_3," *J. Biomed. Mater. Res.*, *10*, 231–247, 1976.

V. Wright (ed.), *Lubrication and Wear in Joints*, J. B. Lippincott Co., New York, 1969.

B. M. Wroblewski, "15–21 Year Results of the Charnley Low-Friction Arthroplasty," *Clin. Orthop. Relat. Res.*, *211*, 30–35, 1986.

CHAPTER 15

TRANSPLANTS

As we have seen in the previous chapters, biomaterials have many uses in aiding healing, restoring a lost form or function, and correcting a deformity. The limitations of artificial materials become apparent when we realize that only the simplest mechanical, structural, optical, and chemical functions can be assumed by nonliving materials. Functions that can only be performed by living tissues can be restored either by transplanting a new tissue or a new organ or by regenerating the tissue or organ that has lost its function.

15.1 OVERVIEW

Transplants may be classified according to the relationship between the donor and the recipient as given in Table 15-1. Transplants are also classified according to their location in the recipient.[1,2] Orthotopic transplants are placed in the same location as the original organ, while heterotopic transplants are placed in a different location.

The historical basis for transplantation can be traced several thousand years ago to Greek myths of human-animal hybrids. In the Middle Ages, there were anecdotal reports in stories of successful transplants of teeth, noses, and even whole limbs. At that time, noses were lost to sword cuts or from advanced syphilis. Instances were reported of autografts (transplants from one part of the body to another) of skin and flesh from the arm to the nose. The first such skin flap

Table 15-1. Types and Definitions of Transplants

Type	Definition
Autograft	Within one individual, from one part of the body to another
Isograft	Between genetically identical individuals, i.e., identical twins
Homograft, allograft	Between different individuals of the same species
Heterograft, xenograft	Between members of different species

technique was reported by Tagliacozzi in 1596. Such techniques are currently used in modern plastic surgery. There were also reports of transplants of noses and teeth from slaves. It was thought, incorrectly, that the graft would survive only as long as the donor lived.

Modern transplantation is based on the technique of vascular anastomosis, developed in 1902, and the understanding and control of the immunological basis of rejection, which has evolved only in the past 50 years. Kidney transplants were tried in the early 1950s and these failed because of rejection resulting from a lack of immunosuppression. In 1954 a kidney transplant between two identical twins (isograft) was successful. Later isografts were successful, but allografts (transplants between two unrelated individuals of the same species) were generally unsuccessful despite the use of whole body irradiation for immunosuppression. The period from 1962 to 1975 saw the development of immunosuppressive drugs, which substantially improved the success of transplants. Since the development of cyclosporin, an immunosuppressive drug, in 1975, further improvements have been realized, and organs such as the liver and heart have been successfully transplanted.

As for the donor, both living and cadaver donors are used in transplantation. Living donors are appropriate in the case of blood or bone marrow, which the body can replace, or in the case of the kidney, of which each person has two but can function perfectly well with one. The concept of a "cadaver" donor is based on the fact that the tissues and organs of the body do not die at the same rate. Obviously, an organ containing only dead cells is useless as a transplant, unless it is performing only a passive function. The requirement for a cadaver donor is that of *brain death.* Brain death involves (1) unreceptivity and unresponsivity, (2) absence of spontaneous respiration or movement, (3) absence of reflexes, and (4) a flat (zero signal) electroencephalogram (EEG). Patients with hypothermia or with severe endocrine or metabolic disturbances are excluded from consideration as brain-dead donors since they can at times be resuscitated from a very unresponsive state. The distinction between brain death and earlier concepts of death as the irreversible stoppage of heartbeat, blood circulation, and breathing is a relatively recent one. Until recently, the heart, not the brain, has been considered by most people to be the central organ of life and human existence. The brain stem controls breathing, and its death results in a cessation of breathing; more than a few minutes of this results in the permanent demise of the entire organism. The role of the brain stem is relevant since machinery to maintain respiration in the absence of brain stem viability has been available only in relatively recent years. Cadaver donors are usually those who have suffered a massive head injury or a cerebral vascular problem sufficient to kill the brain.

At body temperature, irreversible brain damage occurs about 5 minutes after cessation of blood circulation. The kidney, however, remains viable for about 1 to 2 hours without blood circulation. This amount of time permits transplantation if the donor and recipient are nearby and surgery can be done rapidly. Cooling of the kidney extends the viability time to 6–8 hours with useful function retained at 24 hours; cooling combined with perfusion extends viability to 3 days. Cooling of the liver and heart extend viability to 5–8 hours. These preservation procedures are useful in that they increase the availability of organs for transplantation.

15.2. IMMUNOLOGICAL CONSIDERATIONS

Most of the early attempts at organ transplantation failed for reasons that became understood only recently. It is now known that the death of a transplanted tissue or organ is a result of an immunological response known as *rejection*, by the recipient to the transplant. In animal experiments in the 1940s it was found that skin grafts between two different animals of the same species survived for about 7 days before being rejected. A second transplant between the same donor and recipient was found to be rejected twice as fast. This so-called "second set response" became an experimental model for study of the immunology of transplantation.

Rejection is currently prevented by the typing of tissues according to *histocompatibility*, and by the use of immunosuppressive drugs. Histocompatibility refers to the compatibility of different tissues in connection with immunological response. By contrast, *biocompatibility*, discussed in earlier chapters, refers to the compatibility of nonliving materials with living tissues and organisms. The destruction associated with the rejection of a transplant occurs from the effects of antibody-activated macrophages and cytotoxic T cells. Hyperacute rejection, occurring within 24 hours following transplantation, is antibody mediated and is seen in recipients presensitized against donor antigens. Accelerated rejection, which occurs within 5 days, is considered a second set reaction and may be mediated by antibodies or cells. Acute rejection, which occurs in the first few weeks after transplantation, is a T-cell-mediated immune response of the recipient against the graft. Finally, chronic rejection is associated with a gradual decline of function of the graft and results from humoral factors and possibly a low-grade cellular attack. Rejection is viewed not as an all-or-none phenomenon but as a matter of degree. Incipient rejection episodes can be detected and controlled with an increase of drug dosage or use of additional drugs. Drugs currently used include azathioprine (Imuran®) and corticosteroids, as well as the more recently developed cyclosporin. Cyclosporin has revolutionized the transplantation of hearts and livers (see Figure 15-1).

Other immunosuppressive techniques that have been tried include the following. Splenectomy has been performed based on the fact that the spleen is the largest lymphoid organ. Removal of the spleen increases transplant survival in the first 2 years but reduces patient survival after that as a result of a higher risk of infection. The procedure, in conjunction with antibiotic therapy, is still under investigation. Irradiation of the transplant has been tried; the procedure yields equivocal results and is being studied. Thoracic duct drainage is a short-term procedure in which lymph is removed from the main thoracic lymph duct, the lymphocytes extracted by centrifugation, and the remaining fluid returned to the body. This is no longer widely practiced. Whole body irradiation was also used

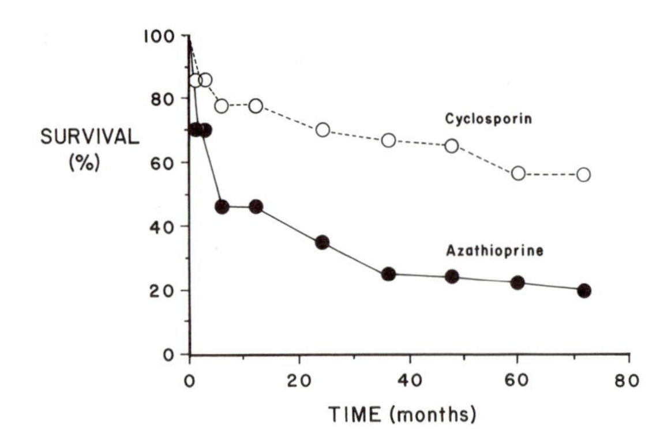

Figure 15-1. Comparison of survival of orthotopic liver transplant recipients under immunosuppressive therapy with cyclosporin and azathioprine. From G. J. Cerilli, *Organ Transplantation and Replacement*, J. B. Lippincott Co., New York, 1988.

in early transplants, but is nonspecific and nontitratable, and has been supplanted by the use of drugs.

Immunosuppressive therapy carries a variety of risks, including a susceptibility to infection, and a significantly increased incidence of a variety of cancers, including Kaposi's sarcoma, other skin malignancies, and cancer of the lymphatic system and other organs. Because the incidence of cancer increases with time of follow-up, transplant patients are expected to be monitored for cancer development indefinitely. The risks are to be balanced by the fact that many recipients of vital organ transplants would have faced a more rapid death without them.

Histocompatibility is associated principally with the human lymphocyte antigen system. The ABO blood group antigens are considered to be important antigens in the context of transplantation. Almost all transplants are between ABO-compatible people; however, recently sporadic attempts have been made with additional immunosuppression to cross the ABO blood group barrier in an effort to make more organs available. Testing for histocompatibility is currently done by serologic tests in which mixed lymphocyte cultures are incubated and examined for reactions.

Example 15-1

Discuss transplantation of the stomach and of the internal ear.

Answer

As we have seen, avoidance of rejection is a challenge in transplantation surgery. Since it is possible to live reasonably well without a stomach by modification of the diet and by use of supplementary enzymes, there is probably not sufficient motivation to transplant the stomach. As for the internal ear, it would be desirable to cure deafness caused by inner ear disease or damage. However, establishment of the appropriate nerve connections is at present an insurmountable problem.

15.3. BLOOD TRANSFUSIONS

The transfer of blood from one individual to another is a transplant of blood tissue. The task of the donor is made easy by the fact that blood is a liquid and is easily removed, and any lost blood is rapidly replenished by the body. Of great importance is the fact that only a limited number of histocompatibility antigens are relevant to the success of blood transfusions. These are the A, B, AB, and O blood groups, and the rh factor, positive or negative. Matching of blood groups is immunologically sufficient to ensure a successful acceptance of the transfusion. Blood transfusion is commonly done during major surgery and it can save lives in cases of severe bleeding from trauma or disease. An issue of current concern is the prevention of transmission of infectious disease such as acquired immune deficiency syndrome (AIDS) and hepatitis via transfused blood. In addition, some people oppose blood transfusion on religious grounds. Consequently, there is increasing interest in artificial blood substitutes. Perfluorocarbon liquids have been used as temporary artificial blood in some emergency patients for whom transfusions cannot be found. Oxygen is highly soluble in these materials. By contrast, in natural blood, oxygen is reversibly bound in the hemoglobin within red blood corpuscles. Another approach to the problem of disease transmission by blood transfusion is the use of autologous blood, i.e., blood from the same person who is to receive it. A person anticipating elective surgery can denote several pints of blood, which is refrigerated, and wait a month or so for the surgery during which time the lost blood is replenished by the body. Obviously such a procedure will not help in emergency cases.

15.4. INDIVIDUAL ORGANS

15.4.1. Kidney

Kidneys are transplanted more often than any other vital organ despite the availability of renal dialysis as indicated in Table 15-2. The artificial kidney, described in Chapter 12, can sustain life in a patient lacking renal function, but there are problems associated with physical and mental health, maintenance of the percutaneous device, and quality of life for the dialysis patient. Physical complications of dialysis include hypotension, cramps, malaise, headache, air

Table 15-2. Transplant Statistics, United States[a]

	Number of cases			
Organ	1983	1984	Number waiting	Potential benefit
Cornea	21,500	23,500	3,500	250,000
Kidney	6,112	6,730	12,000	25,000
Heart	280	400	100	14,000
Liver	168	308	330	5,000
Pancreas	61	87	50	10,000
Heart-lung	13	17	50	1,000

[a] From G. J. Cerilli, *Organ Transplantation and Replacement*, J. B. Lippincott Co., Philadelphia, 1988.

embolism, aluminum toxicity, neuropathy, osteomalacia, dementia, and nausea. Peritoneal dialysis makes use of the diffusion properties of the natural abdominal membrane, which permits passage of larger molecules up to 30,000 daltons. Patients treated with this technique do not suffer severe complications discussed above such as neuropathy, but there is a risk of infection.

In view of the above considerations, a transplanted natural kidney is a more desirable solution than artificial kidney dialysis. Both living and cadaver donors are used for kidney transplants; cadaveric sources account for 60–70% of kidney donors. Living donors are either blood relatives or spouses. In removing a kidney from a living donor, a diagonal incision is made paralleling the 11th rib as shown in Figure 15-2. The rib is removed, the kidney is mobilized from the peritoneum, adrenal gland, and fat, and the renal vein and artery and the ureter are identified, clamped, and cut. The kidney is then removed and immediately perfused with a chilled saline solution containing procaine and heparin. As for cadaver donors, it is necessary to ascertain that the donor is free of communicable disease or cancer. Both kidneys as well as a segment of aorta and vena cava are removed from the cadaver as shown in Figure 15-3 and are perfused and cooled as described above. The suturing procedure for reconnection of the vessels (anastomosis) is illustrated in Figure 15-4.

Immunosuppressive treatment for kidney transplant patients currently involves Imuran® and low-dose steroids in the case of a living human donor, and cyclosporin combined with Imuran® and low-dose steroids in the case of a

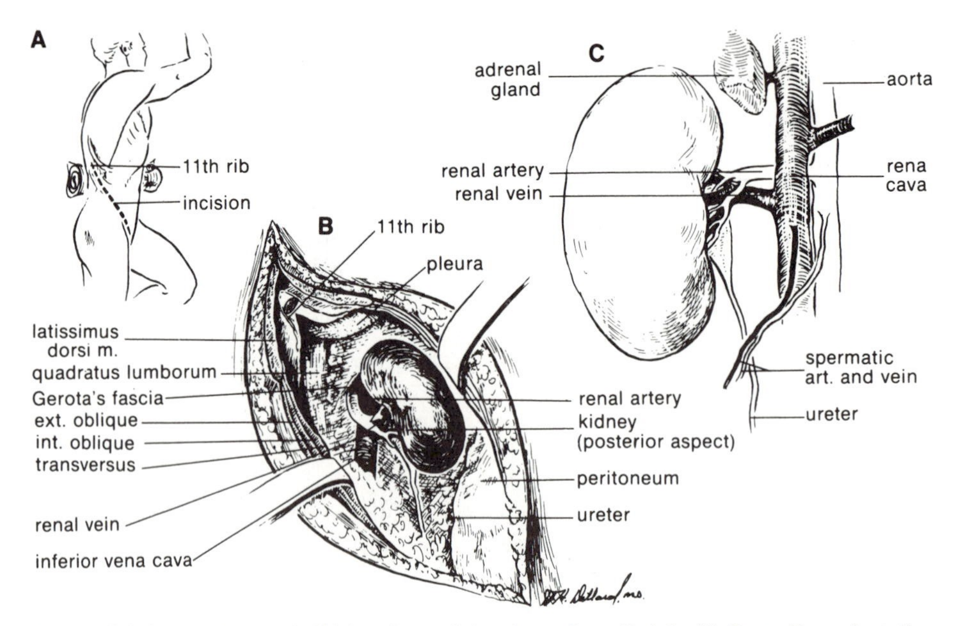

Figure 15-2. Surgical removal of kidney from a living donor. From G. J. Cerilli, *Organ Transplantation and Replacement*, J. B. Lippincott Co., New York, 1988.

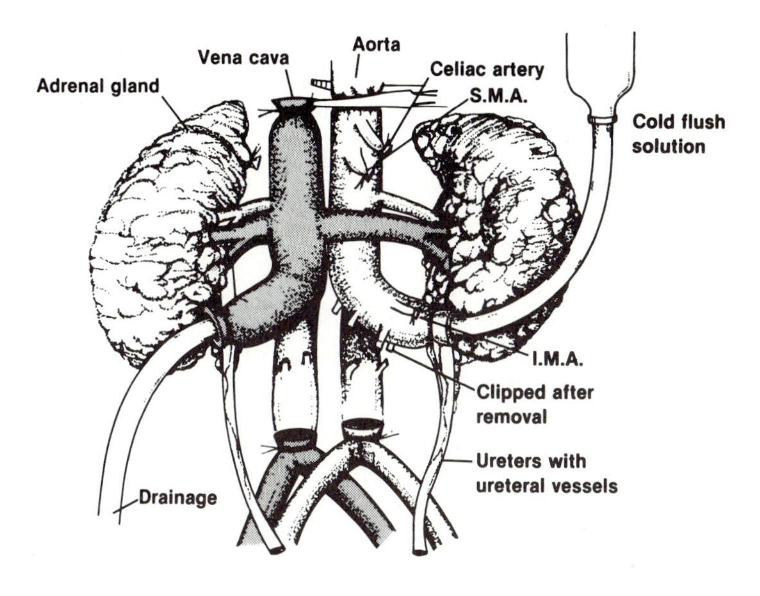

Figure 15-3. Surgical removal of kidney from a cadaver donor. From G. J. Cerilli, *Organ Transplantation and Replacement*, J. B. Lippincott Co., New York, 1988.

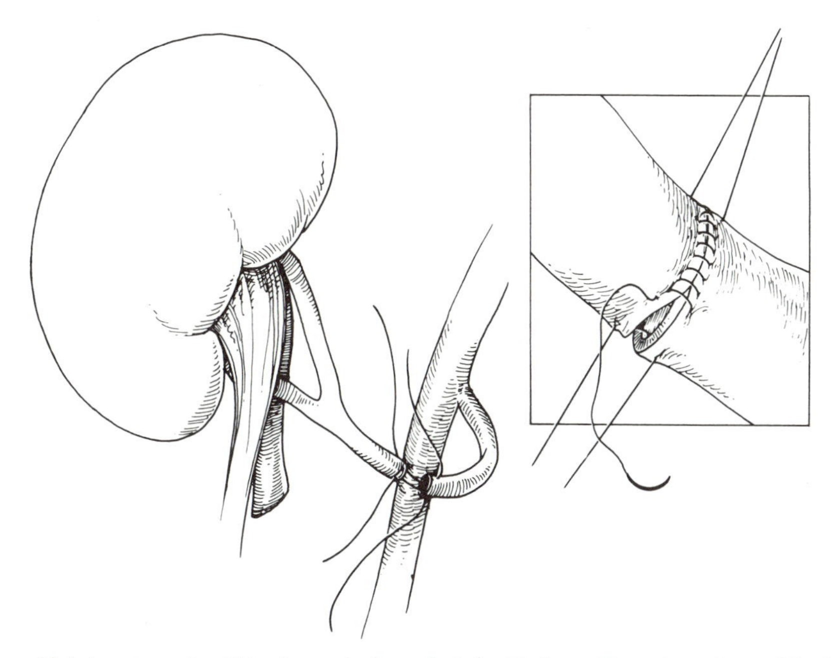

Figure 15-4. Anastomosis of blood vessels. From G. J. Cerilli, *Organ Transplantation and Replacement*, J. B. Lippincott Co., New York, 1988.

cadaver donor. A rejection episode may occur in spite of the above precautions. Rejection is manifested in fever, tenderness of the kidney, weight gain, reduction in urinary output, and change in urinary composition. Rejection is treated with drugs; however, if the treatment fails during acute rejection, the transplanted kidney is removed.

15.4.2. Liver

The liver, as its name suggests, is essential to life and its functions cannot be taken over, even temporarily, by any artificial device or material. The functions of the liver include active ones such as secreting bile, metabolism of carbohydrates, fats, and proteins, and the detoxification of many toxic substances. Patients with active chronic hepatitis, severe cirrhosis, inborn metabolic errors, and primary liver tumors are potential candidates for liver transplantation.

Following the early success of kidney transplantation in the 1960s, clinical liver transplants were attempted. However, since none of the patients survived longer than 1 month, the procedure was abandoned for some time. Liver transplants were tried again as an experimental procedure with improved immunosuppressive drugs in the 1970s, but more than half of these patients died within the first year. The introduction of cyclosporin as an immunosuppressive drug combined with improved surgical technique has led to dramatic improvements in the success of liver transplantation (Figure 15-1), and its acceptance as a clinical procedure.

Transplantation of the liver is rendered difficult by immunological considerations, and by the fact that the liver's blood supply is very complex, much more so than the kidney. Liver transplants require interruption of the inferior vena cava, the main return to the heart. Heterotopic transplants (i.e., placement of the organ in a location other than the anatomically correct one) of the liver were tried initially to simplify the circulatory connections. Interruption of liver function is not tolerable in contrast to the case of the kidney. Consequently, marginal viability problems and rejection episodes are very serious in liver transplants. Nevertheless, modern liver transplants have been successful, with 60–80% of recipients surviving 5 years. Recently, living donor liver transplants to children have been attempted, with a *piece* of a parent's liver used for the transplant. The loss of a piece of liver is not a problem since the liver, like bone, is capable of true regeneration.

15.4.3. Heart and Lung

The first heart transplant performed by Dr. Christian Barnard in 1967 had a dramatic impact on society in view of the central role of the heart in maintaining life, and the psychological perception of the role of the heart. Dr. Barnard's patient survived 17 days, and heart transplants in the years immediately following did not last very long. Greater success in heart transplantation resulted from the introduction of cyclosporin, increasing clinical experience with immunosuppressive drugs, and improved patient selection. Recipients of heart transplants tend to be relatively young people with end-stage cardiac disease. The survival rate of transplant recipients is now about 70–80% at 1 year after surgery. A major problem in heart transplantation is that too few donors are available for the number of potential recipients. Attempts to circumvent this problem by using baboon hearts (in children) have not been successful. Since donated organs cannot be stored for long periods, a role is seen for cardiac assist devices and artificial hearts as a "bridge" to transplantation. Such devices can be used to support the patient for a period of days to weeks until an appropriate donor can be found. As described elsewhere in this book, the total artificial heart is experimental and has not been successful as a long-term solution to end-stage heart disease.

Heart valve transplants, discussed elsewhere in this book, are routinely performed. These are based on porcine xenografts (transplants from another species) that have been fixed in glutaraldehyde. They do not contain any living cells. Consequently, rejection is not expected to be a problem.

Transplants of the lung or of the heart and lung are more difficult owing to the complexity of the surgical technique and are less commonly done than the heart alone. The first successful combined heart and lung transplant was done in 1981. Heart and lung transplant patients suffer a transient defect in lung function within the first month, and this is thought to result from lung denervation, interruption of the lymphatics, and surgical trauma. The nerve supply of the lung is important to its function. Transplanted lungs are always denervated and the nerves do not regrow. Heart and lung differ immunologically, and rejection of each can occur independently.

15.4.4. Bone

Bone grafts, used widely in orthopedic surgery, serve the roles of structural support, of providing a framework for new bone tissue from the recipient to grow into, and a stimulus for new bone growth. The bone cells need not be living for a bone graft to succeed. Therefore, frozen bone tissue is suitable for this purpose. Autografts are commonly performed in which bone is removed from a location from which it can be spared, and used for repair. Bone from the rib, iliac crest, and fibula may be implanted in a viable form, by microsurgical anastomosis of the vascular supply. It is possible to walk relatively normally with a portion of the fibula removed. Allograft replacement of a large segment of bone removed for tumor surgery is also done, as shown in Figure 15-5.

Bone *marrow* is transplanted to correct severe immunological deficiency, aplastic anemia, and damage to the marrow caused by radiation poisoning or chemotherapy for leukemia. Histocompatibility matching is performed, and marrow is extracted from the iliac crest or sternum of a living donor under spinal or general anesthesia. The marrow is screened, heparinized, and injected intravenously into the recipient.

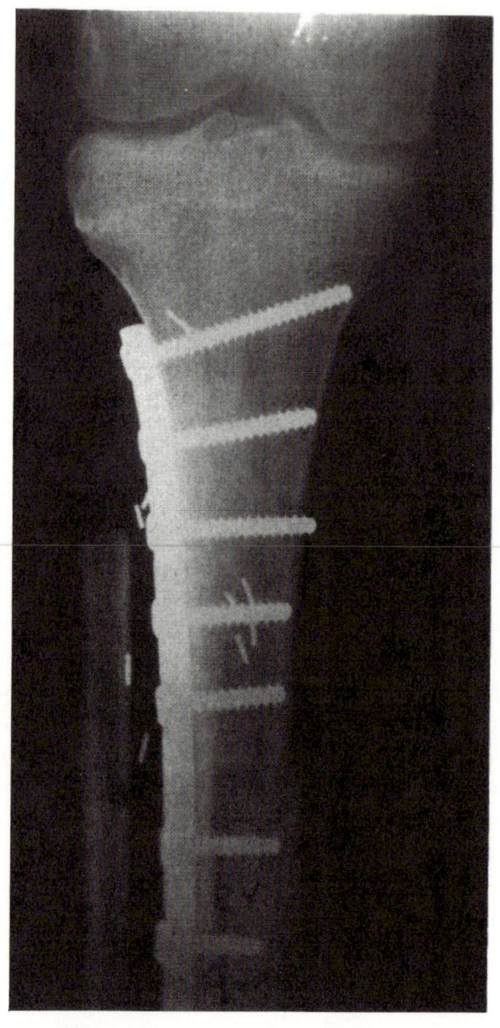

Figure 15-5. Large bone graft. Allograft replacement of tibial segment removed to treat tumor. From G. J. Cerilli, *Organ Transplantation and Replacement*, J. B. Lippincott Co., New York, 1988.

15.4.5. Skin and Hair

Reconstructive surgery of the skin is performed to correct defects resulting from trauma, removal of skin tumors, or vascular problems such as pressure sores. It is sometimes necessary to provide "new" skin to replace the portion that was lost or damaged. The best source for such skin, if a relatively small amount is needed, is elsewhere on the patient's body—an autograft. If the skin is simply rearranged in the vicinity of the defect, it is called a *flap*; if it is moved from elsewhere on the body, it is called a *graft.* Rearrangement of skin in a representative technique is shown in Figure 15-6. Grafts differ from flaps in that

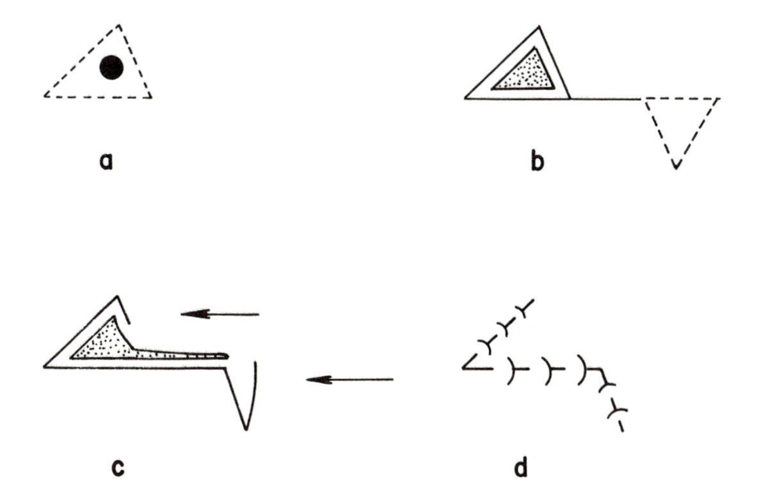

Figure 15-6. Surgical skin flap procedure. (a) Tumor is identified and a triangular segment of skin marked. (b) Segment with tumor is removed and incisions are made. (c) Skin is moved. (d) Wound is closed and sutured.

they are totally separated from their original blood supply and must develop a new supply from the location where they are placed. Consequently, the vascularity of the recipient bed is crucial to the success of a skin graft.

Hair transplants are a treatment for male pattern baldness. The procedure is done for cosmetic reasons. Hair transplants are always autografts, so rejection is not a problem. A local anesthetic is used to numb the scalp and saline is injected into the scalp to increase its turgor. Plugs of skin, including the hair follicles, are removed from the back part of the head, which resists balding. The plugs, about 4 mm in diameter, are harvested using a powered circular punch. Holes in the recipient area are cut using a slightly smaller punch and the plugs are reimplanted at the top and front of the head.

15.4.6. Nerve and Brain

The brain is an immunologically privileged site. The blood-brain barrier prevents cells of the immune system from reaching the brain, so that transplants into the brain are at low risk of rejection. Whole brain transplants are not to be expected since (1) the recipient would presumably lose his memories with his brain, (2) volunteer living donors could not be found, (3) "cadaver" donors, by definition brain-dead, are unsuitable, and (4) nobody knows how to establish new nerve connections with the spinal cord. However, transplants of nerve tissue or neuroendocrine tissue into the brain have been considered for the correction of Parkinson's disease and senile dementia. In the case of Parkinson's disease, autografts of tissue from the patient's own adrenal glands have been tried. Transplants of embryonic brain cells in rats will form new connections in the recipient rat, and improve the rat's brain function. As for humans, the use of human fetal tissue has dramatic ethical implications and would generate considerable controversy if it were attempted. Transplants of nerve and brain tissue are considered experimental at this time.

15.4.7. Pancreas

In insulin-dependent diabetes, the endocrine function of the islands of Langerhans in regulating blood glucose is lost. Injections of insulin allow the patient to substitute an artificial or animal-based hormone for the natural one that has been lost. There remain many complications associated with insulin-treated diabetes. Many of these are related to the bolus character of the injected insulin and the associated fluctuations in blood sugar levels. Experimental treatment has been conducted with "artificial pancreas" devices consisting of a glucose sensor, a feedback amplifier, and a controlled insulin injector. Problems with these devices have resulted from a lack of biocompatibility of available glucose sensors, tissue growth around the sensor, and aggregation of the stored insulin. Pancreatic transplantation is a possible solution, although transplantation of the whole pancreas is made difficult by the corrosive nature of its exocrine secretion, a digestive enzyme; and by the vulnerability of pancreatic islet cells to rejection. Islet cell allografts have been tried but have not been successful as a result of rejection. Islet cell autografts into the portal vein have been done for patients undergoing removal of the pancreas for painful chronic pancreatitis, and some of these have been successful. An interesting and novel approach combining transplantation techniques with biomaterials is to implant only the islet cells in semipermeable artificial tubules or between semipermeable artificial membranes. Glucose and insulin can diffuse through the membranes but immune cells and substances of large molecular weight cannot. Tubule-based islet cell transplants have been successful in controlling glucose levels in diabetes for about 12 months, after which the tubules become overgrown with recipient cells, slowing the diffusion process. Methods for pancreas support are still under investigation.

15.4.8. Cornea

Corneal transplantation (keratoplasty) is the most commonly done solid transplant procedure, as shown in Table 15-2. Conditions that impair the clarity of the cornea sufficiently to obscure vision include trauma, scarring following an infection, congenital abnormality, and degeneration. Success of corneal transplants is aided by the fact that the cornea is avascular. Diffusion to the blood system is very slow as is the turnover of corneal proteins. Rejection, therefore, is not normally a problem. It occurs in a small number of cases and is treated by topical steroids. Donated corneas, owing to their slow metabolism, may be removed within 4 to 12 hours following death. They may be preserved for 96 hours to 3 weeks in a culture medium, or frozen in a cryoprotectant solution for as long as 18 months. The surgical procedure for the recipient involves great care in aligning the new cornea with surrounding structures. Suturing is done with 10-0 or 11-0 monofilament nylon suture under an operating microscope.

Example 15-2

Why and how would an "autograft" of a human leg be performed?

Answer

Autografts are usually performed with biological material that can be moved from one part of the body to another, as with bone and skin. This makes no sense in the case of a leg. However, a leg, arm, or finger that has been accidentally cut off can sometimes be reattached, a procedure that is called "replantation" rather than transplantation. For the procedure to be successful, most of the cells in the limb must remain alive. Therefore, the replantation surgery must be done quickly; it is very helpful to cool (but not freeze) the limb during the time prior to surgery.

Replantation of a limb involves joining the bone by the same techniques used for broken bones, e.g., bone plates. The muscles are joined individually by sutures. The blood vessels are also sutured together; connection of the smaller ones involves microsurgical techniques. The nerve fibers in the severed limb will die; however, if the nerve sheath is properly joined and aligned, new fibers can grow from the nerve cell bodies in the patient's body. In the case of a leg, this regrowth may take a year.

15.5. REGENERATION

Transplantation of organs and tissues has proven successful in the treatment of many serious maladies. The lack of sufficient donors is likely to remain a major constraint on such procedures. An alternative approach is to consider the possibility of regenerating the diseased, damaged, or lost tissue or organ. The use of electrical stimulation to regrow bone across nonunions has been discussed elsewhere in this book. In a related development, attempts have been made to stimulate regeneration of amputated limbs. The background of this idea is the fact that animals of low order have remarkable regenerative capacity. For example, an earthworm cut in half can regrow into two new worms; some amphibians can regenerate a limb that has been amputated. In higher animals and in humans, regeneration is very limited, although young children can regrow an amputated fingertip. The regeneration in the lower animals is accompanied by a distinctive pattern of electrical potential in the injured limb, a potential that is opposite in polarity to that seen in a nonregenerating higher animal or human. Researchers have stimulated partial regeneration of limbs in amputated mammals such as rats or in a nonregenerating amphibian such as a frog, by applying an electrical potential pattern similar to that in a regenerating amphibian. These procedures are under investigation. While bone tissue has been regrown in humans in a clinical setting, whole limbs have not.

15.6. ETHICAL CONSIDERATIONS

Organ transplantation raises ethical, moral, and sociological issues not to be found in inert biomaterials. Issues that are of concern include the decision-making process as to which patients in need of transplants receive them, and the means by which donations are obtained. The high cost of transplant surgery tends to lead to a patient selection favoring the wealthy, although medical insurance covers some procedures. On the donor side, we have described in Section 15.1 historical involuntary transplants that would be unacceptable in civilized societies today. Fiction author Larry Niven has created a hypothetical future society in which technical obstacles to transplantation of all organs have been overcome. This fictional society uses capital punishment subjects as forced transplant donors. To extend their lives by a ready supply of transplants, individuals in this society then voted progressively more minor crimes, including traffic citations, to be capital offenses.

In the modern world, there is evidence that organs are being sold in countries outside the United States. This practice is problematical in that it is viewed as exploiting poor people who are coerced into donation by economic forces. Although the voluntary donation of an organ to a family member can be a very positive altruistic experience, there have been cases of people who have been pressured into donating organs to their sick family members. In view of these issues, the American Society of Transplant Surgeons has prepared a set of guidelines for the procurement and use of transplant organs in the United States.

PROBLEMS

15-1. Although transplants perform functions that cannot be carried out by inert biomaterials, various biomaterials are used in transplantation surgery. What are these biomaterials? Is their role any different in the case of transplants than in applications not involving transplants?

15-2. Discuss transplantation of (1) the external ear, (2) the entire brain, (3) tonsils, (4) the entire arm, (5) the whole eye, (6) whole teeth. Include in your discussion the questions of need, technical feasibility, ethics, economic and social aspects.

15-3. Discuss the relative merits of heart valve replacement by porcine xenograft, by a tilting disk valve design using pyrolytic carbon, and by a silicone ball in cage valve design. Include in your discussion the issues of hemodynamic flow, likely failure mechanisms, use of anticoagulant drugs, and expected performance.

15-4. Discuss the differences between replacement heart valves based on porcine xenograft valves and valves made from bovine pericardium (see Chapter 12). What are their pros and cons?

15-5. Many people object to the use of animals for testing biomaterials. Under what circumstances are *in vivo* tests required for evaluating a particular device such as a heart valve? Can you suggest ways of avoiding animal tests entirely? Keep in mind the fact that testing with human patients is *in vivo* but does not involve animals. Under what circumstances do you think it is ethical to test an unproven biomaterial in humans? If you are in favor of animal experimentation, would you accept being used for invasive experiments by a group of people who are more intelligent and technically sophisticated than you are? If you oppose animal experimentation, are you also vegetarian? Would you refuse treatment for a major disease if that treatment were developed with the aid of animal experiments?

DEFINITIONS

Allograft: Transplant between unrelated individuals of the same species.
Anastomosis: Joining two blood vessels or other tubular structures in the body.
Antigen: A foreign protein that elicits an immune system response.
Autograft: Transplant within an individual, from one part of the body to another.
Biocompatibility: Compatibility of nonliving implant materials with the body.
Brain death: Death of the brain including the brain stem resulting in unreceptivity and unresponsivity, absence of spontaneous respiration or movement, absence of reflexes, and a flat EEG.
Cyclosporin: An immunosuppressive drug.
EEG: Electroencephalogram, which is a plot of the electrical activity of the brain as measured via electrodes on the scalp.
Graft: A transplant.
Heterograft, xenograft: Transplant between members of different species.
Heterotopic: Transplanted in a location different from the anatomically correct one.
Histocompatibility: Compatibility of transplant tissue with the recipient's body in relation to immune response.
Homograft, allograft: Transplant between different individuals of the same species.
Hypothermia: Abnormally low body temperature. It can cause an unresponsive condition that can mimic brain death, so transplants are not taken from hypothermic patients.
Iliac: Referring to the ilium, the large flat bone of the pelvis.
Island of Langerhans: A cluster of endocrine cells found in the pancreas; they secrete insulin.
Isograft: Transplant between genetically identical individuals, i.e., identical twins.
Kaposi's sarcoma: A type of cancer now largely associated with immune system deficiencies such as those produced in AIDS; it produces bluish skin lesions.
Keratoplasty: Replacement of the cornea.
Orthotopic: Transplanted in the anatomically correct location.
Parkinson's disease: A neurological disorder characterized by tremor and paralysis. Attempts have been made to cure it by transplantation of brain or adrenal cells into the brain.
Porcine: From a pig.
Replantation: Surgical reattachment of a severed body part.
T cells: Immunologically competent cells derived from the thymus.
Transplantation: Transfer of a tissue or organ from one body to another, or from one location in a body to another.
Tubule: A small tube.

REFERENCES

1. F. T. Rapaport, J. Dausset, *Human Transplantation*, Grune and Stratton, New York, 1968.
2. G. J. Cerilli, *Organ Transplantation and Replacement*, J. B. Lippincott, Philadelphia, 1988.

BIBLIOGRAPHY

R. O. Becker, "Stimulation of Partial Limb Regeneration in Rats," *Nature*, *235*, 109–111, 1972.

F. H. Gage, S. B. Dunnett, V. Stenevi, and A. Björkerd, "Aged Rats: Recovery of Motor Impairments by Intrastriatal Nigral Grafts," *Science*, *221*, 966–969, 1983.

G. Kolata, "Grafts Correct Brain Damage," *Science*, *217*, 342–344, 1982.

L. Niven, *The Long Arm of Gil Hamilton*, Ballantine Books, Inc., New York, 1974.

O. T. Norwood and R. C. Shiell, *Hair Transplant Surgery*, Charles C. Thomas Publ., Springfield, Ill., 1984.

T. A. Tramovich, S. J. Stegman, and R. G. Glogau, *Flaps and Grafts in Dermatologic Surgery*, Year Book Medical Publ., Inc., Chicago, 1989.

APPENDIXES

APPENDIX I: PHYSICAL CONSTANTS AND CONVERSIONS

Physical Constants

Name	Symbol	Units
Atomic mass unit	amu	1.66×10^{-24} g
Avogadro constant	N	6.023×10^{23}/mol (or molecules/g mol)
Boltzmann constant	k	1.381×10^{-23} J/K or 8.63×10^{-5} eV/K
Permittivity (vacuum)	ε	8.85×10^{-12} C/V · m
Electron charge	q	1.602×10^{-19} C
Electron volt	eV	1.60×10^{-19} J
Faraday constant	F	0.649×10^{4} C/mol
Gas constant	R	8.314 J/mol · K or 1.987 cal/mol · K
Planck's constant	h	6.62×10^{-34} J · s
Standard gravity	g	9.807 m/s^2

Conversions

1 angstron (Å)	= 10^{-10} m
1 inch (in)	= 0.0254 m
1 calorie (cal)	= 4.1868 J
1 erg	= 10^{-7} J
1 dyne	= 10^{-5} N
1 joule (J)	= 1 N · m
1 kg force (kgf)	= 9.807 N
1 pound force (lb)	= 4.448 N
1 pound (mass)	= 0.4536 kg
1 atmosphere (standard)	= 0.1 Pa
1 inch of Hg (60°F)	= 3.37685×10^{3} Pa
1 dyne/cm^2	= 0.1 Pa
1 kgf/mm^2	= 9.807×10^{6} Pa
1 lb/in^2 (psi)	= 6.895×10^{3} Pa
1 MPa	= 145 psi
1 poise	= 0.1 Pa · s
1 rad	= 0.01 Gy

APPENDIX II: SI UNITS

The International System of Units or SI (Le Système International d'Unités) defines the base units as follows:

Base Units

Quantity	Unit	Symbol
length	meter	m
mass	kilogram	kg
time	second	s
electric current	ampere	A
temperature	kelvin	K
amount of substance	mole	mol

Derived Units

Quantity	Unit	Symbol	Formula
frequency	hertz	Hz	1/s
force	newton	N	$kg \cdot m/s^2$
pressure, stress	pascal	Pa	N/m^2
energy, work, quantity of heat	joule	J	$N \cdot m$
power	watt	W	J/s
absorbed dose	gray	Gy	J/kg

APPENDIX III: COMMON PREFIXES

Multiplication factor	Prefix	Symbol
10^{18}	peta	P
10^{12}	tera	T
10^{9}	giga	G
10^{6}	mega	M
10^{3}	kilo	k
10^{-2}	centi	c
10^{-3}	milli	m
10^{-6}	micro	μ
10^{-9}	nano	n
10^{-12}	pico	p
10^{-15}	femto	f
10^{-18}	atto	a

APPENDIX IV: PROPERTIES OF SELECTED ELEMENTS

Element	Symbol	Atomic number	Atomic weight (amu)	T_m (°C)	Solid density[a] ρ (g/cm^3)	Crystal structure[a]	Atomic radius (nm)
Hydrogen	H	1	1.008	−259.14	—	—	0.046
Lithium	Li	3	6.94	180	0.534	bcc	0.1519
Beryllium	Be	4	9.01	1289	1.85	hcp	0.114
Boron	B	5	10.81	2103	2.34	—	0.046
Carbon	C	6	12.011	>3500	2.25	hex	0.077
Nitrogen	N	7	14.007	−210	—	—	0.071
Oxygen	O	8	15.999	−218.4	—	—	0.060
Fluorine	F	9	19.0	−220	—	—	0.06
Neon	Ne	10	20.18	−248.7	—	fcc	0.160
Sodium	Na	11	22.99	97.8	0.97	bcc	0.1857
Magnesium	Mg	12	24.31	649	1.74	hcp	0.161
Aluminum	Al	13	26.98	660.4	2.70	fcc	0.14315
Silicon	Si	14	28.09	1414	2.33	diamond cubic	0.1176
Phosphorus	P	15	30.97	44	1.8	—	0.11
Sulfur	S	16	32.06	112.8	2.07	—	0.106
Chlorine	Cl	17	35.45	−101	—	—	0.0905
Argon	Ar	18	39.95	−189.2	—	fcc	0.192
Potassium	K	19	39.10	63	0.86	bcc	0.2312
Calcium	Ca	20	40.08	840	1.54	fcc	0.1969
Titanium	Ti	22	47.88	1672	4.51	hcp	0.146
Chromium	Cr	24	52.00	1863	7.20	bcc	0.1249
Manganese	Mn	25	54.94	1246	7.2	—	0.112
Iron	Fe	26	55.85	1538	7.88	bcc	0.1241
Cobalt	Co	27	58.93	1494	8.9	hcp	0.125
Nickel	Ni	28	58.69	1455	8.90	fcc	0.1246
Copper	Cu	29	64.54	1084.5	8.92	fcc	0.1278
Zinc	Zn	30	65.38	419.6	7.14	hcp	0.139
Silver	Ag	47	107.87	961.9	10.5	fcc	0.1444
Tin	Sn	50	118.69	232	7.3	bcc	0.1509
Antimony	Sb	51	121.75	630.7	6.7	—	0.1452
Cesium	Cs	55	132.9	28.4	1.9	bcc	0.262
Tungsten	W	74	183.85	3387	19.4	bcc	0.1367
Gold	Au	79	197.0	1064.4	19.32	bcc	0.1441
Mercury	Hg	80	200.59	−38.86	—	—	0.155
Lead	Pb	82	207.2	327.5	11.34	fcc	0.1750

[a] Solid density and crystal structure are for room temperature.

APPENDIX V: PROPERTIES OF SELECTED ENGINEERING MATERIALS (20°C)[a]

Material	Density (g/cm^3)	Thermal conductivity ($W/m \cdot °C$)	Linear expansion ($°C^{-1}$)($\times 10^{-6}$)	Electrical resistivity, ρ ($ohm \cdot m$)	Average modulus of elasticity, E	
					MPa	psi
Metals						
Aluminum (99.9+)	2.7	0.22	22.5	29×10^{-9}	70,000	10×10^6
Copper (99.9+)	8.9	0.40	17	17×10^{-9}	110,000	16×10^6
Iron (99.9+)	7.88	0.072	11.7	98×10^{-9}	205,000	30×10^6
Monel (70 Ni–30 Cu)	8.8	0.025	15	482×10^{-9}	180,000	26×10^6
Silver (sterling)	10.4	0.41	18	18×10^{-9}	75,000	11×10^6
Steel (1020)	7.86	0.050	11.7	169×10^{-9}	205,000	30×10^6
Steel (1040)	7.85	0.048	11.3	171×10^{-9}	205,000	30×10^6
Steel (1080)	7.84	0.046	10.8	180×10^{-9}	205,000	30×10^6
Steel (18Cr–8Ni stainless)	7.93	0.015	9	700×10^{-9}	205,000	30×10^6
Ceramics						
Al_2O_3	3.8	0.029	9	$> 10^{12}$	350,000	50×10^6
Graphite (bulk)	1.9	—	5	10^{-5}	7,000	1×10^6
MgO	3.6	0.03	9	10^3 (1100°C)	205,000	30×10^6
Quartz (SiO_2)	2.65	0.012	0.05[b]	10^{12}	310,000	45×10^6

SiC	3.17	0.012	4.5	0.025 (1100°C)	—	—
TiC	4.5	0.030	7	50×10^{-8}	350,000	50×10^{6}
Glass						
Plate	2.5	0.00075	9	10^{12}	70,000	10×10^{6}
Borosilicate	2.4	0.0010	2.7	$> 10^{15}$	70,000	10×10^{6}
Silica	2.2	0.0012	0.5	10^{18}	70,000	10×10^{6}
Polymers						
Melamine-formaldehyde	1.5	0.00030	27	10^{11}	9,000	1.3×10^{6}
Phenol-formaldehyde	1.3	0.00016	72	10^{10}	3,500	0.5×10^{6}
Urea-formaldehyde	1.5	0.00030	27	10^{10}	10,300	1.5×10^{6}
Rubber (synthetic)	1.5	0.00012	120	—	4–75	600–11,000
Rubber (vulcanized)	1.2	0.00012	81	10^{12}	3,500[c]	0.5×10^{6}
Polyethylene (L.D.)	0.92	0.00034	180	10^{13}–10^{16}	100–350	14,000–50,000
Polyethylene (H.D.)	0.96	0.00052	120	10^{12}–10^{16}	350–1250	50,000–180,000
Polystyrene	1.05	0.00008	63	10^{16}	2,800	0.4×10^{6}
Polyvinylidene chloride	1.7	0.00012	190	10^{11}	350	0.05×10^{6}
Polytetrafluoroethylene	2.2	0.00020	100	10^{14}	350–700	50,000–100,000
Polymethyl methacrylate	1.2	0.00020	90	10^{14}	3,500	0.5×10^{6}
Nylon	1.15	0.00025	100	10^{12}	2,800	0.4×10^{6}

[a] Various sources.
[b] Fused quartz.
[c] Highly vulcanized and filled.

NAME INDEX

SUBJECT INDEX